Kister Distillation Symposium 2013

Topical Conference at the 2013 AIChE Spring Meeting and 9th Global Congress on Process Safety

San Antonio, Texas, USA
28 April – 2 May 2013

ISBN: 978-1-62748-456-5

Printed from e-media with permission by:

Curran Associates, Inc.
57 Morehouse Lane
Red Hook, NY 12571

Some format issues inherent in the e-media version may also appear in this print version.

Printed by Curran Associates, Inc. (2013)

For permission requests, please contact AIChE
at the address below.

AIChE
3 Park Avenue
New York, NY 10016-5991

Phone: (203) 702-7660
Fax: (203) 775-5177

www.aiche.org

Additional copies of this publication are available from:

Curran Associates, Inc.
57 Morehouse Lane
Red Hook, NY 12571 USA
Phone: 845-758-0400
Fax: 845-758-2634
Email: curran@proceedings.com
Web: www.proceedings.com

TABLE OF CONTENTS

(4a) Mr. Tak Yanagi – Ahead of His Time

Mike Resetarits, Tony Cai and Simon Chambers, Fractionation Research, Inc., Stillwater, OK

While working for FRI from 1957 to 1989, Mr. Tak Yanagi already answered many of the questions that today's distillation engineers ask repeatedly. Initially, the results of most of Tak's work were only available to FRI members. Subsequently, however, many of the associated FRI reports were released to the public.

From 1960 until 1980, 60% of the work performed by Tak and his colleagues at FRI was directly related to tray hydraulics. He coauthored articles regarding high-pressure downcomer operation, downcomer biphase densities, vacuum operation, RTD, liquid mixing effects on efficiencies and the impacts of sieve deck open areas on efficiencies. His primer on trays, "Film A", is still shown to hundreds of engineers and technicians annually. Around 1980, however, the packing revolution began – and Tak was at the frontier. He coauthored articles regarding ceramic rings, liquid distributions, liquid maldistributions, distributor fabrication/installation, Marangoni effects, and scale-up. Roughly speaking, about half of the world's towers are now packed – and Tak's work set the stage.

On a regular basis, FRI staff members contact Tak and ask for his remembrances regarding projects that he worked on many years ago. His memory is incredible. He never forgets. FRI and the global distillation community has not forgotten either. We have not forgotten Mr. Tak Yanagi's contributions to today's distillation world.

(4b) A 100 Year Search for the "Ideal" Mass Transfer Device

Christoph Ender, Koch Glitsch, Wichita, KS and Izak Nieuwoudt, Koch-Glitsch, Wichita, KS

With evidence of distillation dating back to the 1st century AD, the evolution of the distillation process and mass transfer equipment has a long history. Around 1810, a French inventor was already using sieve trays, and glass balls were in use in the 1820s. Over the last 100 years, Koch-Glitsch has explored various innovations for enhanced trays and packings used in distillation columns. Various mass transfer devices have moved in and out of favor, for example, in the 40s and 50s, trays were the equipment of choice. The 60s and 70s saw a return to packing, with sheet metal structured packing revolutionizing low pressure distillation in the 70s only to have trays rebound in the form of high performance devices in the 1990s. This paper explores the factors that drove various changes in mass transfer equipment with the continual search for improvements in efficiency, capacity, and reliability through the development of the ideal mass transfer device.

(4c) Gamma Scan and CAT-Scan Data From a Distillation Column At Various Loadings

Simon Chambers[1], Lowell Pless[2], Ron Carlson[2] and Michael Schultes[3], (1)Fractionation Research, Inc., Stillwater, OK, (2)Tracerco, Pasadena, TX, (3)Raschig Ring Department, RASCHIG GMBH, Ludwigshafen, Germany

Rare, and possibly first-of-a-kind, gamma scanning of a distillation column produced some interesting and important results. While FRI's 4-foot diameter High Pressure Test Column contained a 10 foot bed of Raschig-Jaeger random packing with a 2-stage liquid distributor and operating with the C6/C7 system at 23.5psia, the column was conventionally scanned several times from top to bottom. The column was baseline scanned when it was dry and was scanned at 45, 79, 100, and 102% of the Maximum Useful Capacity of the packing. Distributor liquid levels derived from the scan data were compared against liquid level data from FRI bubblers. Additionally, CAT-scans were performed at two different column elevations. At each elevation, 81 data points through the bed of packing were recorded: 9 scanning source positions with 9 different detector positions for each source position around the column circumference. Mathematical processing of the CAT-scan data was one of the challenges of the project. Appreciable new knowledge was gained regarding liquid holdup and liquid flow patterns in random packed columns.

There is evidence in the open literature that volatile components (with a high activity coefficient) can concentrate at a liquid interface and allow for very efficient separations beyond the traditional limitation of VLE. The paper discusses the phenomenon and ponders if it can be used for large scale separation applications

Troubleshooting and Solving a Sour Water Stripper Problem

By

Jeremy Ponting,

Fluor Ltd., Farnborough, Surrey, UK

and

Henry Z. Kister, and Richard B. Nielsen

Fluor, Aliso Viejo, CA, USA

Presented at the Distillation Topical Conference, AIChE Spring Meeting, San Antonio, Texas, April, 2013

Troubleshooting and Solving a Sour Water Stripper Problem

Summary

A Sour Water Stripper in a gas processing plant was designed to strip 35 m3/h (154 gpm) of sour water containing approximately 650 ppmw H_2S to a bottoms purity of less than 5 ppmw. Upon start-up, the column could only achieve a feed flow rate of 20 m3/h (88 gpm). Any increase beyond this would initiate flooding. Several theories were proposed to explain the premature flood.

A systematic investigation combining field testing and hydraulic analysis eliminated about half the theories, leaving foaming and damage as the frontrunners. A plant test was performed in which the liquid loads were raised at a constant vapor load. The test showed the flood to be vapor-sensitive and liquid-insensitive. When combined with the hydraulic analysis, the test provided strong evidence that supported the column internals damage theory.

The tower was shut down and inspected, and one of the upper trays was found to be damaged. The damage unsealed one of the upper downcomers. Interference of rising vapor with the descending liquid in the unsealed downcomer was proposed as the explanation to the flood. This explanation is consistent with a vapor-sensitive flood. The damage was repaired and the tower achieved design performance. The most likely cause of the damage was thought to be entry of cold water during start-up.

One lesson learned is the importance of using hot feed during start-up of steam-water systems. If cold feed is used, any sudden feed surges or feed temperature drop may cause rapid condensation near the column inlet and consequent column internals damage. Another lesson learned is that contrary to some literature references, Sour Water Strippers are not always foaming. The column investigated was not adequately designed to handle foaming, yet achieved the design capacity after the damage was repaired

Introduction

A recent book by Lieberman (1) stresses that to combat the environmental crisis that our planet is facing it is essential to make existing units work adequately and efficiently. Equipment that malfunctions or limps along limits production and does not permit good energy and equipment utilization. This, in turn, leads to higher environmental degradation, not to mention plant downtime and lost production costs. Lieberman concludes that good troubleshooting and adequate repair of malfunctioning units is central for efficiently utilizing our resources.

Towers that strip components out of water (so this water can be re-utilized or disposed of) are common in the process industries. In refineries and gas plants, the main components stripped out are hydrogen sulphide and ammonia, and the tower is referred to as a "Sour Water Stripper". The technology for stripping sour water has been established for decades. It was a Fluor engineer, Beychok (2), who wrote the leading text on the subject as far back as 1967. Yet almost five decades later, there are many Sour Water Strippers that do not operate adequately and efficiently. Many "war stories" of Sour Water Strippers that failed to achieve their design objectives are presented in Kister's "Distillation Troubleshooting" book (3).

Here we present another experience with a Sour Water Stripper that initially failed to achieve its design goal due to start-up damage. The problem was correctly identified in a short, systematic and effective investigation that combined field measurements and testing with good hydraulic analysis. The unit was subsequently successfully fixed. Our paper presents the experience and the lessons learned.

Process Description

A Sour Water Stripper unit in a gas processing plant was designed to process 35 m3/h (154 gpm) of sour water containing approximately 650 ppmw H_2S. The column was fed from two sour water storage tanks with an oil skimming facility (Figure 1). The tanks were lined up such that while one tank was receiving sour water from upstream separation units, the other tank was feeding the Sour Water Stripper column (Figure 2).

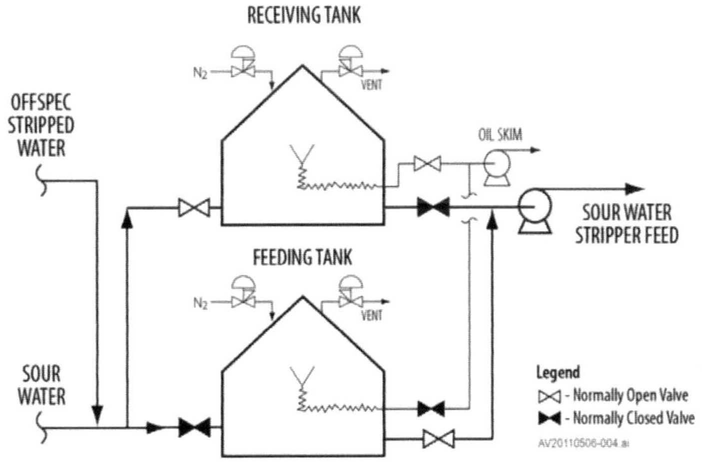

Figure 1. Feed to Sour Water Stripper Unit

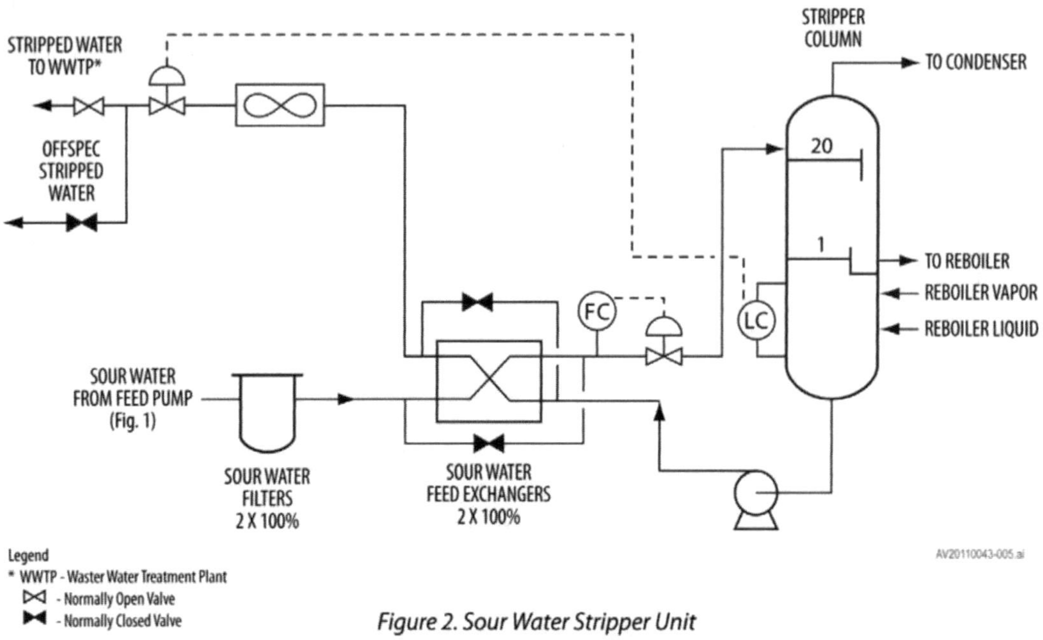

Figure 2. Sour Water Stripper Unit

The column used heat integration in the form of 2 x 100% plate and frame exchangers to exchange heat between the column feed and bottoms (stripped water). The exchanger was designed to preheat the sour water from 55 °C to about 110 °C (131 °F to 230 °F). At 110 °C and at the normal tower inlet pressure the sour water was calculated to be all liquid close to its boiling point.

The column (Figure 3) was equipped with 20 single pass fixed valve trays numbered from the bottom tray up. Feed entered at the top tray. The column was designed to operate with an overhead pressure of 1.4 barg (20 psig) corresponding to a top temperature of 126 C (259 °F). These conditions were sufficient to achieve a bottoms specification of less than 5 ppmw H_2S. The bottoms product was routed to the waste water treatment plant.

Reflux was produced by the tower air condenser, and entered a small reflux drum from which it was pumped back to the tower on drum level control. Uncondensed sour gas was vented from the reflux drum via the tower pressure controller (mounted on the vent line) to the acid gas header. There was no liquid product. A temperature controller at the condenser outlet adjusted the speed of the condenser fan motor to set the split between the gas and reflux.

Reboil to the tower was supplied by a kettle reboiler that was gravity-fed by liquid drawn from a trap-out tray just beneath the bottom tray. Liquid from the kettle overflowed a weir into the kettle draw compartment, from where it gravity-flowed to the column bottom sump. Stripped sour water from the column bottom sump was cooled in the plate and frame feed-bottom exchangers (Figure 2) and then flowed to the waste treatment plant.

There was a bypass that permitted bypassing tower feed around the plate and frame exchangers. If the exchangers were bypassed, the column feed temperature was close to ambient temperature. This no-feed-preheat case set the design surface area of the tower kettle reboiler. In case of reboiler fouling the column was also equipped with a live steam injection line.

During start-up column feed preheat was essential to minimise the temperature difference between the incoming feed and the column top temperature. Column feed preheat minimizes the chance of sudden condensation in the top section of the column, which could lead to tray damage, similar to two cases described in detail in Reference 3.

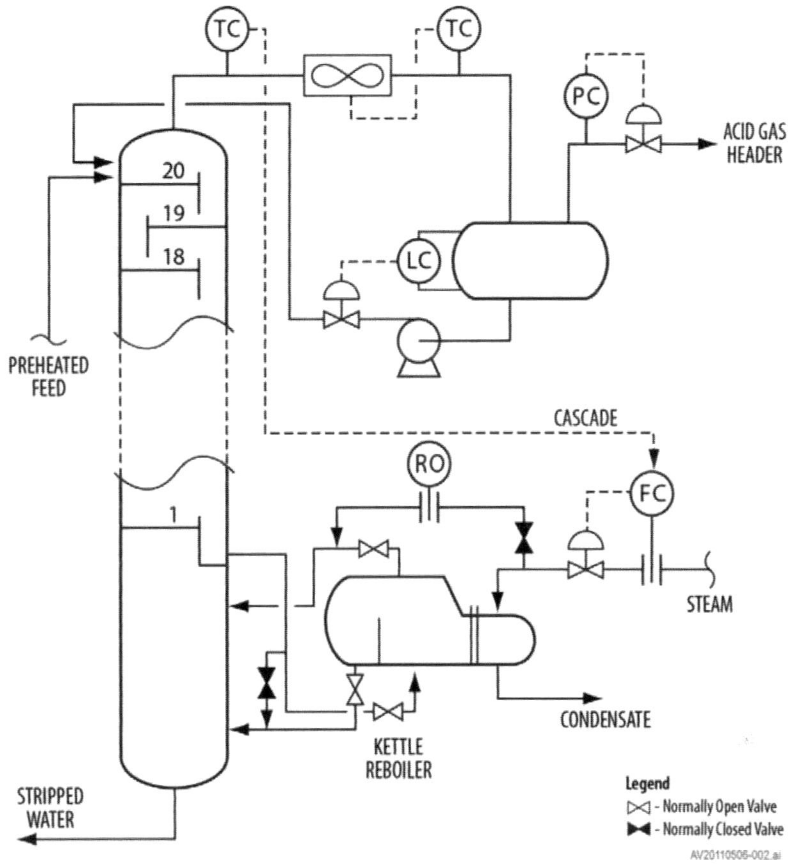

Figure 3. Sour Water Stripper Column and Auxiliaries

Initial Operation

During start-up the column feed flow rate was gradually increased towards the 35 m3/h (154 gpm) design rate with the feed preheat exchangers lined up. At approximately 60% of design feed rate the reflux drum flooded. The reflux level control valve was subsequently driven wide open increasing the reflux to the column. The top temperature fell dramatically and the column pressure fell from its set point of 1.4 barg (20 psig). The only way of bringing the tower back under control was by reducing the column feed rate. The reflux was then brought back into line and the column gradually brought back up to design temperature and pressure.

There were several attempts (with small variations) at increasing the feed to the design flowrate. Each attempt resulted in the same scenario as described above. It made no difference whether the column was operating with the kettle reboiler or with live steam injection.

The operating behaviour and DCS data were closely reviewed in order to establish the root cause of the problem. At feed flow rates below 16 m3/h (70 gpm) the column ran without problems with all parameters within their correct operating range. As the column feed rate approached 20 m3/h (88 gpm), there was a sudden increase in the pressure drop across the trays. Initially the operators observed a rapid increase in pressure drop followed by an equally rapid fall (Figure 4). These peaks became more dramatic as the column feed increased until the reflux drum flooded.

Figure 4

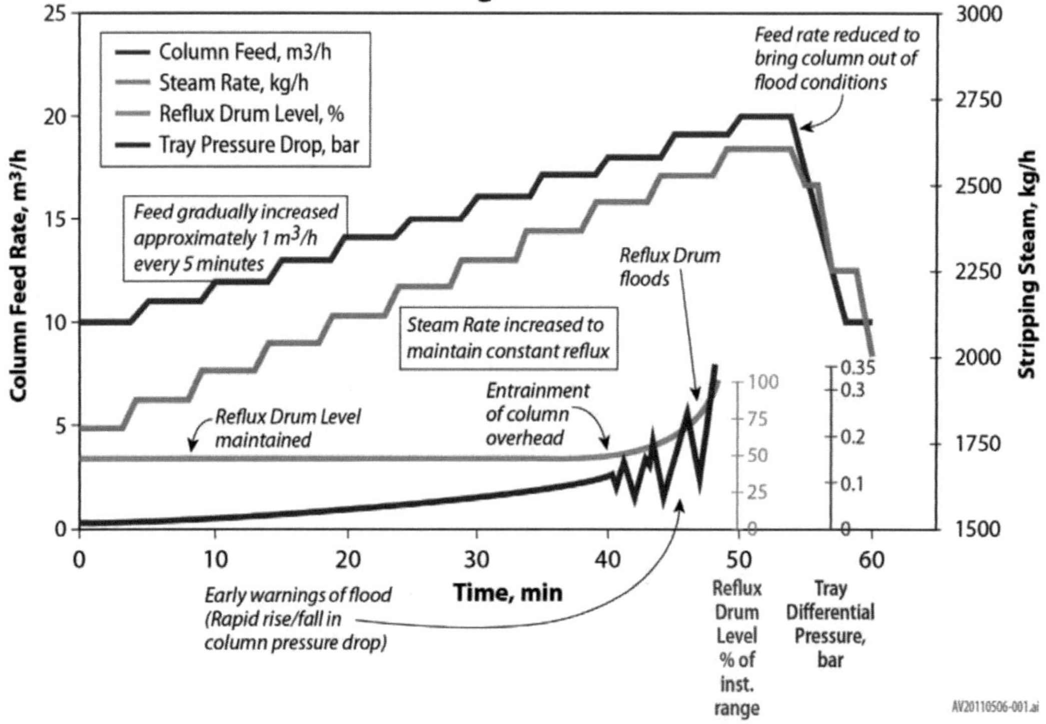

The sudden rise in column pressure drop is a primary symptom of flood. Entrainment from the top of the tower, as evidenced from the overfilling of the reflux drum and the full opening of the reflux control valve, is also a primary symptom of flood. Pressure fluctuations and drop of the overhead temperature are also symptoms of flood. Fluctuations in the column sump level, which followed the pressure drop fluctuations, were also noticed, and provide one additional symptom of flood. It was concluded that above a certain feed flow rate, the column became flooded.

At this stage the cause of flooding was unknown since the column was designed to operate under a much higher feed flow rate.

The tray vendor was consulted by communicating the symptoms of:

i) Rapid increase in tray differential pressure
ii) Fluctuating overhead pressure
iii) Fluctuating column sump level

iv) Reflux drum flooding

The vendor concurred that the cause of these symptoms was flooding. They concluded that it was downcomer flooding.

Theories

As in a typical troubleshooting investigation, many theories were proposed by various members of the start-up team (including the vendor and outsiders) to explain the premature flood. Some could be quickly eliminated, others took more testing.

1. **Tray plugging** is a common experience in refinery Sour Water Strippers (4). However, in this tower there was little presence of fouling compounds such as heavy hydrocarbons and asphaltenes. The feed was filtered down to 2000 micron particles to protect the feed/bottoms exchanger (plate and frame type). If fouling was an issue, the filters would have needed frequent cleaning and this was not the case. In an endeavor to address fouling, the filter screen was changed from 2000 micron to 250 micron. Even after changing, the filters blocked only once, indicating little solids in the feed. Further, the exchanger should have fouled before the trays and this was not observed. Also, the problem was first observed after a short run, much shorter than it takes to plug the trays of a typical Sour Water Stripper even under heavy fouling conditions (e.g., Ref,. 4). Based on these observations, plugging was branded a highly-unlikely root cause for flooding in this column.

2. **Trap-out kettle reboilers** are notoriously troublesome, usually due to poor hydraulic design of the lines leading to or from the kettle, incorrect force balance on the kettle, or liquid weep from the tray feeding the kettle (5).

   In this column, an issue with the kettle was easily ruled out by operating the column with direct injection of stripping steam instead of the kettle reboiler. The flood occurred at the same rates irrespective of whether the reboiler or the stripping steam injection was used.

3. **Incorrect inlet/outlet nozzle designs** and velocities were closely checked. The possibility of flashing at the feed inlet was addressed by reviewing operating data. The preheater was designed to preheat the feed from 55°C to 110°C, but during the initial runs the feed was colder, coming in at 40°C, and the preheat temperature was consequently lower at 90°C - 95 °C (131 °F - 230 °F), so flashing would have been highly unlikely.

4. **Incorrect tray design**. One concern was that the tray open slot area was low, about 7.3% of the active area, and this could result in premature jet flood or excessive downcomer backup. This is addressed further in the next section.

5. **Foaming**. Sour Water Strippers are generally considered foaming systems. The industry's practice for system factors, reflected by the factors listed in Table 14-9 in Perry's Handbook (8th Ed., 2008, p.14-41, Reference 6), lists Sour Water Strippers as "Heavy Foaming" and assigns them a system factor of 0.5 to 0.7.

Further support for possible foaming came from the observation that there was a higher than anticipated hydrocarbon content (about 20,000 ppm oil) in the feed to the sour water feed tanks (Figure 1) and an increased quantity of particulates in the feed compared to the very original design. The filtration specification on the Sour Water Stripper feed filter was relaxed after the vendor of the feed/bottoms plate and frame exchanger stated that a 2000 micron filtration level was sufficient to prevent excessive fouling without the requirement for the operator to clean this filter frequently. The original specification on the filter of 5 microns was expected to result in frequent cleaning of this filter. Particulates and heavy hydrocarbons are well-known to promote foaming (3, 7).

Our experience in general has been that most Sour Water Strippers do not foam. The times we experienced foaming in Sour Water Strippers were when significant oil entered the system. Solids may also catalyze foams, but we are not aware of cause-and-effect experiences in which solids by themselves catalyzed foams in Sour Water Strippers. Among more than 60 case studies of foaming experiences surveyed from the open literature and abstracted in Kister's "*Distillation Troubleshooting*" book (3), not one came from a Sour Water Stripper.

The theory of foaming emerged as a front-runner, having significant support, which was by no means conclusive.

6. **An installation error**. There was not much evidence supporting that an installation error was responsible for flooding. The tower was well-inspected prior to start-up and no problems were found. Nonetheless, there were some concerns and we therefore kept on installation errors on the troubleshooting list. Specifically, there was a suspicion of incorrect downcomer clearance (which could be manually adjusted (with a tolerance of +/- 2mm)). The inspection records were checked and the inspection report only indicated that the recommended clearances had been met with but with no values recorded.

7. **Tray damage**. Damage is not uncommon in steam-water operations (8), such as in Sour Water Strippers. Steam rapidly condenses when it comes into contact with a large quantity of cold water. The pressure in the contact zone plunges and induces a rush of steam from above and from below into the contact zone. This rush imposes a large downward-acting differential pressure across the upper trays, and a large upward-acting differential pressure across the lower trays. Tray damage may result if any of these differential pressures exceeds the allowable stress of the trays, Trays above the contact zone will be deformed downwards, and those below the contact zone will be deformed upwards. This damage has been previously reported in Sour Water Strippers (3).

   The above theories hinge on cold water entering the stripper. In this column, the cold water needed to enter during start-up, because the column never reached the design feed rates. The feed to the tower was hot, 90 to 95 °C (194 to 203 °F). At these feed temperatures damage is unlikely. At start-up, live steam was first brought in at a slow rate, the reflux and condenser were commissioned, and then the tower bottom was circulated through the preheater back to storage. When feed eventually entered the tower it was hot.

Nonetheless, there were short periods in which the preheater was bypassed, dropping the tower feed temperature from 90°C to 40°C (194 °F to 104 °F) within a few minutes. Stepping up the feed rate at the lower temperature, or a sudden drop of the feed temperature (e.g., by opening the bypass valve too fast), or a sudden drop in steam flow rate while the feed was cold, could have led to tray damage. Also, there were some teething problems with the stripping steam piping and with leaking gaskets on the plate and frame exchangers and the column went through a number of start-up/shutdown cycles as well as some feed-bottom exchanger switches. Each of these could have led to feed changes and dropping of feed temperature over short periods, possibly resulting in the type of damage described above. So while there was no reason to suspect damage, it would have been very conceivable.

With the first three theories above ruled out or considered highly unlikely, four theories remained, with foaming being a slight front-runner, and the installation error bottoming the list. Further work on narrowing-in on the root cause, the most likely explanation, and the path forward are described below.

Hydraulic Analysis

Our hydraulic analysis looked at the design case as well as the operation bottleneck, and considered the possibility of foaming. Our calculations were based on methods by FRI and Perry's Handbook (6). We compared our findings to values from the vendor rating sheets.

A. At Design Loads

At design loads, the vendor calculated 76% percent of jet flood, while we calculated 70 to 76% of jet flood from the FRI valve tray and equivalent sieve tray correlations as well as from published methods (6). The numbers are the same, but there was a huge difference: the vendor's calculation applies the 0.7 system factor specified in the column data sheet. Our calculations did not apply a system factor. If we applied the specified system factor to our calculations, the trays would rate at 100% to 109% of jet flood.

The FRI jet flood correlation is known as the best in the field, gives predictions within 10-15% accuracy, and concurs with published methods with which we have had excellent experience. Also, the FRI calculations as well as the published method reflect the industry's experience that sieve and valve trays of equivalent open area give much the same jet flood capacity. We concluded that our calculations are closer to reality than the vendor's.

The practical implication of this discrepancy is that the current design is suitable for a non-foaming system, but does not follow the industry's practice of using a system factor of 0.5 to 0.7 for Sour Water Strippers (6).

The downcomer backup we calculated is 264 mm (10.4") compared to the 199 mm (7.8 ") calculated by the vendor. With a total height of the downcomer plus outlet weir of 660 mm (26.0"), both numbers are adequate for a non-foaming system. The larger of the two numbers (264 mm) is too tall for a foaming system for which an aeration factor of 0.4 (or even less) has been recommended (6).

The downcomer inlet velocity is simply the downcomer inlet clear liquid flow rate divided by the downcomer inlet (top) area. Our calculation gave a downcomer inlet velocity of 0.122 m/s

(0.4 ft/s). This velocity is adequate for a non-foaming system. For a foaming system, the maximum recommended downcomer inlet velocity is 0.061 - 0.076 m/s (0.20 - 0.25 ft/s) (6). It can be concluded that the downcomer design is adequate for a non-foaming system, but grossly inadequate for a foaming system. The downcomer inlet areas are too small to accommodate the liquid rate.

Our design case evaluation therefore confirmed that the trays/downcomers were not adequately designed for foaming conditions. The downcomer velocities are far too high and the system factor application in the design is questionable.

B. At Operating Loads

We also hydraulically rated the column based on operating data at the loads just before flooding was observed. It is important to recognize that this occurred at about half the design loads. Our calculations show that for these conditions, the trays operated at 37% of jet flood, and at 170 mm (6.7") downcomer backup, both of which are very far from the limit even for a foaming system. The only parameter close to a limit for a foaming system was the downcomer inlet velocity, which was 0.075 m/s (0.25 ft/s), right at the upper end of the recommended range for foaming systems. So based on this calculation, it is possible that foaming bottlenecked the column and caused the premature flood.

This finding kept the foaming theory alive, possibly providing slight additional support. Based on the hydraulic analysis, if foaming were the bottleneck, it would limit the tower by a downcomer choke ("downcomer velocity flood") mechanism. At the operating conditions where the column flooded, no other flood mechanism was near a limit, even under foaming conditions.

The operating load analysis also ruled out the possibility of an incorrect tray design (in the absence of foaming) causing the premature flood.

Foaming or No Foaming?

We were left with three theories:

- Foaming
- Installation error (less likely)
- Damage

The solution for an installation error and damage would have been much the same: shutting down the tower and fixing the problem. The solution for foaming would have been redesign and reinstallation of all (or at least some) of the 20 trays. So there was a need for further differentiation between foaming and the other theories.

Gamma scanning is a common column diagnostic technique that can provide useful insight into tray damage and/or foaming. However, gamma scanning would have required transporting radioactive materials across international borders and security systems, which would have been a forbiddingly expensive exercise. A column shutdown was a more economical option here. Gamma scanning was therefore ruled out.

One technique invaluable for testing theories is the vapor/ liquid sensitivity test. A limitation can be sensitive to vapor load, liquid load, or both. In a flood test, feed (or reflux) flow rate is often raised while the column products are kept on-spec until flood occurs. Raising the feed liquid or reflux flow raises the column liquid load, but keeping the products on-spec increases reboil or stripping steam rate. When flood occurs, there is no telling whether the flood is due to the initial increase in liquid load or the subsequent increase in vapor load due to the higher reboil. This test therefore cannot tell whether a limit is due to excessive vapor load or excessive liquid load.

In a liquid-sensitivity test, liquid feed or reflux flow rate is raised at constant boilup. If flood occurs, it is liquid-sensitive, since the vapor load is unchanged. During such a test the bottom product usually goes off spec on lights, as the increases in feed or reflux flow rate are unmatched by increases in boilup. In a vapor sensitivity test, boilup is raised at a constant liquid load. If flood occurs, it is vapor-sensitive. During a vapor-sensitivity test, the top product usually goes off-spec on heavies because the increase in vapor load is unmatched by higher reflux. So such tests can be performed only if off-spec operation can be tolerated for the duration of the test, typically 2-3 hours.

To test the validity of the remaining three theories, we decided to proceed with a "High Liquid Test" in which we raised the stripper feed flow rate to near the design rates (35 m3/h or 154 gpm)) without largely increasing the steam flow rate. The test was accompanied by a hydraulic evaluation based on the field measurements. A comparison of the test with the operation is provided in Table 1.

Table 1. High-Liquid Rate Test

	OPERATION	HIGH-LIQUID TEST
Feed Flow Rate, m^3/h	20	35
Reflux Flow Rate, m^3/h	1.5	0
Steam Flow Rate, T/H	2.8	3.1
Feed Temperature, °C	90 - 95	70
Bottom Temperature, °C	128	115
ppm H_2S in Bottom	<10	100 - 200
CALCULATED VARIABLE		
Internal Liquid Load, m^3/h	27	41
Internal Vapor Load, m^3/s	0.54	0.60
% Jet Flood	37	48
Downcomer Backup, mm	170	210
Downcomer Inlet Velocity, m/s	0.075	0.115

AV20110043-001.ai

Compared to the operating case, the tower internal liquid load in the High Liquid Test was 53% higher, and the internal vapor load was 13% higher. During the test, the tower operated under non-flooded conditions for 10 hours. The tower then flooded. The tower was taken out of flood by a large reduction in steam flow rate and only a 14% reduction in feed flow rate. This verifies that the flood was insensitive to an increase in liquid rate, but sensitive to an increase in vapor rate. This strongly argues against a flood by excessive downcomer velocity, which is always primarily a liquid flood.

Our hydraulic evaluation shows that at the High Liquid Test, the trays operated at 48% of jet flood, and at 210 mm (8.3") downcomer backup, both of which are far from the limit even for a foaming system. The only parameter that exceeded the limit for a foaming system was the downcomer inlet velocity. The velocity of 0.115 m/s (0.38 ft/s) was well in excess of the maximum recommended 0.061 - 0.076 m/s (0.20 - 0.25 ft/s) (for a foaming system (above).

The hydraulic evaluation above suggests that at the operating case, the only hydraulic limit that was approached under foaming conditions is an excessive downcomer velocity. The fact that in the test the liquid rate could have been increased by 53% beyond this limit without flooding strongly argues against a foaming bottleneck.

So the High Liquid Test produced key evidence denying the likelihood of foaming causing the bottleneck in the stripper. Foam-floods due to excessive downcomer velocities are liquid floods, not vapor floods. And the trial produced strong evidence to argue against a liquid flood.

Mysteries Explained

This left us with the damage theory in the forefront, with an installation issue as a second possibility. Both were addressed by shutting down the column and performing a full inspection.

The inspection found that tray 18 (two trays beneath the top tray and column feed) was badly damaged (Figure 5). The tray was bowed downwards and had become detached from its support clips on all 3 sides. Other trays in the tower top section were also bowed but remained in place.

All downcomer clearances were checked and recorded. Without exception all the clearances were within the manufacturer's tolerance.

It was observed that the seal from the Tray 19 downcomer had been lost as the side section of tray 18 (underneath the tray 19 downcomer) was also bent downwards. This loss of seal offered an explanation to the observed flood. It was postulated that the steam flow from tray 17 in the top section was routed preferentially through the tray 19 downcomer rather than through the fixed valves of tray 19. This explained why the top section of the column quickly flooded (evident from a sudden increase in tray differential pressure) and flooded the reflux drum.

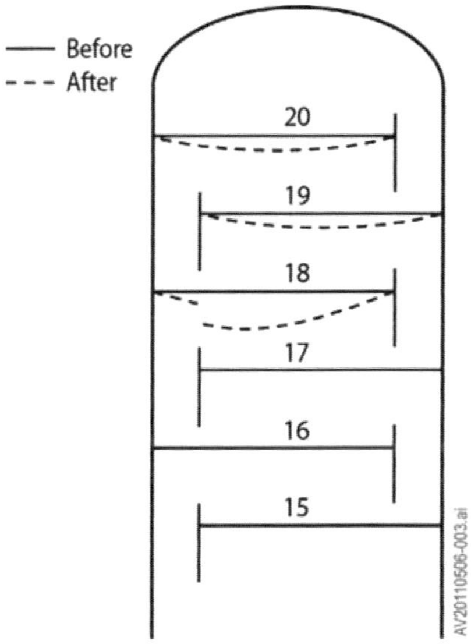

Figure 5. Observed Tray Damage

The observation that the flood was vapor-sensitive and liquid-insensitive is very consistent with a sealing-loss flood theory. When the vapor rate is high enough, it will break through the downcomer's broken seal and severely interfere with liquid drainage, inducing flood on the tray above. In this case, the small hole area (7.3% of the active area) will aggravate the sealing problem by resisting flow through the tray holes and acting to divert vapor to the downcomer. Sealing loss leads to a vapor flood, just as observed in the High Liquid Test.

The damage was repaired and the column returned to service. During a performance test, it has had no problem handling the feed flow rate of 35 m3/h (154 gpm) at the corresponding steam flow rate with the reboiler online and is producing a water bottom stream of <5 ppmw, as per design

Lessons Learnt

The key for avoiding damage in water strippers is to keep the feed hot. This lesson was also advocated in two other cases of column internals damage in Sour Water Strippers reported in Reference 3. Awareness of and understanding the mechanism of damage when cold feed enters a column containing a steam-water system may lead to many additional means of minimizing its incidence, as advocated in one of the two case studies reported in Reference 3.

The case confirms our experiences that Sour Water Strippers generally do not foam. As stated earlier, our design case evaluation concluded that the trays/downcomers were not adequately designed for foaming conditions. The downcomer velocities are far too high and the system factor application in the design is questionable. Yet, now that the damage has been fixed, the tower has been operating well above design rates, suggesting absence of significant foaming in this column.

References

1. Lieberman, N. P., “Process Engineering for a Small Planet”, Wiley, NJ, 2010

2. Beychok, M. R., “Aqueous Wastes”, John Wiley and Sons, London, 1967.

3. Kister, H.Z., “Distillation Troubleshooting”, Wiley-Interscience, NJ, 2006.

4. Hauser, R., and R. T. Kirkey, "Refinery Tests Demonstrate Fixed Valve Trays Improve Performance in Sour Water Stripper", in "Distillation 2003: on the Path to High Capacity, Efficient Splits", Topical Conference Proceedings, p. 163, AIChE Spring National Meeting, New Orleans, LA, March 31-April 3, 2003.

5. The FRI Design Practices Committee “Reboiler Circuits for Trayed Columns”, Chemical Engineering, p. 26, January 2011.

6. Kister, H. Z., P. M. Mathias, D. E. Steinmeyer, W. R. Penney, B. B. Crocker, and J. R. Fair, “Equipment for Distillation, Gas Absorption, Phase Dispersion and Phase Separation”, in D. W. Green and R. H. Perry “Perry’s Chemical Engineers’ Handbook” 8th Ed., Section 14, McGraw-Hill, New York, 2008

7. Lieberman, N. P., "Troubleshooting Process Operations", 3rd Ed., PennWell Books, Tulsa, OK, 1991.

8. Kister, H. Z. “Distillation Operation”, McGraw-Hill, NY, 1990.

NAGAOKA PACKING "Super H pack"

Tadayoshi Nagaoka
Laboratory of structured packing for distillation

Abstract

In the distillation and absorption processes of chemical plant, the structured packing was introduced instead of historical random packing and increased separation efficiency.

The structured packing, however, could not increase treating volume. When treating volume increases, speed of upward vapor also increases. In this circumstance, downward liquid, specially free liquid droplet at the bottom of distributor and structured packing can not go down due to high speed of upward vapor, and is accumulated in the column, resulting in start of flooding. These phenomena came from layout of all internals, because of separated layout, distributor, and structured packing.(Fig.2)

NAGAOKA PACKING "Super H pack" has connected a distributor, structured packing and collector by changing layout in the column. Super H pack realized increased speed operation of vapor and reached two-times larger treatment volume with high efficient separation.(Fig.1)

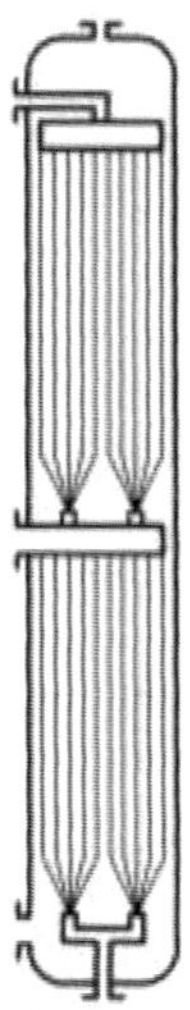

Fig. 1 NAGAOKA PACKING Super H Pack

Fig. 2 Conventional Structured Packing

Structure and feature

NAGAOKA PACKING "Super H pack" has five main features of unique structure as follows.(Fig.1)

1. Epoch-making connected structure with distributor, structured packing and collector pushes away the flooding point because of avoiding free liquid-droplet at the bottom of distributor and structured packing and expand operating range up to a higher F-Factor which enables high speed operation.
2. Super H pack is constructed with multiple vertical capillary wires which run with equal interval and in parallel in the full length.
   Multiple linear liquid flow along the vertical capillary wires and upward organized vapor flow produce efficient vapor-liquid counter-contact.
   Downward liquid flow which is led by the capillary wires can go down against high speed of upward vapor. Therefore high efficient separation and high speed operation which means larger treatment volume are achieved simultaneously.
3. Trough distributor is overflow type and has many equally spaced liquid ports and all capillary wires of structured packing hang from each of all liquid ports. Number of liquid drip points of distributor over 50,000/m2 can be available.
   Combination of many equally spaced liquid ports and hanging capillary wires produces uniform liquid distribution over the entire column.
4. Many equal liquid distribution starts at top of structured packing.
   Elimination of the mal-distribution results from the constitution of vertical and parallel wires in the full length.
   Therefore minimum specific surface area of Super H pack produces high efficient separation.
5. Constitution of vertical and parallel wires and minimum specific surface area achieve minimum pressure drop in operation.

Test result of distillation and valuation

Vacuum batch distillation.
(Photo 1)
Head pressure 100mmHg.
Fluid; Chlorobenzene and Ethyl benzene.
Separation efficiency ;NTS/m (Fig.3)
Pressure drop; mbar/NTS (Fig.4)

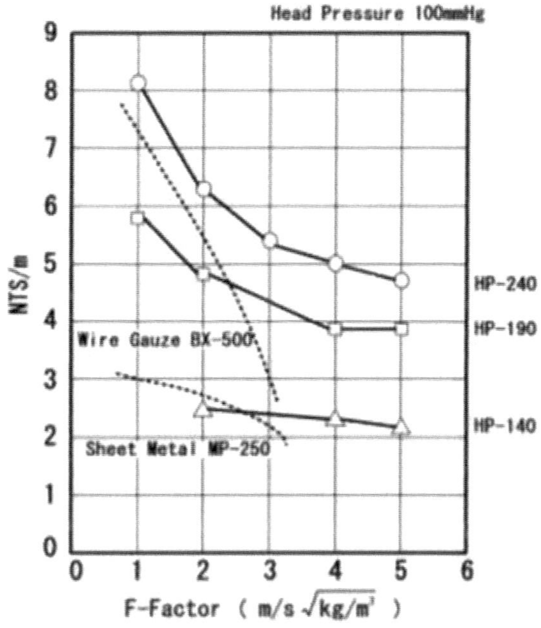

Fig. 3 Number of Theoretical Stages per Metre /F-Factor

Photo1

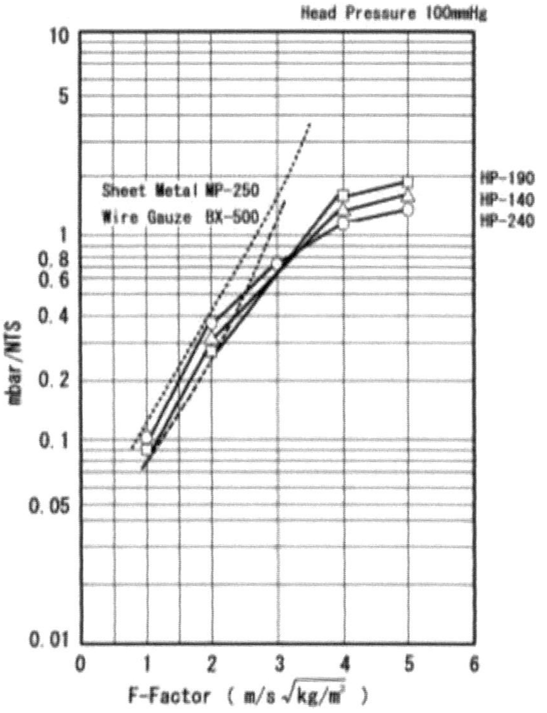

Fig. 4 Pressure Drop per NTS (mbar /NTS)

Indication	Remarks	Specific Surface Area m2/m3
HP-140	Super H pack	140m2/m3
HP-190	Super H pack	190m2/m3
HP-240	Super H pack	240m2/m3
MP-250	Conventional Sheet Metal	250m2/m3
BX-500	Conventional Wire Gauze	500m2/m3

◆ Comparison of vapor velocity, F-Factor(Load) m/s√kg/m3 (Fig.3 and 4)

Super H pack; -------------------------Max. F-Factor=5

Conventional structured packing

Sheet Metal and Wire Gauze; Max. F-Factor=3

After F-Factor=3,

Conventional structured packing----Pressure drop going up sharply
Flooding start. Impossible operation

Super H pack------------------------------Pressure drop going up slowly
Stabilizing operation until F-Factor=5.

◆ Comparison of separation efficiency NTS/m (Fig.3)

After F-Factor=3

Conventional structured packing------Separation efficiency going down sharply
Impossible operation

Super H pack------------------------------Separation efficiency going down slowly
Stabilizing operation until F-Factor=5

◆ Stabilizing operation of Super H pack in wide range is a main feature which separation efficiency(Fig.3) and pressure drop(Fig.4) graphs show .

◆ Separation efficiency of Super H pack in minimum specific surface area is superior to conventional structured packing with large specific surface area which show in Fig.3.

Conclusion

Super H pack is epoch-making structured packing which realized two-times larger treatment volume with high efficient separation simultaneously in condition of minimal specific surface area and minimal pressure drop.

Above features are proved in the distillation tests

In the revamping case, separation efficiency will be raised and treating volume will be increased two-times.

In the new plant case, column size will be reduced and cost-down will be realized in initial stage.

Column height-----------Shortening by higher NTS/m

Column diameter--------Shortening by higher F-Factor/NTS

Running cost down will be added by higher separation efficiency, larger treating volume and lower pressure drop.

Large economic impact of initial stage and running cost down will be expected in the actual chemical plant.

Super H pack which is of a simple structure with multiple vertical and parallel wires will be applicable to processing of fluid which will condense in the column or processing of dirty materials.

Application field of Super H pack are not only distillation but also absorption, stripping, cooling and aeration etc.

Especially, application of Super H pack for absorption and stripping of CO_2 which will contribute to the earth will be highly expected.

.

(28a) Mitigating Chloride Content On Vacuum Diesel

Ana Lucia M. T. L. Malheiros[1], **Andre C. Bellote**[2], Arlan Lucas de Souza[1], Braz Marino Zanutto[3], Diego Telles Fernandes[1], Glaucia A. Torres[4], Helton Siqueira Maciel[1] and **Silvia Waintraub**[1], (1)CENPES – Petrobras Research and Development Center, Petrobras, Rio de Janeiro, Brazil, (2)REPAR - Presidente Getulio Vargas Refinery, Petrobras, Araucaria/PR, Brazil, (3)SIX, Petrobras, Brazil, (4)Headquarters, Petrobras, Rio de Janeiro, Brazil

Traditionally, diesel is recovered as direct product from the atmospheric tower. With the growing demand and tighter specifications, vacuum diesel recovery has becoming an attractive complementary production strategy for many refiners.

The chloride content tends to be higher in vacuum diesel than in the direct atmospheric diesel. Although in low concentration (ppm level), chloride can drastically affect the downstream diesel hydrotreating unit, including plugging, corrosion and costly unscheduled shutdown.

Pilot and field tests have been conducted to determine how chloride content in vacuum diesel is related to vacuum tower operating conditions.

The results show that a higher fractionation reflux rate, besides the expected sulphur and end boiling point reductions, has the benefit of mitigating the chloride content in vacuum diesel.

(28b) Side Reboiler Redesign Improves Alky Isostripper Operation

Daryl Hanson, Valero Energy Corporation, San Antonio, TX and Kannan Subramanian, Valero Texas City Refinery, Texas City, TX

In 2004, the Valero Texas City Alkylate Isostripper was revamped with high capacity trays to improve the tower capacity. Results from the revamp were less than expected and the tower was revamped again in 2006. This revamp also proved to be less than successful.

In 2007, troubleshooting was initiated in order to locate and fix the problems. It was thought that the trays were not the problem, but a detailed troubleshooting strategy was put together. Troubleshooting revealed that the tower was flooding due to design issues in the side reboiler circuit. There were several design issues with the reboiler circuit design that resulted in limitations and premature tower flooding.

Reboiler draw tray and side reboiler circuit modifications were initiated to improve the tower capacity. After almost 4 years of operation, post-revamp data indicates that the tower is meeting all expectations. Changes increased the side reboiler duty and improved product quality, while meeting all the capacity expectations for the revamp.

Troubleshooting jet-fuel production on crude distillation unit in Pemex Cadereyta Refinery

Guillermo Trujillo Rendón
Luis Alberto Quintero Rivera (*)
Jorge Arturo Recio Espinosa
Ricardo Alberto Loubet Guzmán

Pemex Refinación, Bahía de San Hipólito No. 56, Col. Petróleos Mexicanos, Delegación Miguel Hidalgo, D.F., 11311, México

(*) Corresponding author. E-mail: luis.alberto.quintero@pemex.com

---

Troubleshooting jet-fuel production on crude distillation unit in Pemex Cadereyta Refinery

Guillermo Trujillo Rendón
Luis Alberto Quintero Rivera (*)
Jorge Arturo Recio Espinosa
Ricardo Alberto Loubet Guzmán

Pemex Refinación, Bahía de San Hipólito No. 56, Col. Petróleos Mexicanos, Delegación Miguel Hidalgo, D.F., 11311, México

(*) Corresponding Author. E-mail: luis.alberto.quintero@pemex.com

Abstract

This paper shows the troubleshooting and analysis that led to a successful modification on the jet-fuel draw-off tray of a crude distillation tower. The modification solved a problem of yield reduction and difficulties to maintain product specifications.

The problem was reviewed qualitatively and quantitatively using different methods proposed by famous authors in this field like Kister and Lieberman. We used process simulators, heat and material balances, mechanical drawings of tower internal construction, and flow sketches through sections inside the tower. By the use of these tools and the recommended bibliography we solved the problem and evaluated the modification after plant turndown.

The modification of the jet-fuel chimney tray allowed an enhanced performance, a more stable operation for tower profile control, and an increase in the naphtha yield by 1.2% and jet-fuel by 8.3%. All this led to an improvement on plant preheat train thermal stability which gave energy savings of around 5.4 MMUSD per year.

Due to the improved crude preheat resulting from the modification, a better desalting is achieved, the pre-flash tower takes more lights off the crude, which allow a better pressure control on atmospheric tower. A large reduction in fuel gas in fired heaters was achieved because of the increase in heater inlet temperature.

Yield increase of naphtha allows earnings of around 54.3 MMUSD per year. In addition, a potential jet-fuel earning could be achieved because of the production increase, which represents 55.3 MMUSD per year.

Key words: *chimney tray, heat balance, flow pattern analysis, draw-off stability, jet-fuel yield improvement.*

Introduction

Pemex Cadereyta Refinery is located at Nuevo Leon, Mexico. Its facilities are strategically located to supply the north side of the country fuel requirements.

The Refinery has a 270,000 BPD crude oil total capacity divided into two crude distillation units of 120,000 BPD and 150,000 BPD respectively. Formerly Combinada 1 plant (CDU No. 1) had a capacity of 100,000 BPD. During the revamp of the Refinery in 2000 its capacity was increased to 120,000 BPD.

The Combinada 1 plant has an atmospheric distillation tower (101-CO). Before the revamp, the tower contained 36 two-pass trays. Products from this tower include straight-run naphtha, jet-fuel, kerosene, diesel, atmospheric gas oil and atmospheric residue which is sent to a vacuum distillation tower. Side products draws are made via four total draw chimney trays. Each of these side draws is split into (i) a side-stream product which feeds a steam stripper, and (ii) a reflux to the tray below which is pumped back the tower after cooling, rather than overflowing internally from the draw tray to the tray below. This is a kind of "Type R" column as Watkins classified.

According to Watkins this tower can be regarded as five towers operating in series, where the bottoms of each tower splits into a product stream and a reflux to the tower below for product rectification. (1)

Figure 1 shows a simple sketch of the tower arrangement. The main modifications made during the revamp were the installation of high efficiency trays in section 1 and section 2, and the change from trays to packing on sections 3, 4 and 5. The bottoms stripping section remains the same with 7 two-pass trays.

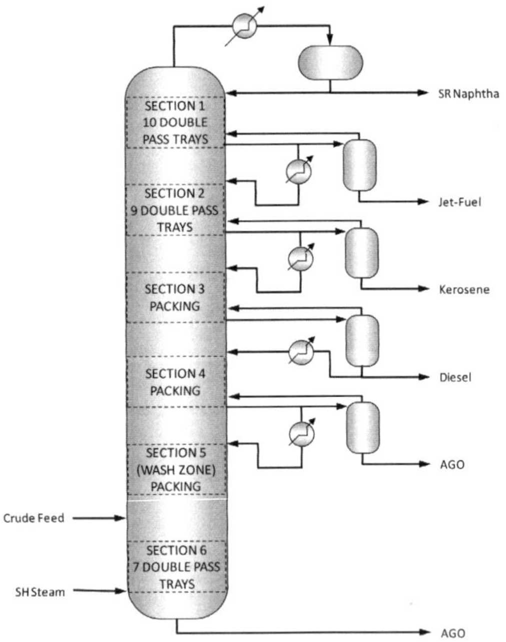

Figure 1. General sketch after revamp.

Following the modifications described above, the tower experienced problems controlling quality specifications and quantity of products draws.

The main problem was due to the level on the jet-fuel chimney tray, which couldn't be maintained. When the level dropped, a major portion of heavier vapors passes through to the naphtha section above due to loss on the reflux flow control. The jet-fuel level control valve, which sets the product draw rate, closes to restore the level on the tray. This product sub-cools because of the long residence time that it remains on the pipe to the stripper, and therefore couldn't be stripped, so light material remained in the jet-fuel product affecting flash point specification.

In order to improve the tower control and operation, several modifications were made to the top section trays during plant turndowns after the revamp. Main modifications are described below:

(a) Central down-comers have a width distance of 10 inches. Anti-jump baffles were installed in order to avoid premature flooding which may cause tray damage.

(b) The bottom tray of Section 1 did not have a seal pan. This allowed vapor up-flow through the downcomer which would affect tray performance. In order to solve the problem a seal pan was installed as recommended. (2), (3)

Even after these modifications were carried out, the instability and loss of level on chimney tray continued, so the quality, specification and yield problems continued.

Figure 2 shows operating charts illustrating the problem. Reflux rates would drop from as high as 140 m^3/h to 20 m^3/h, while product draws would drop from 70 m^3/h to zero. These changes were accompanied by temperature changes as high as 30-40° C.

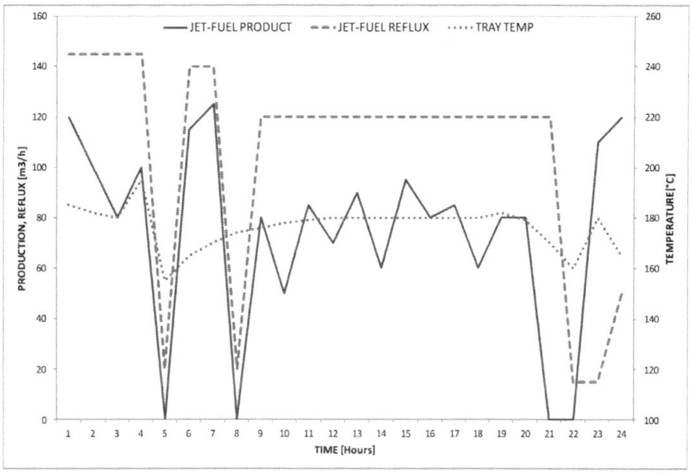

Figure 2. Jet-fuel draw-off and reflux, typical operating chart.

The situation described affected the jet-fuel yield which was lower than expected around 5.3%, equivalent to 42 m^3/h. The temperature profile swings also and considerably affected the crude pre-heat exchange, as illustrated in Figure 3. This diagram compares the actual to expected temperature profile in the tower, showing lower than expected temperatures in most of the tower.

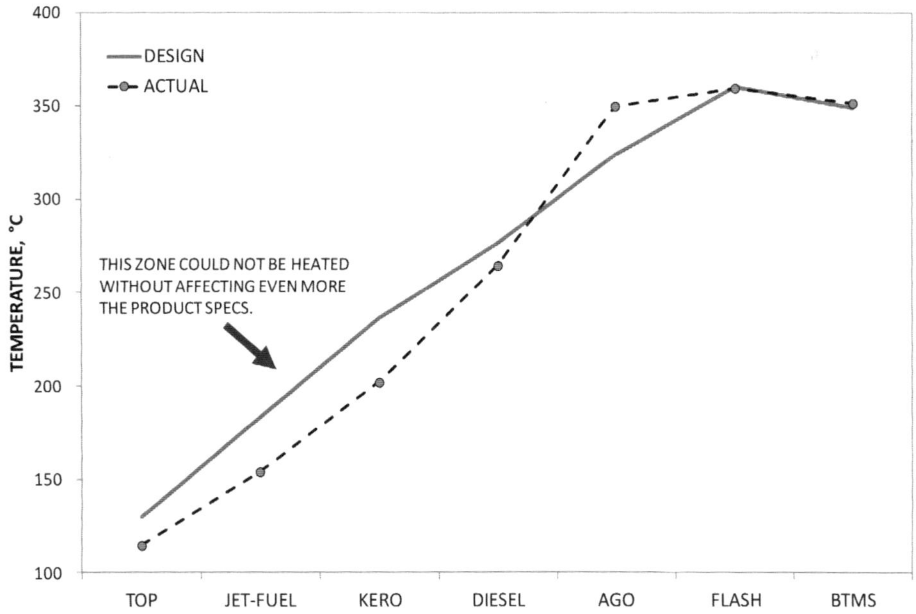

Figure 3. Performance of atmospheric distillation tower before modifications, July 2010 average data.

Problem analysis and initial theory

According to field operators, jet-fuel draw-off losses occurred continuously since the tower was revamped. The only way they could maintain liquid level on chimney tray was to substantially raise top and jet-fuel reflux. However, this reduced the jet-fuel production rate, since all the jet-fuel that has been drawn off the tower was now used as reflux. The yield of the very valuable jet fuel product was reduced when operating this way.

Figure 4 shows the control loops for the draw-off and reflux. During upsets operators increases the set-point of the reflux which disturbed even more the control of the tower. The draw-off control valve fully closed in order not to lose the tray level. The no-flow product cooled because of liquid stagnation. When the draw-off stopped, the level slowly returns to its set-point value. This allowed the operators to control the tower again, but the temperature profile was cooler than expected.

Design rate for reflux is around 115 m^3/h for a 120,000 BPD design unit feed rate. When level loss occurred, reflux was increased to values around 150 m^3/h, even with a unit feed rate of 90,000 BPD. This represents a 75% over what's really required.

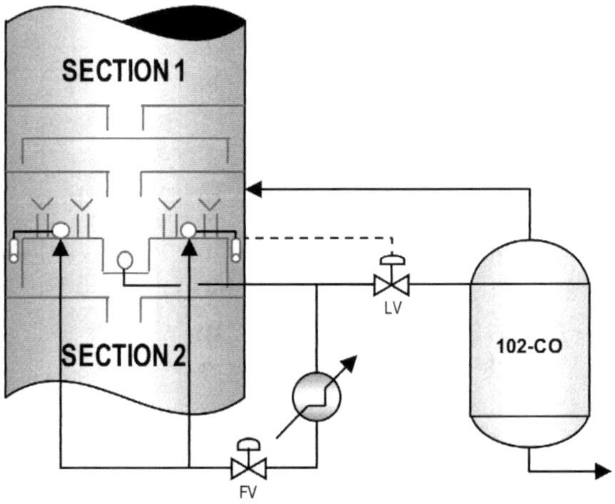

Figure 4. Jet-fuel draw-off and reflux control philosophy.

It was theorized that the problem may have been caused by an obstruction on the jet-fuel reflux feed to the downcomer. Figure 5 shows a detail diagram of the draw-off section and reflux feed to tower.

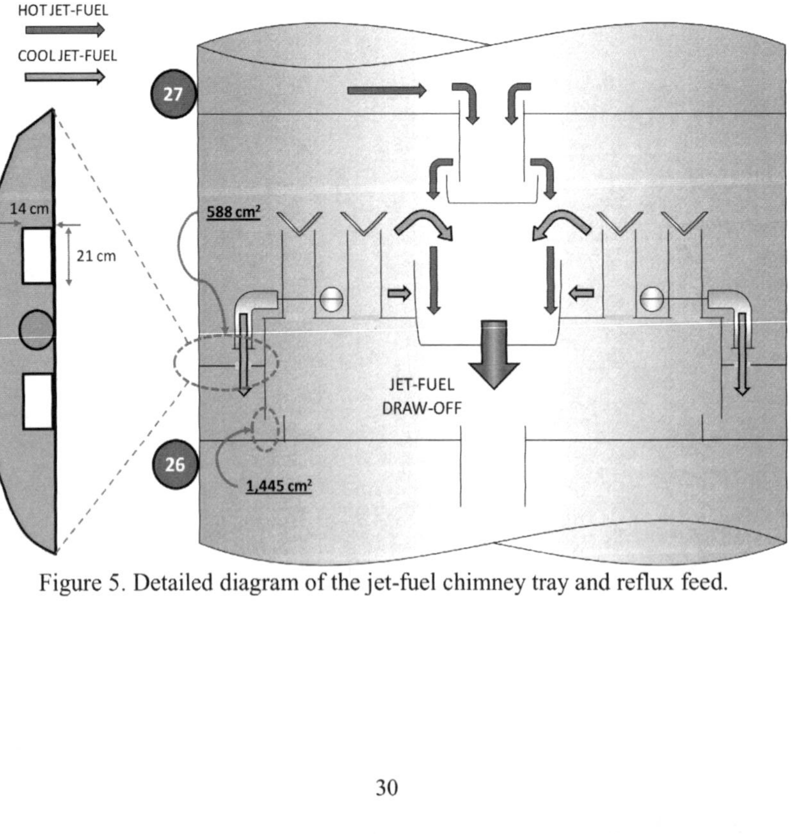

Figure 5. Detailed diagram of the jet-fuel chimney tray and reflux feed.

Figure 5 shows that the jet-fuel draw chimney tray is divided into two compartments. The center compartment received all the hot liquid from tray 27 above. From there, the hot liquid was drawn to the stripper and to the reflux, as described earlier. This is a total draw sump, whose level was controlled by the control valve on the stripper feed (see Figure 4). The second compartment received the cooled reflux going to the jet fuel–kerosene separation section below. This compartment operated with no level control, with the cold reflux return flowing into the pipes which sent it to tray 26 below. The way it was meant to work, there will be no mixing between the hot and cold liquids.

The theory above argued that there was an obstruction in the path of the cold reflux return into the downcomer. Figure 5 shows that the cold reflux is fed by the return pipes onto an impingement plate in each downcomer. Each impingement plate had two 21 cm x 14 cm openings through which the liquid descended onto tray 26 (Figure 6). The theory argued that the free area on the impingement plate may be insufficient for the reflux rate, causing liquid to back up and flood the downcomer above the plates. The backed up liquid would back up on the chimney tray, eventually overflowing into the hot compartment. This would mix cool and hot jet-fuel in the draw-off sump, which will sub-cool the stripper feed. The subcooled feed would cause the stripper to lose efficiency. (4), (5) Figure 5 illustrates the flow patterns theorized to occur.

Figure 6. Chimney tray lateral down-comer. Jet-fuel pump back feed.

Note that in order for the steam stripper to work correctly, product at its bubble point should be fed. Steam reduces the partial pressure of the product which restores liquid-vapor equilibrium. Necessary energy for this action comes directly from the product not the steam, therefore a diminution of the jet-fuel temperature fed to the steam stripper promotes a loss of efficiency of stripper. This loss of performance doesn't allow the stripper to eliminate light material from jet-fuel and flash point of product if affected. (5), (6)

The above theory suggests a jet-fuel contamination with reflux. This theory explained the subcooling of the stripper feed, but did not explain the reduction in product rate. This theory actually suggested that the product flow rate should increase (product + contamination), affecting only product specification.

Another argument against this theory was provided by hydraulic calculations. Flow analysis through orifices shows that free area on each downcomer impingement plate can handle flows greater than 200 m^3/h.(7) The design flow of reflux for maximum capacity was 115 m^3/h and maximum flow reached during upsets was around 150 m^3/h. This ruled out that no reflux contamination was happening.

Heat balance

In order to review this problem a heat and material balance is compiled on jet-fuel draw-off and reflux section, results can be seen on Figure 7 and procedure is described below.

Equation 1 ... $Q_I = Q_{TR} + Q_{BTV} + Q_{SV}$

Equation 2 ... $Q_O = Q_{OHV} + Q_{JFR} + Q_{JFP}$

Where:

Q_I = Total heat going in of tower section,
Q_O = Total heat going out of tower section,
Q_{TR} = Heat of top reflux,
Q_{BTV} = Heat of vapors up-flowing from tray below,
Q_{SV} = Heat of jet-fuel striper vapors,
Q_{OHV} = Heat of over-head vapors,
Q_{JFR} = Heat of jet-fuel reflux,
Q_{JFP} = Heat of jet-fuel product.

Figure 7 shows the heat and material analysis which has been compiled on jet-fuel section, total heat going inside the section was 792 MMBtu/h and heat taking off tower section was 760 MMBtu/h.

The analysis showed that the heat going into the section was much larger than heat going out, by around 32 MMBtu/h. According to bibliography this suggests a product leakage. Kister's case studies 9.1 and 9.2 (4) show that a heat balance indicating more heat going in than coming out means leakage or bypassing of liquid from the collector tray into the section below. Further, when discussing these cases, Kister (4) states that the common symptom of less recovery of an upper product, more of the lower product, and colder temperatures below is a symptom of leakage or other way of liquid bypassing into the lower trays. These coincide with the symptoms observed in the crude tower.

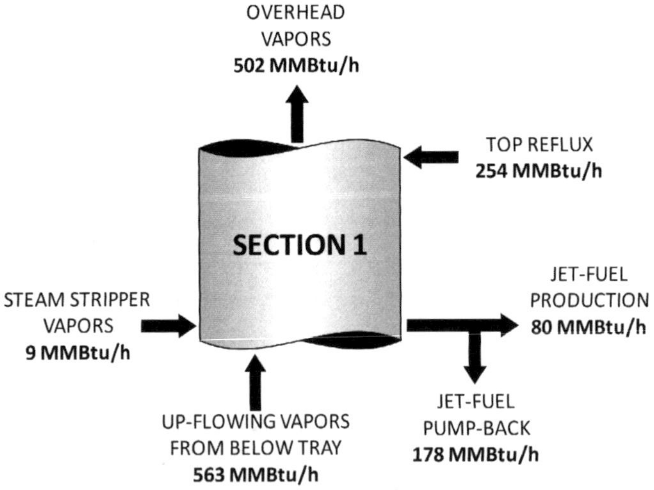

Figure 7. Heat and material balance on the jet-fuel chimney tray.

This leads us to a review of past turnaround reports of tower internals inspections. Leakage tests have been always carried out and showed correct operation of the chimney tray. No leak and no damage report or corrosion was found.

The next plant turnaround was close. During the turnaround an internal inspection was made, the leakage test was repeated, and satisfactory seal was found. This denied the leakage theory.

A detailed analysis of chimney tray geometry was conducted, several sketches with dimensions were drawn. Figure 8 (a) is a simple sketch of jet-fuel chimney tray. Figure 8 (b) adds the expected flow patterns to the sketch.

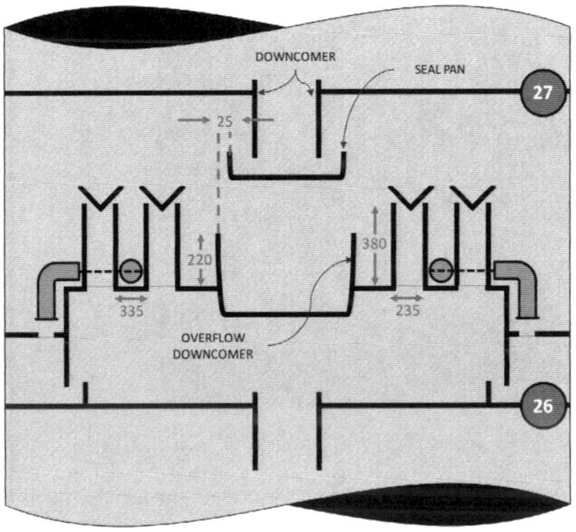

Figure 8 (a). Chimney tray sketch. All data are in millimeters.

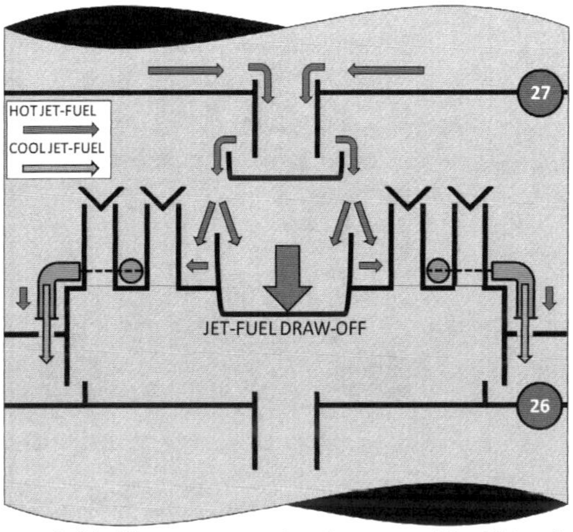

Figure 8 (b). Flow patterns sketch on tray 27 and jet-fuel draw-off tray.

Inspection of the flow patterns in Figure 8 (b) clearly shows that the free area available for jet-fuel catching and directing to the sump is too small, and that not all the jet-fuel can be drawn-off. Part of the jet-fuel drops into the sump, with the remainder dropping outside it (Figure 9). The portion that dropped outside went to the next tray together with cooled pump-back jet-fuel. This jet-fuel was lost to kerosene section.

This situation, observed as a product leakage with heat and material balance, was proved with a simple flows patterns sketch over the chimney tray.

These finds together, with the analysis described above, served as a basis for chimney tray and downcomer modifications. The modifications included replacing the draw overflow weirs from right below the seal pan from tray 27 to the edges of the chimney tray and to increase the free area available on impingement plate to make it equal to the clearance available on the overflow seal. The main purpose of these modifications was to catch all jet-fuel coming from tray 27 and to avoid possible flow restrictions in the downcomers which may cause premature flooding.

Figure 9. Free area for jet-fuel catching to the sump.

Figure 10 shows pictures of chimney tray with red arrows pointing main modifications proposed on tray and flow patterns expected with these modifications proposed is shown on Figure 11.

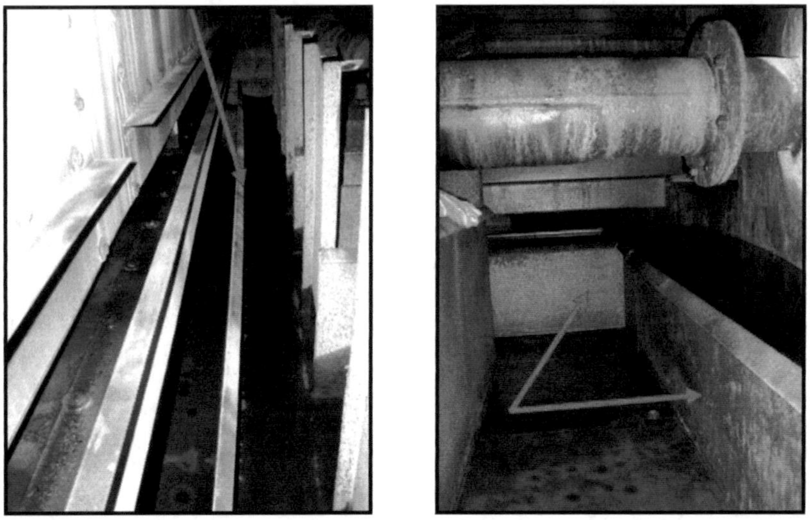

(1) Cut overflow weir in order to increase available area for jet-fuel catching.

(2) Install overflow weirs on the edges of chimney tray for separating hot and cold compartments.

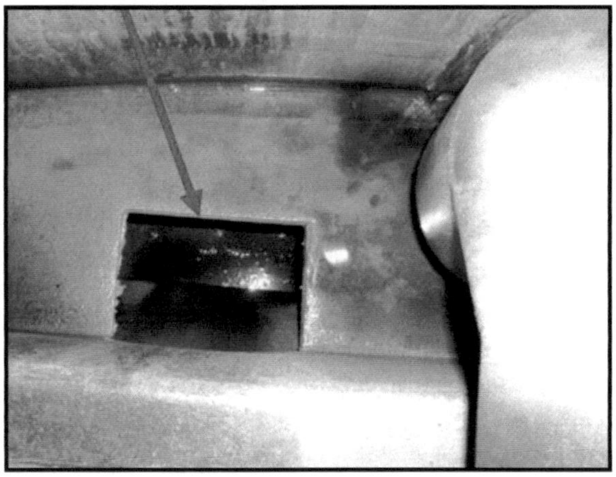

(3) Increase free area on impingement plate for jet-fuel pump back flow.

Figure 10. Modifications proposed on chimney tray.

Results

After modifications described on previous section, which were carried out during 2011 plant turnaround, no more level, production and specification problems have been detected on jet-fuel. See Figure 12.

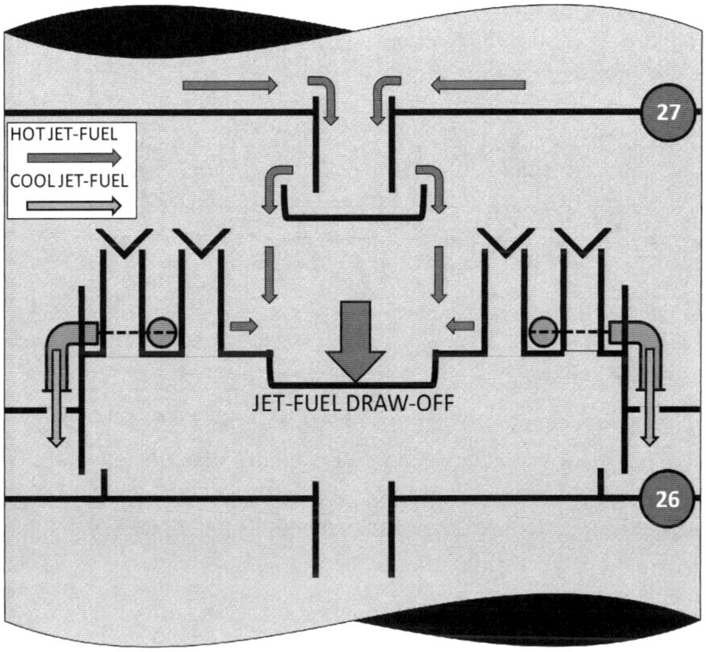

Figure 11. Modification and flow pattern of liquid through tray 27 and chimney tray.

The quality of the products has improved significantly, and most important, the jet-fuel yield largely increased, see Table 1. This has improved tower temperature profile, which is actually hotter than before modification. See Figure 13.

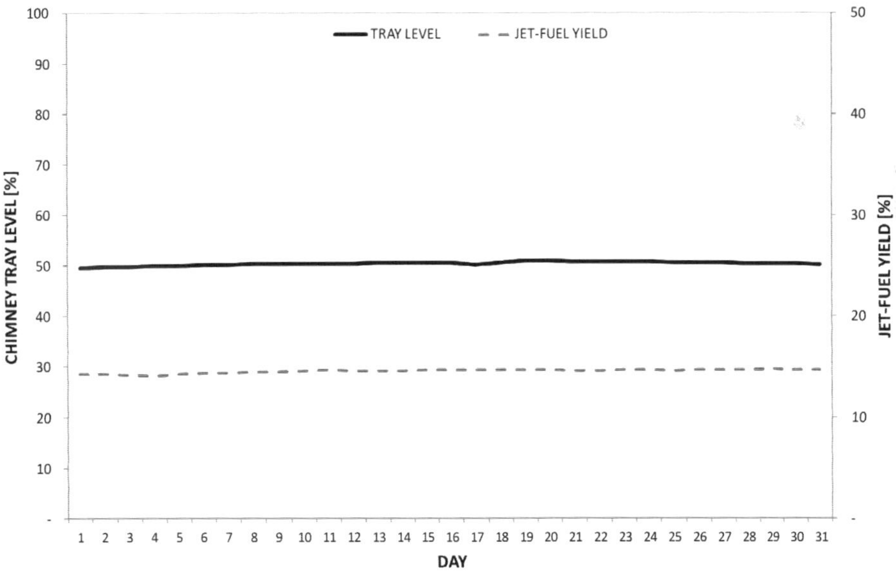

Figure 12. Typical operating chart after modification.

Currently more production and refluxes of products are handled, and these streams exchange heat with crude stream. This increase in product and reflux flows has improve crude temperature for desalting, pre-flashing and fired heater feed temperature, allowing energy savings due to lower fuel gas consumption. The crude temperature going to fired heaters has increased by 15°

C, from 220 to 235° C. This temperature improvement has lowered energy consumption per unit oil barrel processed from 100.6 to 72.6 kBtu per barrel, representing 5.4 MMUSD per year on fuel savings.

Table 1. Result analysis.

SR NAPHTHA		BEFORE	AFTER
YIELD	% vol	16.8	18.0
ASTM D86 @ 95%	°C	149.7	155.8
JET-FUEL			
YIELD	% vol	6.6	14.9
ASTM D86 @ 5%	°C	152.1	170.5
ASTM D86 @ 95%	°C	195.0	241.9
FLASH POINT	°C	35.2	46.0

Increase of gasoline yield represents earnings of around 54.3 MMUSD per year. In addition, a potential jet-fuel earning could be achieved because of the production increase, which represents 55.3 MMUSD per year. This value depends on market because when no jet-fuel is needed this stream is mixed with the kerosene and diesel streams and sold as diesel.

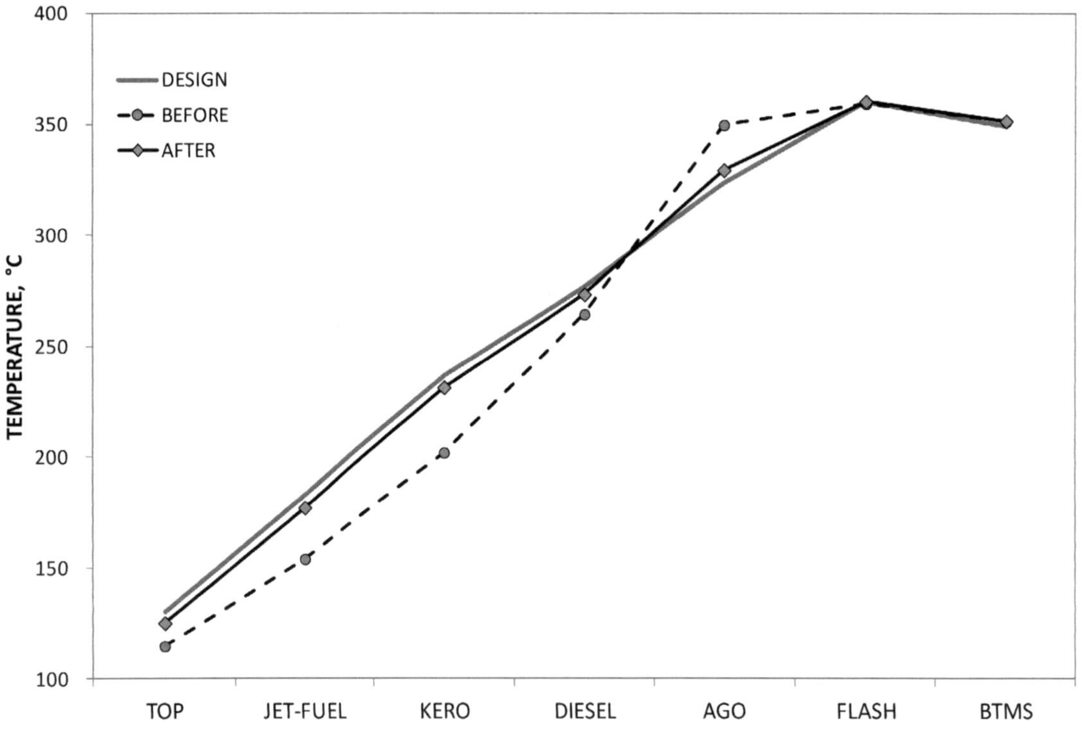

Figure 13. Crude tower temperature profile, design, before and after modification.

Conclusions

For distillation tower troubleshooting it is important the use heat and material balances and graphic analysis of possible flows inside the tower.

Physical tests during turndown periods have same importance. Double checks of tray details and dimensions should be priority in order to avoid flow problems during operation.

We need to assure correct flow of vapor and liquid at the points of transition (e.g., chimney trays, feed regions), as Kister says when dealing with these sections "You don't need an expert you need a sketch". When reviewing draw-off sections, correct installation and geometry should ensure that all the liquid product be directed to the sump. At the same time, it is important to assure that no flow obstructions are present on downcomers. This could be particularly problematic in reflux feeds of "Type R" columns.

In the present case, overflow weir relocation and an increase of the free available area for reflux flow on downcomer were the solution.

Bibliography

(1) **Watkins, R. N.** *Petroleum refinery distillation.* Houston, TX : Gulf publishing company.

(2) **Ludwig, Ernest E.** *Applied Process Design for Chemical and Petrochemical Plants Vol. 2*: Gulf Publishing, Houston, 1979.

(3) **Kister, Henry Z.** *Distillation Operation,*: McGraw-Hill, Inc., 1990.

(4) **Kister, Henry Z.**. *Distillation Troubleshooting.* New Jersey : John Wiley & Sons, Inc., 2006.

(5) **Lieberman, Norman P. and Lieberman, Elizabeth T.** *Working Guide to Process Equipment*: McGraw-Hill, 2008.

(6) **Lieberman, Norman P.** *Troubleshooting Process Operation.* Tulsa : PennWell Books, 1991

(7) **Gieck, Kurt.** *Manual de fórmulas técnicas.* México : Representaciones y servicios de ingeniería, S.A., 1981.

(8) **Reklaitis, G. V. and Schneider, Daniel R.** *Balances de Materia y Energía.* México : Interamericana, 1986.

Refinery Tower Inspections: Discovering Problems and Preventing Malfunctions

By

Rodrigo Cardoso,

CITGO Petroleum Corporation, Lake Charles, LA

and

Henry Z. Kister,

Fluor, Aliso Viejo, CA, USA

Presented at the Distillation Topical Conference, AIChE Spring Meeting, San Antonio, Texas, April, 2013

Refinery Tower Inspections

Discovering Problems and Preventing Malfunctions

By

Rodrigo Cardoso, CITGO Petroleum Corporation, Lake Charles, LA,

And

Henry Z. Kister, Fluor, Aliso Viejo, CA

There are many possible causes for refinery tower malfunctions. Our previous survey (1) identified installation mishaps to be one of the top five causes of tower malfunctions. Such malfunctions lead to poor separation, instability, lost capacity, and higher energy consumption, all with negative economic impact. In some cases, a tower may cease to work forcing a premature outage. Proper inspection following construction and during turnarounds is the best tool to identify installation mishaps, design oversights, fouling, and damage and to correct them before they turn into malfunctions.

This paper presents several case histories of refinery towers where inspection and turnaround testing identified potential and actual bottlenecks caused by improper internal installation, inadequate past inspection, fouling, and internal damage. In each case, the inspection led to problem identification followed by a simple, low-cost solution. The paper demonstrates that thorough and well thought-out inspection often prevents a major malfunction in operation. When it comes to towers, **you get what you inspect (not always what you expect).**

Each case history will be presented with a simple description of related equipment, the problems observed during operation, the findings during inspection after shutdown, and how internals were tested and problems solved.

Case 1: Bad piping for reboiler steam inlet line limited tower operation

Description: A deisobutanizer tower has two horizontal once-through thermosiphon reboilers in parallel. Both use 250 psig steam with two passes on the heating side. Boilup was supposed to be regulated by a bottoms temperature controller that indirectly reset the set-points of both flow control valves (one for each reboiler) located in the steam condensate outlet lines (no condensate pot).

Problem: Due to control problems, console operators would not operate the reset control, leaving the steam condensate valves in either manual or auto (flow control only). To maintain product quality (Reid Vapor Pressure of the bottom product), the operators would manually make adjustments to the condensate control valve settings. The south reboiler always had the process side return line colder, consumed less steam, and showed no steam condensate level on a sight glass located in the channel head. The north reboiler, even though it consumed more steam, always showed a few inches of steam condensate in the channel head.

Cause: One concern was that the shell side of the south reboiler was severely fouled. It was water-washed online, but with no appreciable improvement in heat transfer.

A damaged pass partition was another possible cause. A field observation revealed an unusual arrangement for the steam supply line to the reboiler (Figure 1). A block valve was located near the bottom of the vertical section of the line with no provision to drain the steam condensate that would accumulate above the valve when it was closed during a shutdown.

During the reboiler shutdown, steam condensate would accumulate above the block valve. When the valve was opened, a large slug of condensate accelerated and hit the pass partition causing it to fail. Acceleration of a slug of condensate was much greater during start-up because the cold reboiler (and cold tower bottoms) caused low pressure that increased velocity. Figure 2 shows the damaged pass partition.

Solution: A steam trap and a drain were installed just upstream of the block valve in order to remove liquid condensate before the block valve is opened. There will still be some condensate accumulation between the drain/steam trap and the gate in the block valve (see Figure 1), so operators were trained to open the valve slowly to avoid a slug of condensate hitting the pass partition. Finally, “stiffeners” were installed in the channel head to provide additional strength to the pass partition plate (Figure 3). The ultimate solution though would be the relocation of the block valve and drain/steam trap to the horizontal section of the line leading to the reboiler steam nozzle.

Figure 1: Steam supply line prone to steam condensate accumulation.

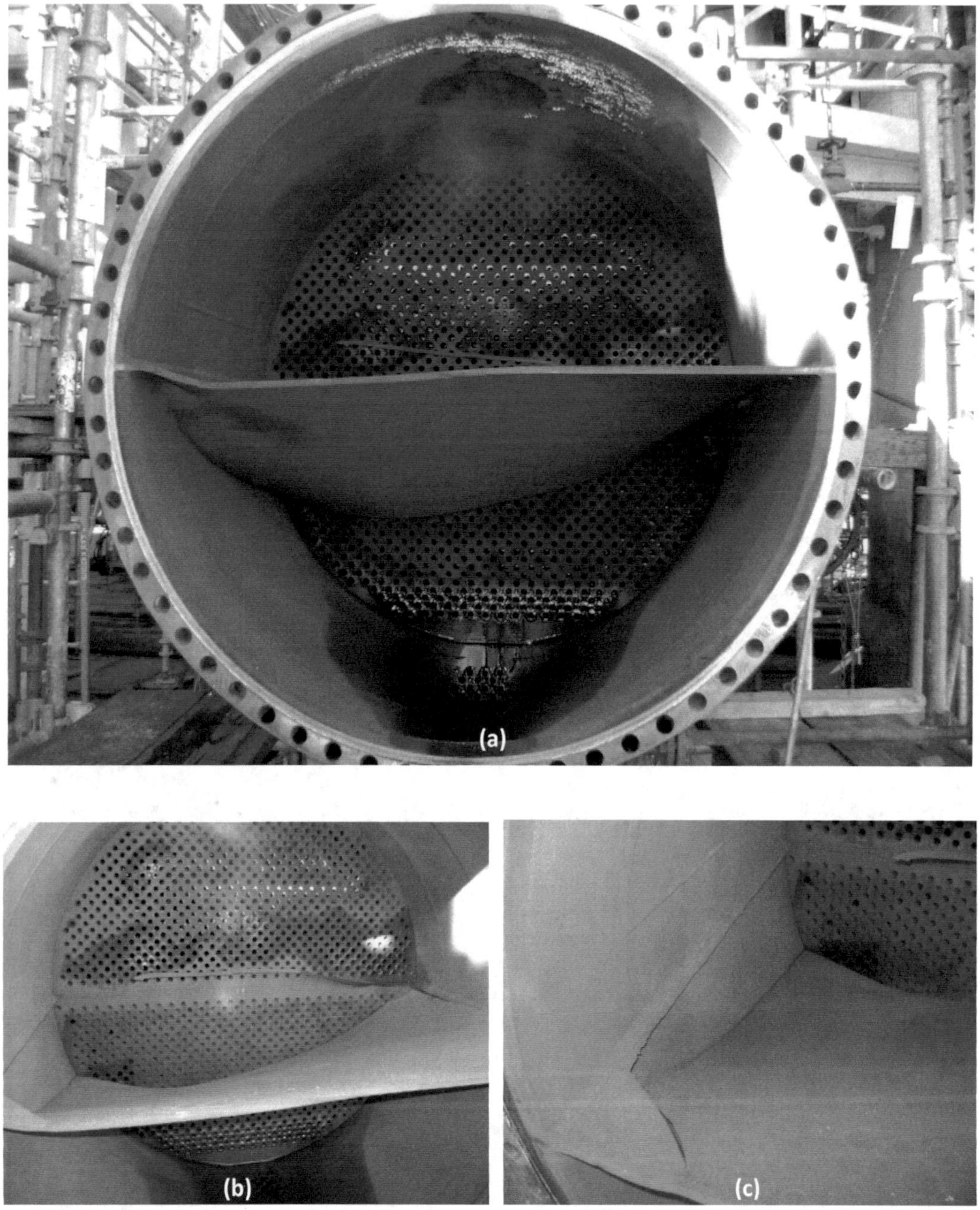

Figure 2: Damaged pass partition. (a) as found once the cover end was removed; (b) and (c) details of damaged pass partition.

Figure 3: “Stiffeners” to avoid future damage to the pass partition plate.

Case 2: Poor installation and faulty inspection can lead to poor tray efficiency

Description: A 11.5 ft ID deisobutanizer tower equipped with 11 three-pass valve trays in the rectifying section is used to make a high purity isobutane stream to be recycled to the reaction section of a sulfuric acid alkylation unit. The reflux distributor is arranged such that flow is split: 1/3 and 2/3 to seal areas on the top tray into which it enters behind inlet weirs. Liquid overflowing the inlet weirs feeds the top tray passes (Figure 4a).

Problem: Process simulation based on plant data indicated that the rectifying trays were 25% efficient, compared to 90% expected. The low efficiency caused increased amounts of normal butane in the isobutane product.

Analysis: Three pass trays are highly prone to maldistribution due to their high degree of non-symmetry. There are many factors that could lead to maldistribution of liquid or vapor on these trays. The following may lead to maldistribution of liquid, vapor, or both, and therefore account for the lower efficiency:

- Severe fouling hindering valve movement;
- Damaged tray panel or collapsed trays;
- Different clearances under downcomers, different outlet weir heights, damaged outlet weirs and/or unleveled trays.
- Damaged inlet weirs on top tray
- Bad reflux distribution to top tray
- Poor tray balancing or design.

Cause: Turnaround inspection revealed the followings issues:

- Pipe scale was found on the top 5 trays (a lot on top tray and diminishing from top to bottom). Some construction debris were also found. About 1% of the valves per tray were missing or dislodged
- There were some missing or not properly installed seal plates
- There were some misaligned anti-jump baffles
- There were some missing or unfastened downcomer braces

However, we believe that the probable causes of the poor tray efficiency were badly leaking flanges in the reflux distributor and/or a bigger clearance under the downcomer:

- Leaking flanges: metallic spacers were found in lieu of gaskets in the flanges (Figure 4a) of the reflux distributor. This had neither been noted nor corrected in earlier tower inspections. The ¼" gap between the flanges caused a large leak which induced poor liquid distribution on top tray. Once maldistribution is generated on the top tray, it propagates down the tower in multipass trays as demonstrated (2).
- Clearance under downcomer: a few trays below top tray, where clearances were supposed to be 1 ½", one of the clearances was almost 3" (outlet weir height is 2 ½") (Figure 5). The uneven clearances are likely to cause liquid maldistribution between the passes (2) and possible loss of downcomer seal (below).
- The bad liquid distribution due to the flange leaks and the high downcomer clearances could have caused vapor to break through the downcomer seal and to flow up the downcomer as it did in some of the FRI tests for 3-pass trays at low liquid loads (2). As demonstrated in the FRI tests (2), breaking the seal can lead to a large reduction in tray efficiency. The top section (above the feed tray) operated at low weir loads, ranging from 1.9 to 2.1, and these low values would have been even lower due to the poor reflux distribution.

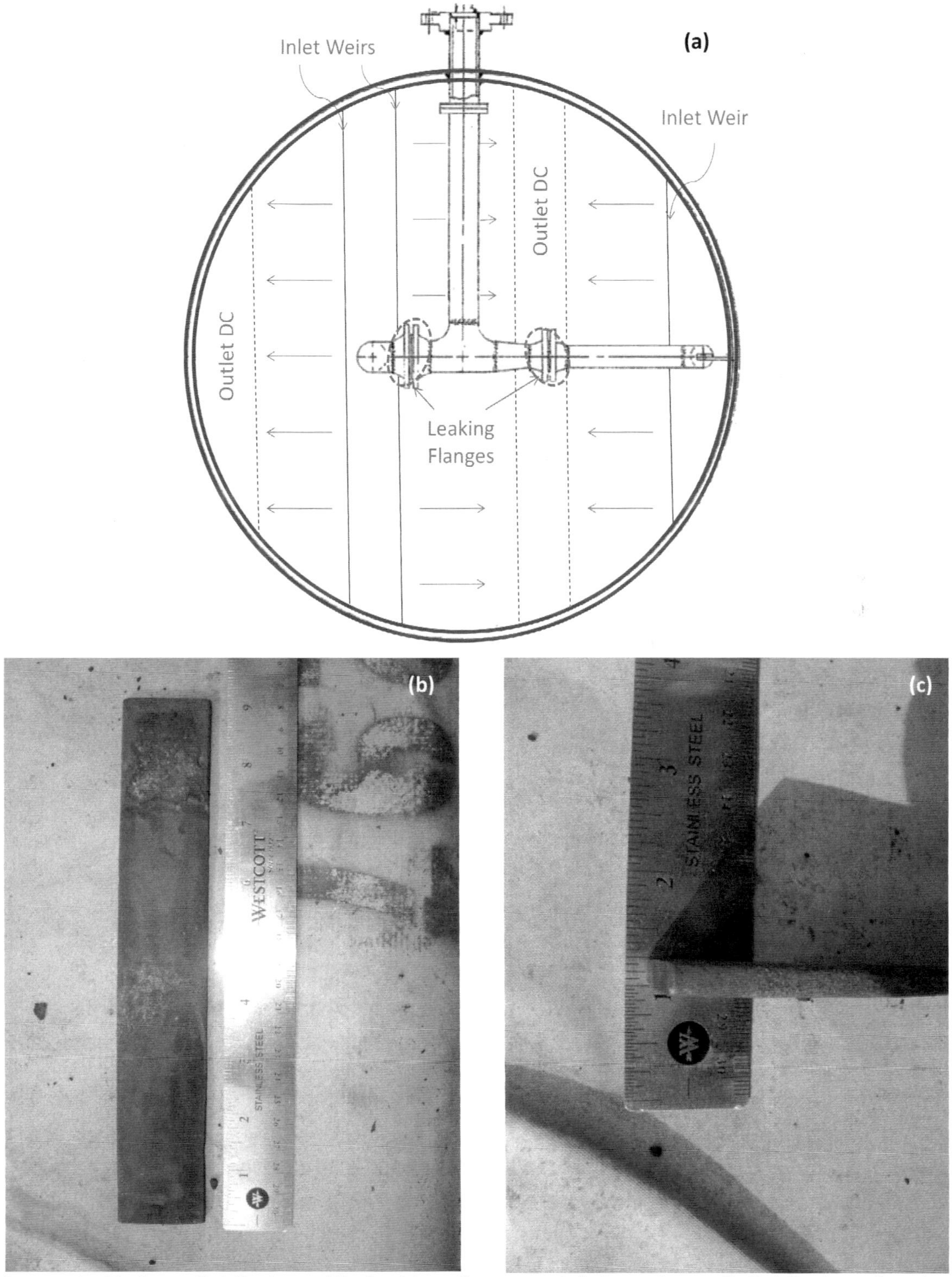

Figure 4: Reflux distributor with leaking flanges: (a) top view of distributor showing leaking flanges; (b) metallic spacer found between flanges (9” long); (c) metallic spacer thickness (1/4”).

Figure 5: Clearance under the downcomer: almost 3" where it was supposed to be 1 ½".

Results: After replacing the flange spacers with gaskets, adjusting downcomer clearance, and taking care of the other minor issues listed above, rectifying section tray efficiency was much higher at 81 percent (compared to 25 percent before the fix), largely increasing the purity of the overhead isobutane product.

Moral: Maldistribution in multipass trays often propagates a long way down the tower, causing a large drop in tray efficiency. It follows that for towers containing multipass trays, careful inspection of internals, especially downcomer clearances and weir heights, is critical to assure proper liquid and vapor distribution. Internal piping should be built to the same standards as outside piping to avoid internal piping leaks or failures and should be carefully inspected. Flanges and gaskets on internal piping should be carefully checked.

Case 3: Leak test revealing non-evident problem

Description: The bottom section of a deisobutanizer tower equipped with three-pass valve trays has two total liquid drawoff pans connected by two 10" balancing lines that feed two once-through thermosiphon steam reboilers. The tower separates isobuane as the top product from a gasoline bottoms with normal butane product recovered as a side draw. The spec on the gasoline is expressed as the Reid Vapor Pressure (RVP). A high RVP signifies a large C4's impurity in the gasoline.

Problem: During normal operation, even at high reboiler steam loads, the butane and isobutane impurity in the gasoline was so high that sometimes the RVP specification was hard to achieve. Higher steam consumption along with the isobutane losses accounted for more than $1.5 MM/year in monetary losses to the refinery.

Cause: The lower temperature of the bottom product compared to the reboiler return, as well as the higher impurity of C4s in the bottom stream (even at high steam flow rates to reboilers), indicated leakage from the bottom tray or drawoff pans. The leaking liquid would bypass the reboilers and raise the lights content of the bottoms stream. Leakage from the bottom tray or draw pans is a common issue with once-through thermosiphon reboilers (1, 3).

At the turnaround, the bottom tray and draw pans were closely inspected visually. The visual inspection did not reveal any integrity problems with the bottom tray panels and/or total liquid drawoff pans.

However, the inspection identified "water marks" indicating leakage from the drawoff pans (Figure 6). A leak test was performed revealing the real extent of the problem: hoses were used to pour water and leakage was so massive that it was not possible to build a liquid level in the drawoff pans (Figure 7).

During operation, the colder liquid from the bottom tray bypassed the reboilers through the leaking total drawoff pans, cooled the liquid returning from the reboilers, and contributed to the high level of C4s in the bottom product stream.

Figure 6: “Water marks” and bottom of drawoff pan identifying the draw pan leaks.

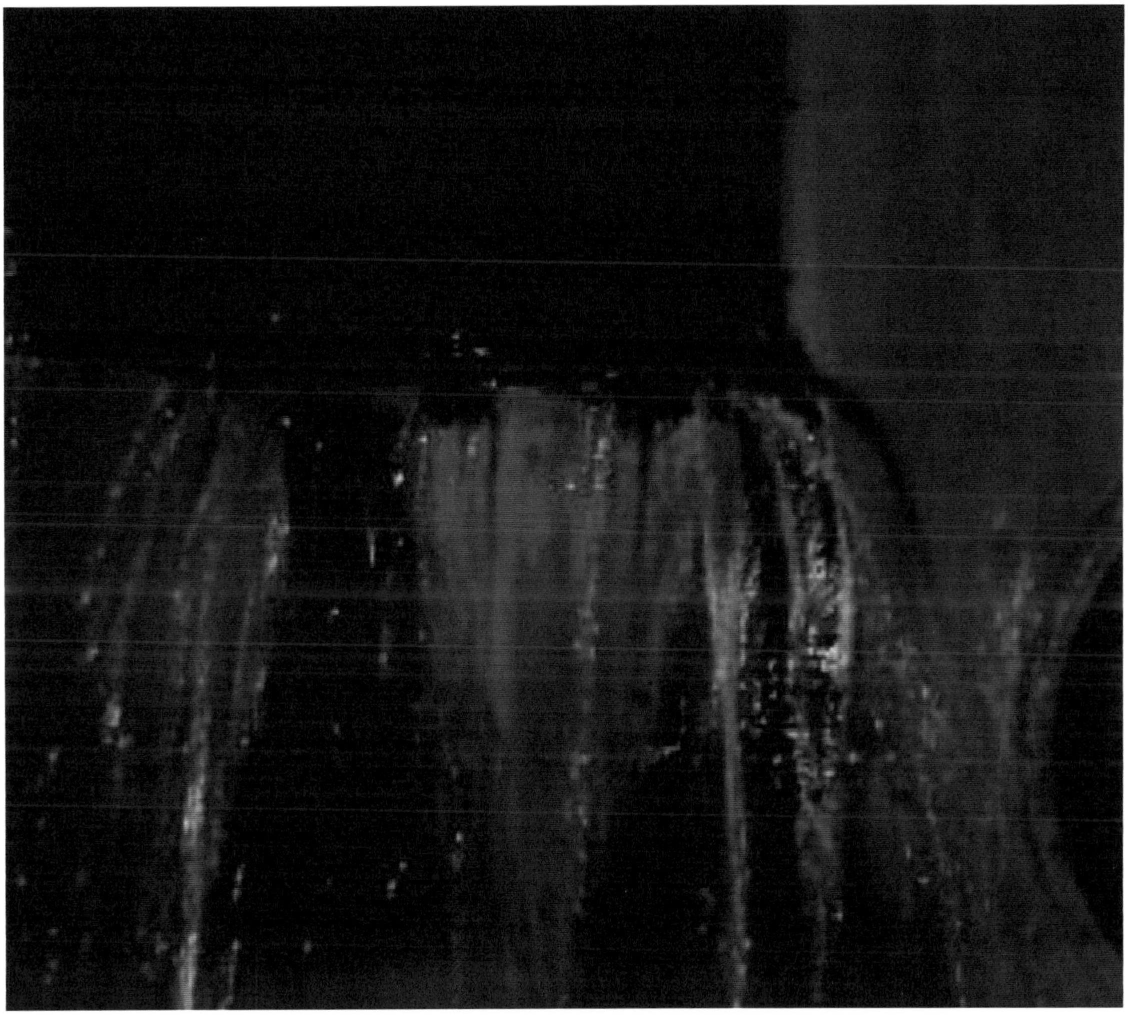

Figure 7: Leak test revealing a massive leak through the draw pan.

Solution: Initially, replacing gaskets was tried with some improvement but still not satisfactory. Seal-welding the whole total drawoff pans was the definitive solution.

Moral: Visual inspection for possible leakage can be misleading. For total drawoff pans, chimney trays, liquid outlet pans, bubble-cap trays, and seal areas behind inlet weirs it is imperative to conduct a leakage test. A leak test is performed by plugging weep holes and then water-filling the pan/ tray/seal area up to the top of the weir and measuring how fast the water leaks down. A leakage rate of 1" per 20 minutes for normal services, or as little as 1" per hour for services where leakage is to be positively avoided, are common criteria used. More details on leakage test can be found in the literature (4).

Case 4: Draw pan responsible for poor fractionation and product recovery

Description: The 14.5 ft lower section of the main fractionator for a hydrocracker unit is equipped with two-pass valve trays. Turbine is drawn from two liquid outlet pans located just below tray #40 downcomers (Figure 8). The remaining liquid is supposed to overflow the draw pan into the four-tray section between the turbine draw nozzles and the two-phase feed inlet.

The turbine draw proceeds from the main fractionator to a stripper, controlled by the stripper bottom level. A flow controller sets the final production rate from the turbine stripper (Figure 9).

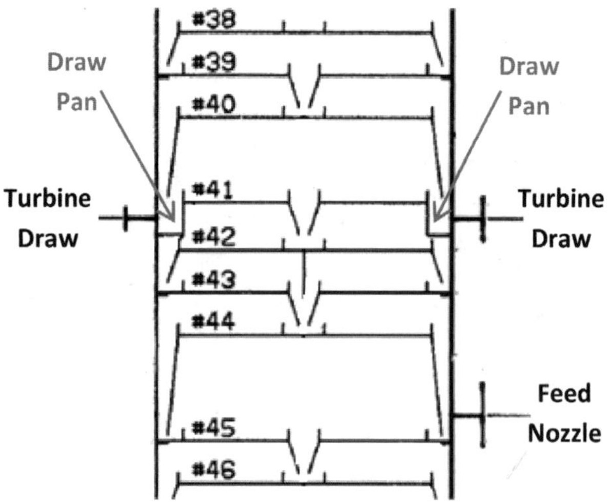

Figure 8: Section of hydrocracker main fractionator.

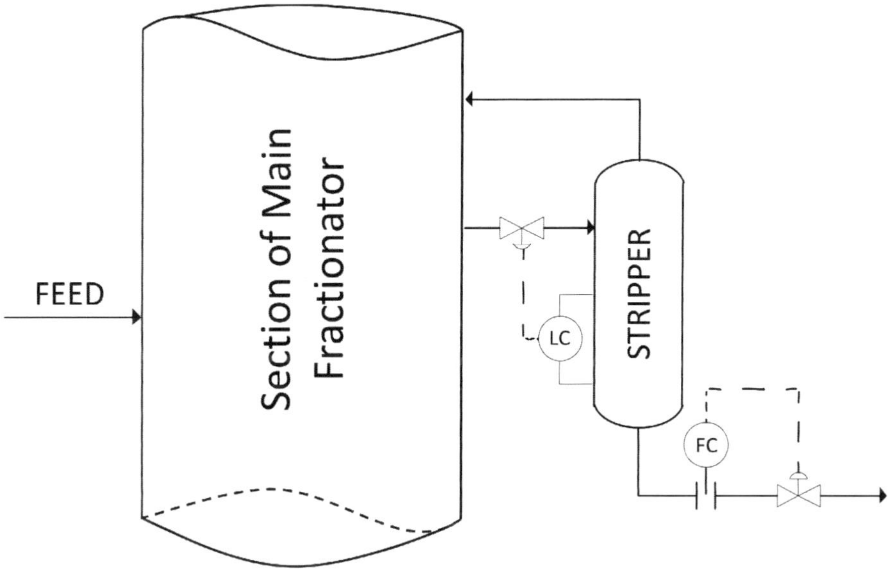

Figure 9: Simplified scheme of the turbine draw system.

Problem: Upon attempts to increase turbine rate above 19-20 MBPD, the stripper would start losing level and the final boiling point of the turbine product would rise considerably. In order to avoid freeze point issues induced by high components in the turbine product, the turbine production would be reduced and a considerable amount of turbine would be lost with the bottom product.

Cause: Visual inspection of the turbine draw sump showed big gaps and no gasket to seal the draw pan (Figure 10a). A leakage test proved that the draw pans were not capable of holding liquid level. White marks on the wall were the references for the water level in the leakage test, but it was not even possible to fill to those marks (Figure 10b).

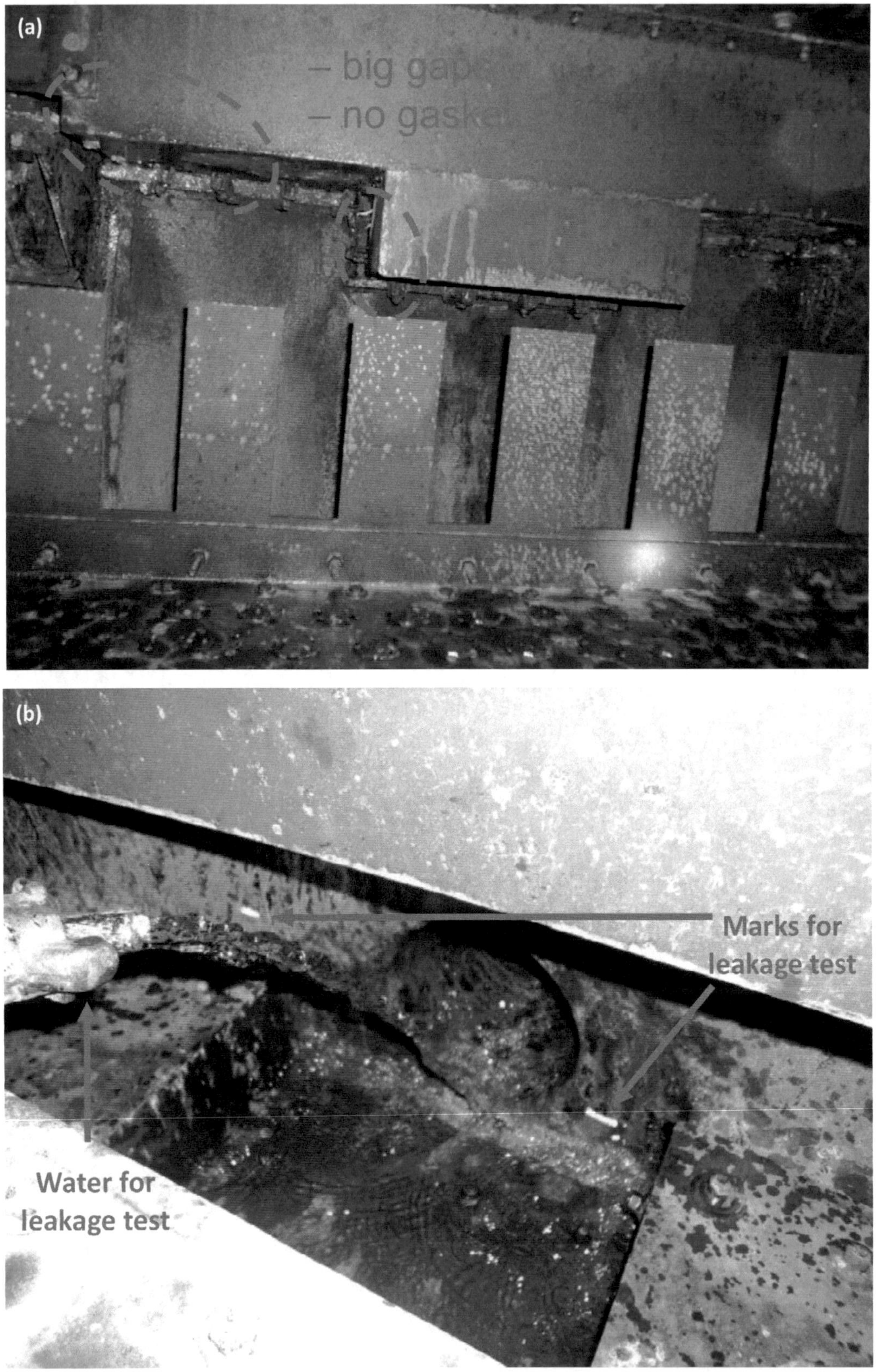
(a)
(b)
Marks for leakage test
Water for leakage test

Figure 10: Turbine draw pan. (a) Visual inspection showing gaps between pan and support; (b) leakage did not allow water level to get to the white marks.

The leaking draw pan together with the high product rate led to dry up trays #41 and #42 (leakage bypassed both trays) and to maldistribute the liquid load to trays #43 and #44. Consequently, the separation on trays #41 to #44 was far worse than design, inducing heavies into the turbine product.

Solution: An attempt was made to seal the draw pans with gaskets, but when the leakage test was repeated no real improvement was observed. The draw pans were then seal-welded, and afterwards passed the leakage test. Eliminating the leakage improved the valuable turbine yield, but also improved the overall efficiency of the four trays between feed inlet and turbine draw, reducing the likelihood of high boiling point components in the turbine product stream.

Case 5: Feed distributor causing tray damage

Description: A feed pipe distributor (Fig 11) containing downward-pointing holes in a hydrocarbon separation tower operating at 95 psig was originally designed for 100% liquid feed. Later, a feed preheater heated by steam condensate from the tower reboilers was added to save energy.

Problem: Most of the valves on the feed tray directly under the distributor holes at its closed end were found missing or severely damaged (Figure 11). The photograph in Figure 11 shows only the last third of the feed distributor pipe near the closed end. There was no damage along the first two thirds of the feed distributor pipe closer to the pipe inlet flange. There was no damage on the tray above or the tray below the feed tray.

Figure 11: High number of valves missing or severely damaged directly under the holes of the feed distributor near the closed end.

Cause: Adding the preheater vaporized some of the feed. At higher heat duties, the feed was up to 32% wt vaporized. This accounts for more than an order of magnitude increase in volumetric flow rate. The high discharge velocity from the pipe distributor perforations caused failure of valves located right below the distributor near its closed end (Figure 12). Weeping and penetration of liquid occurred through the missing valves. Weeping liquid would partially by-pass two trays (feed tray and the tray just below). In this case, minor impact to overall tray efficiency was noticed because the tower had more trays than required for the service.

Figure 12: Closed end of distributor and angle of incidence of feed stream.

Solution: In this particular case it was economical to relocate the feed preheater exchanger so the feed to the tower became 100% liquid.

In a similar experience (5), a downward-pointing high-velocity flashing feed caused turbulence, weeping, and vapor downflow that bottlenecked the capacity of a debutanizer tower. This bottleneck was eliminated by replacing the distributor with a new pipe distributor that discharges the flashing feed at half the previous velocities at a downward angle of 45° towards the downcomer wall. To prevent impingement of the feed on the downcomer wall and possible boiling of the downcomer liquid, an impingement baffle was installed parallel to the downcomer and about 1" away from it. This solution would have also been applicable to the current case too.

Another finding: This same distributor had plate flanges that were found to be leaking (Figure 13). Marks of the liquid spraying from the flange were found on the downcomer plates. Experiences of leaks from warped plate flanges are common (1). The plate flanges were all replaced by standard flanges with standard gaskets.

Moral: Modifications should be critically evaluated considering tower internals configuration. A close review of the tower feed entry is essential when increasing feed preheat. Plate flanges get easily damaged and should be avoided.

Figure 13: Warped plate flanges.

Case 6: Tray installation mistake

Description: A 14 ft ID debutanizer tower equipped with 40 four-pass valve trays removes light components from the hydrocracker reactor effluent stream before further separation in the main fractionator. The trays are numbered so #1 is the top tray. In the 2006 turnaround, tray #21, just above feed inlet, was replaced due to corrosion found during the previous internal inspection. Previously, in the 2004 turnaround, the bottom 19 trays – from tray #22 to #40 – were replaced with duplicates also due to corrosion.

Problem: Upon inspection at a later turnaround, tray #21 was found to have 4" inlet weirs less than 1/2" away from the downcomers exits, and downcomer panels were freely bowing in and out due to absence of braces to hold their bottom edge in place (Figure 14). The inlet weirs were not supposed to be there. No other tray had inlet weirs except for the top tray. Luckily, no operating issues were observed following the installation of the new tray #21 in 2006.

Cause: The cause of this improper installation was a misinterpretation of the tray design drawings and lack of attention to details. The tray drawing for odd trays from #1 to #21 (Figure 15) calls for an inlet weir, but it also specifies that it only applies to tray #1 (top tray for proper distribution of the incoming reflux). The supplier missed that information and the mistake was not noticed during the post-installation inspection in 2006.

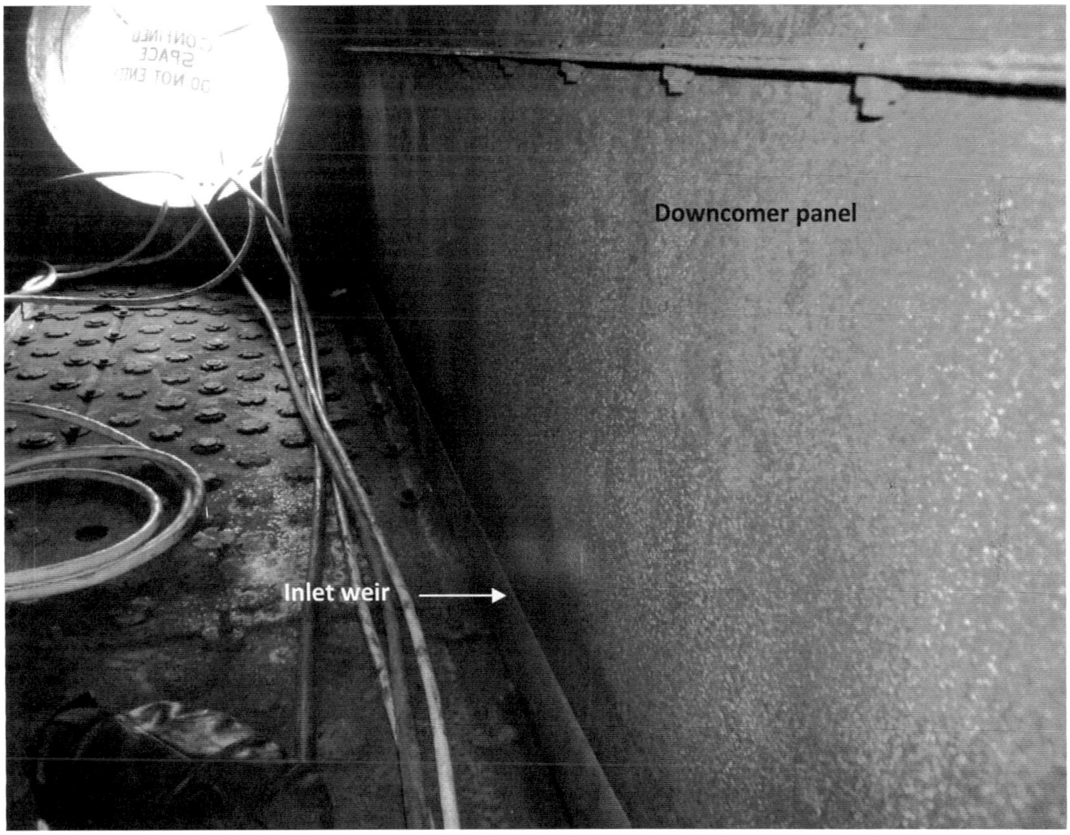

Figure 14: Tray #21 with inlet weir.

No major operational issues were observed because the liquid load for this section of the tower was low: We calculate that at the reduced (1/2") clearance the head loss of exiting liquid was only about 1.25" inches of liquid which is on the high side but still within the normal range for downcomers (5). Also, with a manhole between trays #20 and #21, and consequently a larger 42" tray spacing, the extra height helped prevent downcomer backup flood. We estimate that the downcomers on the feed tray operated at 24% of their maximum downcomer backup capacity. The tower was lucky to "dodge the bullet" of mal-performance in this case.

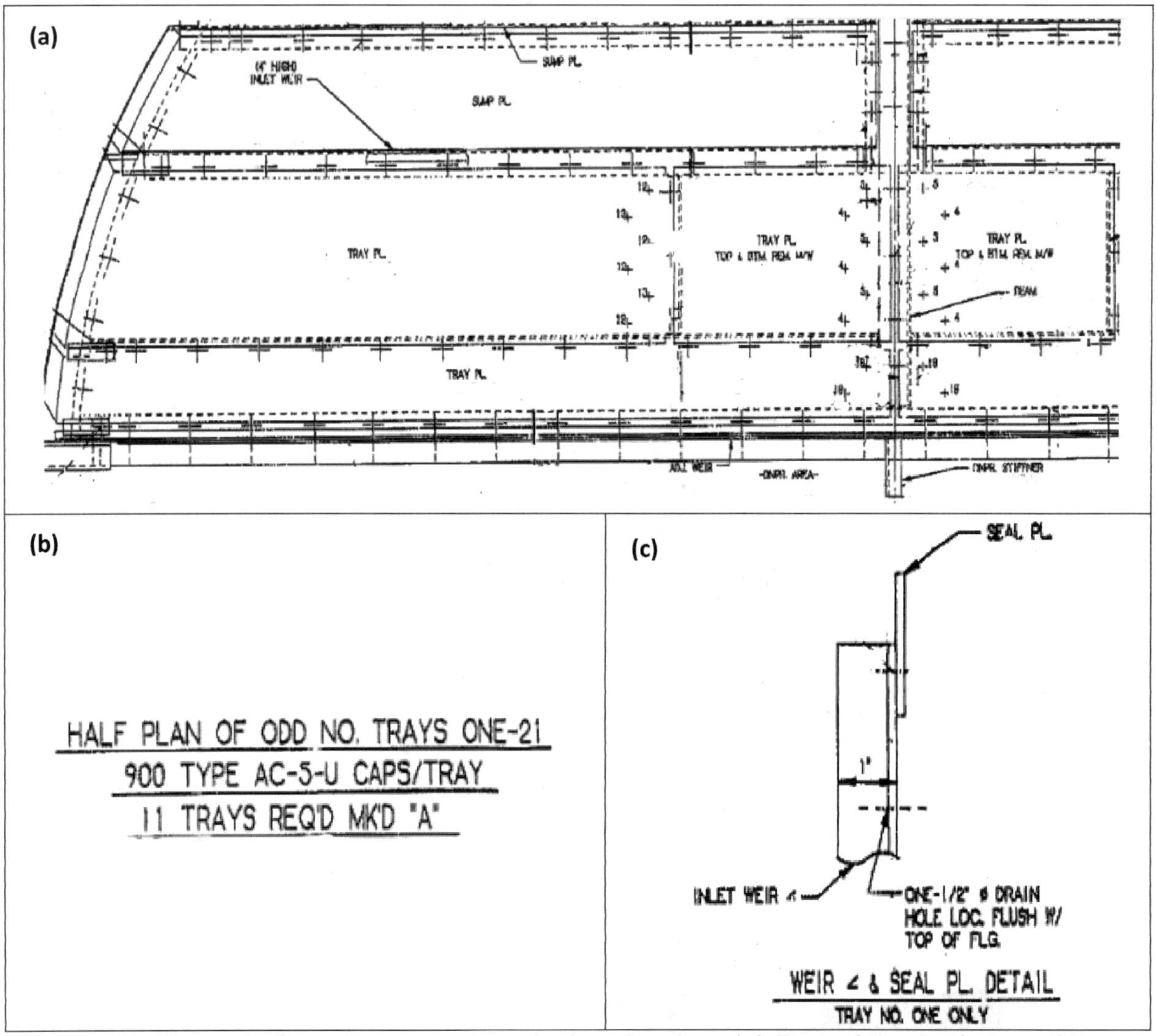

Figure 15: Excerpts from the same tray design drawing. (a) Part of the half plan of the odd trays showing the inlet weir; (b) description of the tray drawing; (c) detail showing that inlet weir applied only to tray #1.

Solution: The inlet weirs were cut and removed from tray #21 and their edges were rounded towards the tray. Braces were installed to attach the downcomer plates to the tray and prevent downcomer bow (Figure 16). The tower performs well .

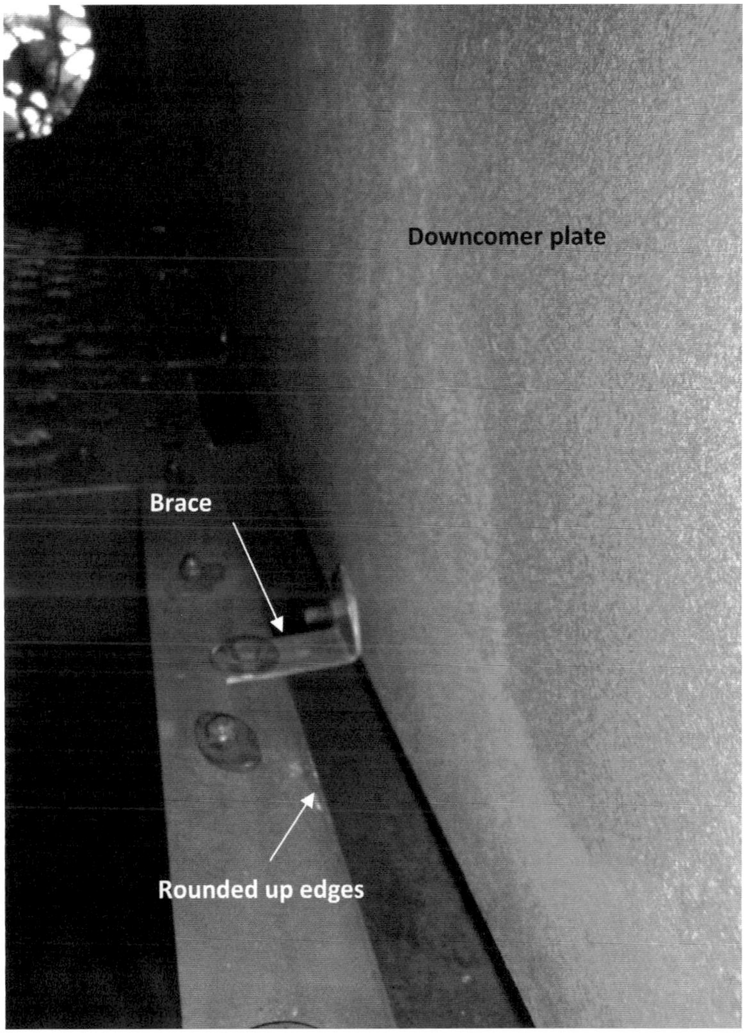

Figure 16: Solution to tray installation problem.

Case 7: How detail of internals in overhead receiver affects the startup procedure – the 6 inches that caused corrosion

Description: In an isomerization unit, the stabilizer tower strips light hydrocarbons and hydrogen chloride from the reactor effluent stream. The overhead system consists of an aircooled condenser, a water cooler and a receiver drum. All condensate is refluxed to the tower. Light hydrocarbons and hydrogen chloride are sent on pressure control to a downstream caustic scrubber (Figure 17). The whole system (except the scrubber) is kept dry in order to avoid hydrochloric acid attack.

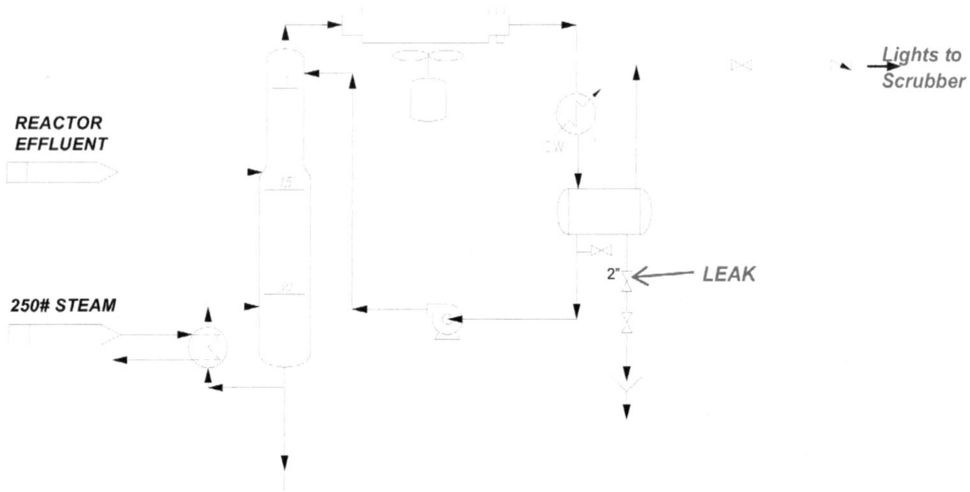

Figure 17: Schematic of the stabilizer tower.

Problem: Corrosion of the 2" drain line and valve on the receiver drum (that was not normally used) took place a few months after a quick outage, leading to hydrocarbon leak to the atmosphere and forcing the unit to have another outage to fix the problem.

Cause: During an outage and in preparation for a planned turnaround, an Internal Rotary Inspection System (IRIS) test was performed to determine if the aircooler finned tubes would need to be replaced. This technique introduces water into the overhead system. The receiver (Figure 18) has a 6" standpipe in the reflux nozzle, allowing free water to accumulate in the drum during startup and be drained (through a different nozzle) instead of circulating with the reflux. This arrangement (Figure 18) speeds startup by separating and allowing easy removal of free water faster from the system.

The dry-out procedure was executed and water was observed neither in the liquid samples from the reflux line, nor in the gas samples from the receiver vent line. The drain line was not operated during the startup to drain water, so water accumulated in the bottom of the drum and in the drain line.

When the unit was put back in service, the water absorbed small amounts of HCl and became acidic. A possible mechanism is that in the condenser, the condensed liquid hydrocarbons absorbed small quantities of HCl gas, some of which would later be extracted into the stagnant layer of water at the hydrocarbon - water interface in the drum.

Some water dissolved into the reflux at the hydrocarbon – water interface in the drum, would re-vaporize upon entry to the fractionator and would be mostly removed in the vent gas, but the remaining water at the bottom of the drum and in the drain pipe was just increasing its acidity and corrosiveness leading to the leak.

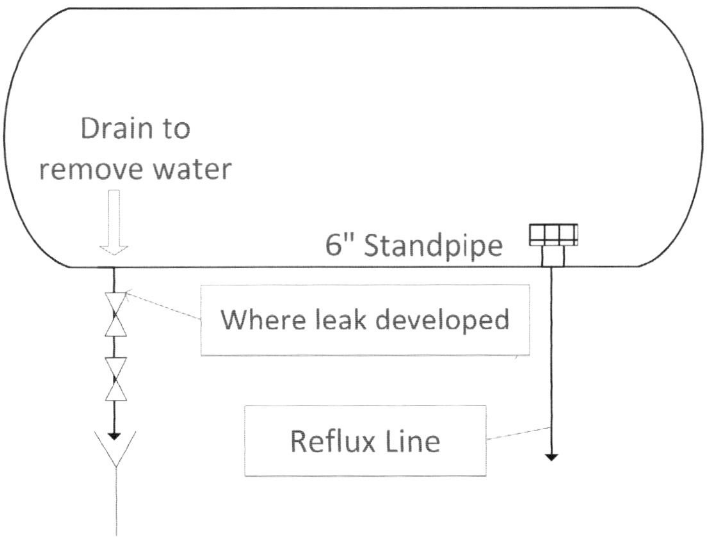

Figure 18: Sketch of the overhead receiver showing the standpipe and drain.

Solution: The dry-out procedure was modified to drain water from the drain line during dryout.

Moral: Details are very important. Not being aware of the standpipe possibly caused the drain line to be left out in the dry-out procedure. Without a standpipe the reflux line could be used for draining water and checking for water presence.

Case 8: Corrosion in the overhead system leading to tower flooding

Description: An 8.5 ft ID rectifying section of a dry depropanizer tower in a Sulfuric Acid Alkylation unit is equipped with carbon steel sieve trays (except for the top 5 stainless steel trays) containing multiple rectangular truncated box downcomers. Liquid from the downcomers was issuing into the active areas below via slots at the bottom of the downcomers (Figure 19a). The overhead product was almost pure propane containing about 8000 ppm of isobutane and about 2000 ppm of SO_2. As long as the system was kept dry, the presence of SO_2 did not pose a problem. However, if a wet stream was diverted to the tower or steam/cooling water leaked into the system during downtime, rapid corrosion would take place.

Problem: The tower flooded near the top. The flood symptom was that as the reflux rate was increased (in order to reduce amount of isobutane in the propane product) there was no corresponding increase in reboiler steam consumption. Since the flooding took place near the top tray, tower differential pressure did not rise significantly to indicate flood.

Cause: At times, the depropanizer would process wet streams (from outside of alkylation unit). Water from the wet feed stream vaporized at the feed, rise, then was condensed in the overhead of the tower. At the colder overhead temperatures, the water often exceeded the solubility limit of the propane stream and appeared as a separate phase. The free water absorbed SO_2 to produce sulfurous acid that corroded steel, generating different iron compound precipitates. These precipitates were recycled back to the tower with the reflux stream and collected on trays and preferentially in the rectangular truncated downcomer boxes.

Turnaround inspection showed a thick layer of corrosion deposits in the downcomer boxes for the top trays (Figure 19). The same deposits were found on all trays, diminishing from top to bottom. At the bottom of the tower, a chimney tray collected a thick layer of this corrosion material.

Figure 19: Corrosion products trapped in the rectangular truncated box downcomers. (a) a truncated downcomer box showing the slots through which liquid flows; (b) a downcomer with a thick layer of deposits covering all slots and causing flooding; (c) bottom view of the downcomer showing the slots completely covered.

Solution: For a dry depropanizer system in a Sulfuric Acid Alkylation Unit, the way to avoid corrosion and the consequences of it (as described in this case) is to keep the system dry. Avoid processing any external wet stream.

In case water gets in the system (cooling water or steam leakage) the following can be done to remove the water and minimize corrosion:

- Drain water from the reflux drum as frequent as possible;
- Raise propane production – it helps the overhead system to eliminate more water (soluble water in that stream). In this overhead system, a bank of overhead condensers would be removed from service – this raises the velocity through the shell side of condensers, increases water solubility in the hydrocarbon phase and consequently reduces the possibility of a water pocket.

Moral: Do not rely only on pressure differential to diagnose flooding near the top of a tower. The energy balance (in this case, a simple trend of reflux and steam consumption rates) gave good indication of tower flooding in the upper section.

References:

1. Kister, H. Z., "What Caused Tower Malfunctions in the Last 50 Years?", Trans. IChemE, Vol 81, Part A, p. 5, January 2003.
2. Kister, H. Z., and M. Olsson, "Understanding Maldistribution in 3-Pass Trays", Chemical Engineering Research and Development, 89, p.1397 – 1404, Elsevier – IChemE, August 2011.
3. The FRI Design Practices Committee, "Reboiler Circuits for Trayed Columns", Chemical Engineering, January 2011.
4. Kister, H. Z., *Distillation Operation*, McGraw-Hill, New York, 1990.
5. Kister, H. Z., D. E. Grich, and R. Yeley "Better Feed Entry Ups Debutanizer Capacity", Petroleum Technology Quarterly, Revamps and Operation Issue, 2003.

(28e) Installation of Secondary Pumparound Resolves Resid Entrainment from Fluid Coker Scrubber

Jill Brown Burns[1], **Daryl Hanson**[2], **Chris Riley**[3] and Cameron Wicklow[3], (1)Valero Energy Corp., San Antonio, TX, (2)Valero Energy Corporation, San Antonio, TX, (3)Valero Benicia Refinery, Benicia, CA

Valero Benicia refinery operates a 28 MBD Fluid Coker. The reactor effluent vapor passes through a single cyclone discharging to a Scrubber that contains 6 rows of 8-pass shed decks. The Scrubber bottoms liquid is pumped both as a recycle stream to the reactor and as pumparound liquid to the top of the sheds fed through a slotted pipe distributor with laterals. The top section of the Scrubber contains a 5 ft bed of structured grid packing washed with heavy gas oil. After the passing through the reactor and scrubber, heavy and light gas oil products are separated in the Fractionator.

While coking and fouling of the Scrubber occurs throughout the 3-year run cycle, at the midpoint of the run heavy gas oil product quality should be maintainable without undercutting. At 1.5 years into the current cycle, significant residue carryover into the heavy gas oil was observed, along with an accelerated growth in Scrubber grid pressure drop. Temperature instruments located at the top shed level indicated that vapor and liquid temperatures were equal, whereas previously the vapor indicator read 50 degF higher than the liquid indicator.

The product quality and operating data suggested an increased entrainment of resid above the shed section, most likely related to a change in distribution quality of the pumparound return. Skin temperature surveys at multiple levels of column elevation supported the data from the TI's, indicating temperature maldistribution focused in one quadrant of the Scrubber. The unit responded positively to increasing pumparound flow through use of the open-tee start-up header located above the normal shed lateral pipe distributor, and thus undercutting the heavy gas oil.

A project was implemented to install piping to feed additional shed pumparound liquid through lances installed in the Scrubber above the shed section. Positioning of the lances and flow have been optimized to target quenching of the hot quadrant. Post-commissioning data indicates that HKGO product quality has decreased from 6-7% resid to 4% resid, and temperatures have dropped by approx. 40degF in the hot zone with grid DP stabilizing. The authors will describe the troubleshooting methodology, discuss the unique piping and lance design, and present post-commissioning data that will show that the new secondary pumparound injection was successful.

Distillation Revamp Project Problems & Prevention
Paper #295774

Doug Bouck, Separation Solutions, Ltd, Aurora, OH

Abstract

An idea turns into a proposal which turns into a project for improving the operation of a distillation unit. The road from the idea to the operating improvement is lined with bumps and potholes. How many things can go wrong? This paper will discuss ten of the many issues that can get in the way of a successful project:

1. Economic Objective Reality Check.
2. Long Term Capacity Plan
3. Change in Engineering Personnel
4. Incorrect Process Simulation Model
5. Incorrect Reading of the Simulation Model Output
6. Failure to Consider Fundamental Heat Transfer Capabilities vs. Hydraulic Capacity.
7. Failure to Test/Define Existing Equipment Limitations.
8. Over Dependence on Vendor/Contractor Design Capability
9. Lack of Specialty Expertise Design Review.
10. Incorrect Installation of New or Revamp Designs.

1. **Economic Objective Reality Check**.

Cyclic Product Values

When setting out to improve the recovery of a product, the average value needs to be accurately computed. When the product value goes through a cycle based on summer/winter or economic swings, the justification must honestly represent the average and projected future. Gasoline and Heating Oil values cycle from summer to winter and many projects have been designed to optimize recovery of the most valuable products of the season.

Supply/Demand

Incremental Supply/Demand cycles can be introduced by startup of a new producing plant. The project for building of Lima OH Aromatics Toluene DeAlkylation plant had been shelved several times due to ups and downs of the chemicals use of benzene. When the project was finally justified and built, the new facility added an incremental 10% to the US benzene production. When the plant was producing the over supply drove the price down. When the plant shut down the price climbed. It is important to determine whether the incremental effects of your changes or competitors changes will have any significant effect on the product value and economics.

When environmental considerations squeezed more benzene out of gasoline, producing benzene out of toluene became unnecessary and Benzene/Toluene splitters became oversized. Redesign projects now revamp aromatics splitters for lower energy use at high turndown rates.

Quantity Requirements (Supply Contract Requirements) vs. Product Values

A recent project was based on maximizing the highest value product stream, however the actual operation was driven by specific product *quantity* demand. As a result the project team and operations team were working in opposite directions.

Regulations Impact

Government or Environmental regulations drive incentives to improve separation or increase capacity to produce the desired product. However, regulations are fickle and can change with a change in administration or environmental concerns. The drive for cleaner gasoline is the obvious case in point. In the 1980s addition of MTBE to oxygenate our gasoline for cleaner air, turned 180 degrees because of environmental concerns that surfaced in the storage and blending of the additive. So by the mid 1990s refiners had to scramble to switch to ethanol as the primary oxygenate additive.

2. Long Term Capacity Plan

Interrelationships with other units

A good long term capacity plan should not be exceeded in the unit operation before the installation of the revamp.

Limitations in upstream or supply units need to be considered in capacity revamps for downstream units. What are the existing upstream or downstream unit limits at current and future operations? Has a long range plan been formulated that includes the project unit, its auxiliaries, and any upstream/downstream units?

Investigation of unit limitations may be as simple as considering what equipment in the system poses the most expensive or impossible to replace or to expand hardware. Management long range planning must also be in the consideration.

Interrelationships with column peripheral equipment

Expanding only one aspect of a unit operation can get the project in trouble. The first item considered may be the column internals. High capacity trays and new packings exceed the capacity of earlier generations which offers opportunity for debottlenecking and expansion of capacity. However higher internals capacity may require a larger condenser which may need more water or reboiler which needs more steam or hot oil. How about new impellers on pumps or new larger pumps requiring new motors, new wiring, new breakers, etc. Is the reboiler large enough or can it be revamped. If so, will the hydraulics work with the higher rates through existing inlet and return piping? Are the flow meters and control valves sufficient? The whole system needs to be considered in any expansion project.

3. Change in Engineering Personnel

A project leader is a critical person in the life of a project. If that person moves on to a greater opportunity, or even dies, the project could end up in limbo, or seriously wounded. The longer the time line for the project from beginning to completion, the higher the risk that the project leader and/or other key persons could change. Documentation is the necessary link to transition between project leaders and maintains the momentum and direction of a project. All files need to be organized and filed. Electronic input and output files as well as calculations need to be saved in safe, redundant company filing systems in addition to residing on personal computers. The project leader's supervisor or associates need to have sufficient understanding of the project so that someone can step in to complete the project or act as bridge to a new project leader.

4. Incorrect Process Simulation Model Input

G(in) = G(out)

Garbage in equals garbage out is still a valid scenario even as computer programs get more sophisticated. Getting the feed and products defined correctly is easier with a propane/butane fractionator than a crude or vacuum column. No matter what the interconversion gurus say, the conversion between TBP and Chromatographic TBP and D-86 or D-1160 is not an exact science. The conversion is dependent on characteristics of the components as well as the boiling points. Elutriation times for an aromatic 300 degree F boiler and a straight chain paraffin are not the same. A coker gas oil and a crude gas oil and a hydrotreated gas oil will not have the same correlation. Subsequent conversion from simulated distillation to D-86 or D-1160 or the reverse conversion can induce simulation errors.

VLE Data

Close behind in importance is the choice of VLE data for calculating the simulated separation. Years ago I saw simulation results run purposefully at every VLE option available in ProII for a de-isobutanizer column. The calculated efficiency represented by the options ran from about 50% to 140%. The extremes would create either a non-functional or non-economic column design. Historical data, plant testing or pilot plant testing can provide data to confirm efficiency for the process simulation model especially where unusual components or mixtures are being separated for the first time. However, the sample taking, lab sample extraction and lab testing must all be of the highest quality or there will be garbage in.

Column Representation

Another important input is the description of the physical set up of the column and auxiliaries. Some columns are very straightforward as in a simple depropanizer: feed, two products, water condenser, reflux drum, steam reboiler, pumps and control systems. Other columns are very complicated as in crude & vacuum columns: cracking in the furnace, partial vaporization in the furnace, additional vaporization in the transfer line, final flash into the column flash zone, entrainment, runback, liquid stripping section, sensitive to specific pressure drop through chimney draw trays and packed wash section, pumparound sections, heat transfer zones and fractionation zones, spray distributors, multiple stage vacuum systems, etc. Add to that: heat integration with feed preheat, desalters, preflash columns, pumparounds, recycles, etc. There are several types of reboilers and condensers and each one must be correctly represented in the simulation model.

Envelope

One very basic part of the simulation procedure is to draw an envelope around the process to be simulated. Make sure that all streams in, out and within are accounted for.

Example:

A recent simulation had a vacuum column with vacuum pumps sealed by a slip stream of light product. The seal oil was cooled by product coolers before being fed to the vacuum pumps, then it was recycled back to the upper pumparound return to the column. However, the recycle return was ignored in the simulation. Less return threw the actual liquid rates off in the top section of the simulated column. But since the seal oil returned cold relative to the rest of the pumparound, there also was a significant heat balance error from the missed cooling effect between the simulation and the real column.

Specifications

Proper specifications to match actual or projected operation is also important.

Examples:

A benzene-toluene fractionator was revamped for new turndown conditions. Control issues persisted after the revamp. Since benzene has a higher value than toluene the column is being operated at virtually zero benzene in the toluene. Temperature profiles are different between the simulation specifications used for the revamp and actual operation. The former temperature control system is inadequate to differentiate control over small changes in the overstripped toluene product and needs to be re-thought and re-designed.

Vacuum column wash zones need to be designed for a specific minimum liquid rate **off** the bottom wash zone stage not the wash zone inlet. Significant vaporization of the wash oil takes place and the wash zone will dry up and coke if significant liquid rate is not maintained at the bottom of the bed.

5. Incorrect Reading of the Simulation Model Output

Very rarely will the model stages of a process simulation match the actual trays in a column. That presents multiple problems of deciding or deciphering the relationship between the theoretical stages and the actual trays or feet of packing. Some of the most common errors in reading simulation outputs:

Stage by stage results are normally shown as liquid FROM the stage and vapor FROM the stage. Typical tray design requires liquid FROM a tray and vapor TO a tray. So information has to be taken from two separate stage outputs and auxiliary stream results. Typical packed bed design requires liquid TO the top and Vapor FROM the top plus Liquid FROM and Vapor TO the bottom of the bed. So values may have to be gathered from several places in the simulation output. Especially when tray efficiency is low, it may be necessary to interpolate rates to calculate the liquid rate from the top tray or other intermediate trays.

Reflux rates especially for designing top distributors and the tops of packed beds must be the liquid rate TO the top simulation stage NOT FROM the top stage.

Liquid rate off the bottom stage of a pumparound with a product draw is usually the sum of three separate numbers in the simulation output and may even not be on the same page of output. Liquid from the draw stage (internal reflux), pumparound or pumpdown return rate (out of the draw stage and returned to another stage), and a product rate.

Knowing exactly how a particular simulator treats a flashing feed is important both to a correct simulation and a correct set of data for the trays or packing above and below the feed. Does the feed flash before entering with the vapor going to the stage above and the liquid entering the feed stage. Or is the liquid **and** vapor mixed into the equilibrium of the feed stage?

Example:

Before installing a new packing distributor it was discovered that the distributor design rates were much higher than the reflux pump capacity. Investigation revealed that somewhere between the simulation and the vendor design, the rate FROM the top column stage rather than the reflux feed TO the top stage had been used in the distributor design. Fortunately the original distributor had been saved and was re-installed.

6. Failure to Consider Fundamental Heat Transfer Capabilities vs. Hydraulic Capacity.

Usually making a change to heat transfer requirements outside a column are logically thought through. If you need more reflux, you probably need a larger condenser, perhaps a larger impeller or larger pump and motor. The higher reflux rate may require a larger reboiler or other means to increase heat into the column. It is easier to forget heat transfer impacts on the inside of columns.

Example:

A revamp plan was to move heat transfer from a lower packed Pumparound to upper packed Pumparound. The intent was to create additional reflux/separation in the middle section of the column and at the same time convert the bottom pumparound bed into an additional fractionation section. The new design loadings were sent to a vendor who sized the beds for hydraulic capacity. Initially no one in the project - owner, contractor or vendor - considered the necessity for checking or increasing the external heat transfer and hydraulics of the upper pumparound system or the heat transfer capability of the upper pumparound internal packed bed as part of the overall project. Subsequently changes were made to increase heat transfer in the external pumparound loop. However, the project will not achieve its full objectives because the internal heat transfer issue should have included a more significant re-ordering of space in the column.

An additional exchanger was provided for more heat exchange on the outside pumparound circuit. However, additional packing volume was needed internally just as additional heat transfer surface was needed in the external pumparound circuit. Adding surface area for additional heat transfer on the inside is more complicated. A timely evaluation of the heat transfer issues as an early part of the project would have revealed the extensive internal changes required to increase the heat transfer of the upper pumparound bed. Early project understanding of the effects of moving the heat transfer would have resulted in a completely different design or determination that the cost of the more extensive changes required were not justified by the economics.

7. Failure to Test/Define Existing Equipment Limitations.

Sometimes there is a temptation to “fix” something without really understanding everything about the problem or the repercussions of changes to an element of a system without considering the whole of the system.

Example 1:

A project sought to increase heat transfer in a pumparound circuit. The design included increasing a pump impeller size, changing the control valve system and adding an exchanger in series with an existing exchanger. All the hydraulic and heat transfer calculations were made using normal engineering methods including seeking confirmation from the exchanger manufacturer. Upon implementation of the new design the circuit failed to meet either design pumping rates or expected heat duty. There were two limitations in the new system:

A. The primary limitation in the circuit was extra pressure drop significantly greater than calculated. The history of the project showed that the extra pressure drop had been in the circuit prior to the revamp and its cause and location had never been determined. No one considered the possibility of a two phase flow, generated at the junction of two streams, creating the limiting pressure drop.
B. The heat transfer achieved in the exchangers was also significantly below design calculations. The original active heat exchanger had never been tested and compared to its design basis.

The above example is a reminder that operating reality is always right. If reality does not match calculations, either something is wrong with the calculations or our understanding of the reality is faulty. To continue forward with a design based on calculations that do not match reality is a process doomed to failure.

Example 2:

Maintenance had a problem side reboiler on a naphtha fractionator. There were repeated leaks in the reboiler requiring shutting down the column for repairs. Someone in maintenance decided to add block valves to isolate the side reboiler from the tower and to allow on-line repairs. The valves were installed during a turnaround without process engineering review. Upon startup, the column exhibited flooding at below maximum normal rates. A Gamma scan revealed flooding of trays starting above the side reboiler draw. The extra pressure drop from the new gate valves in the reboiler circuit was enough to cause backup in the column. There were at least two errors in planning and execution:

A. There was no process pre-review to consider or calculate the effect of the addition of the extra gate valves to the reboiler hydraulics.
B. The reboiler was a total draw and there would have been no way for the liquid to continue down the column with the new block valves closed. So, even if the isolation valves had not been a pressure drop problem, the block valves could not have been used to isolate the side reboiler for repairs during operation of the column. So the original purpose for the change was not consistent with the internal design of the column.

8. Over Dependence on Vendor/Contractor Design Capability

Vendors and contractors are like most companies, they have a range of experience represented in their staffs. A simple hydraulic design based on loads provided by a customer can be created by a new employee that has been taught to run the company's rating/design program. However, a more complicated design requiring interpretation of simulation output to create the loadings takes a higher degree of expertise. If the design includes heat transfer, the requirements go higher. If simulation is included in the design requirements, plant process experience is almost necessary unless a plant engineer is working directly with the simulation engineer. If the simulation engineer has no plant process experience there may be limited ability to recognize suspect output of complicated systems.

9. Lack of Specialty Expertise Design Review.

Experience counts. Younger engineers may be very academically trained right out of University, but experience comes with time and exposure to various situations. Younger engineers often get the simulation jobs because they are a wiz at computer technology. Putting the simulation together and getting the simulation to converge quickly is important, but getting the feeds, products, internal/external column and auxiliaries representation correct is of primary importance. Once the simulation is operating, next is making sure that the output achieves the design purpose and makes sense. Here is where experience can spot inconsistencies either with the plant operating expectations or general process rationale.

Vendor quotes need to be reviewed to make sure that any issues beyond hydraulics are addressed.

Examples:

A packed pumparound bed had packing that was undersized for capacity. The vendor solution fixed the inlet (bottom) capacity with a more open packing without considering the vapor rate and heat transfer profile through the bed. In the proposed split bed the low heat transfer efficiency of the high capacity packing would guarantee flooding the inlet to the upper higher efficiency section. To optimize the heat transfer and capacity the bed needed to be divided into three sections of progressively higher efficiency packing.

Once upon a time a young engineer was given a design problem. A compressor lube oil system needed to have a nitrogen stripping column added to a slip stream of the lube oil circulating system. The young engineer researched for a suitable packing and found a mini ring used in laboratory sized equipment. He diligently did the calculations and a six inch diameter several feet tall should do the job. The equipment was built and added to the lube oil system. Upon startup only a fraction of the design nitrogen rate was achievable before the column flooded. The designer forgot to allow for the relatively thick film of oil on the small packing elements which reduced the open area of the packing significantly. That was my rookie mistake. It could have been prevented by a design review with the company specialist who would have spotted the design error, or at least asked the right questions about the design considerations.

10. Incorrect Installation of New or Revamp Designs.

The final part of any project is to make the installation of the change and start it up successfully. One unknown is always: "what is the experience level of the installation crew?" Whether it is a simple replacement of tray decks or extensive cutting and welding or change out of multiple pieces of equipment, someone has to take final responsibility for correct installation.

Examples of installation problems:

Installation of a new set of high capacity trays was about to begin. The process engineer decided to check the rise of the valves on the various tray sections in the lay-down area. One of the tray sections was inconsistent with the others. A call to the vendor determined that the differences were beyond a tolerance issue and the one set of tray parts had been punched wrong. Quick replacements corrected a problem that could have restricted the capacity by up to 25%.

Installation of a replacement chimney tray. A side reboiler draw tray was re-designed to capture water from periodic water wash. The re-design added a water draw box and weirs to collect and channel the water into the draw box. The drawings had been reviewed by experienced engineers. However, when the new draw tray was laid out in the field prior to installation, the field process engineer noted that the tray deck was set upside down. The installation foreman protested that it was according to the drawings. He was right about the drawings, but no one had noticed that the original deck strengthening ribs were on the top in the vertical position. That was fine for the original deck for feeding the reboiler draw, however the vertical ribs would have prevented water from reaching the water draw box. In other words as designed, the water draw would not work. Fortunately with some minor modification the deck was able to be installed flat side up and functional. This is an example of why it is always best practice to lay out installations in the field prior to installation. It gives a final look to confirm that the design will fulfill both process and mechanical purposes.

<u>An example where process field inspection failed</u>. The top FCC Main Fractionation zone was revamped from 24 inch to 18 inch tray spacing including replacing a chimney draw tray. Upon startup approaching full rate operation, the naphtha fraction overhead turned yellow with obvious entrained light gas oil. A Gamma scan of the section showed that the chimney draw tray was full to as much as several inches above the top of the chimneys. The liquid was flowing down the chimneys and was being entrained in the vapor causing flooding on the trays above which carried the Light Gas Oil into the overhead Naphtha fraction. Further analysis of the drawings indicated that there were downcomers built into the original chimney tray, but **no** downcomers were provided in the new chimney tray design. No one caught that error. Process inspection **should** have realized the downcomers were missing at drawing reviews. Lastly the opportunity to see the process problem as the internals were installed in the field failed as well.

<u>Installation of a re-designed feed pipe</u>. The ~18 inch diameter feed pipe for a large reformate splitter was designed and the pipe had been installed and welded into the vessel wall when the field process engineer arrived. To his horror it had been installed upside down. Neither the plant management or the contractor was interested in carving the pipe and its nozzle out of the column. The baffles inside had to be re-designed on the spot to compensate for the incorrect orientation. The moral here is that project/process supervision needs to be involved at every significant step of installation.

<u>Downcomer dimension</u>. A gamma scan of a 3 pass absorber stripper column showed the initiation of flood at an unusual place that would not be expected based on the column loading profile. The flood began several trays above a side reboiler in the stripping section. When the column was opened at the next turnaround, dimension checks showed that the outside downcomer on that tray was 3/4 inch narrower than design. The column was downcomer inlet limited and correcting the downcomer gained 5% in column capacity which was the calculated value of the extra downcomer inlet area.

<u>General installation quality</u>. ANY new installation of trays or other internals has to be checked for correctness and tightness of hardware. ANY opening and closing of internal manways must have the bolting loosened and/or replaced as necessary before re-bolting for closure. Incorrect closure of internal manways has caused cracking of trays due to vibration of loose parts.

In Conclusion:

Beware of these and other project potholes and pitfalls. Keep your process engineering eyes open while on the road from project conception and completion through installation and startup.

(28g) PROFLUX™ Severe Service Grid – an Alternative Solution for Fouling Applications

Izak Nieuwoudt, Koch-Glitsch, Wichita, KS, Pat Quotson, Koch-Glitsch, LP, Wichita, KS and Dennis J. Schmude, Koch-Glitsch LP, Houston, TX

Severe operating conditions require the use of unique equipment to maintain acceptable performance and reliability. A typical example of this situation is the wash bed in a refinery vacuum crude tower. The wash bed removes entrainment from the flash zone and provides some fractionation of impurities such as unwanted metals. This is currently best handled with a combination of grid packing and structured packing. Data for this superior new grid packing is presented that compares de-entrainment and fractionation performance with the current arrangement. The fouling resistance and structural integrity of the product is studied and compared with conventional grid packing as well. Higher vapor capacities are possible, which are shown in typical design scenarios using complete and combo beds.

ON MODELING OF GAS-PHASE MASS-TRANSFER STRUCTURED PACKINGS

F.J.Rejl[†], *L. Valenz, J.Haidl, V.Linek*
Institute of Chemical Technology, Prague

Introduction

Structured packings are now widely utilized due to their low pressure drop per unit of separation efficiency. Since their introduction number of models predicting their mass-transfer performance has been published. Comprehensive review of mass-transfer models for packed columns has been provided by Wang [4] in 2005. While new and new models are created (some of them are incorporated into the simulation programs) their reliability is still not broadly acknowledged [17].

The models serve often for HETP calculation and their predictions are usually reasonable. Physical background of the models focuses on fundamental mass-transfer data, however, (mass-transfer coefficients and effective interfacial area) out of which the HETP is only calculated. Predictions of the models with respect to these fundamental quantities differ often in multiples and also their trends with phase flows differ considerably. This situation results from verification of the models predictions predominantly through HETP data. HETP is a complex quantity, affected by the equilibrium, phase hydrodynamics and two phase mass-transfer phenomena. Separation of these phenomena in order to determine fundamental quantities often involves application of strong assumptions, as is negligible mass-transfer resistance in the liquid phase or fully wetted packing surface. Application of the plug flow model is another usual assumption performed almost automatically. Verification of the models with respect to the prediction of fundamental quantities through HETP is therefore much more uncertain.

To ease the situation, data from standardized absorption experiments shall be examined, as there is possible to apply systems which have mass-transfer resistance limited to the either liquid or gas phase [5] and other systems for the effective area measurement. Also a method for determination of both phase mass-transfer data under distillation conditions has been suggested [8] and could help in the process.

Only limited absorption gas-phase mass-transfer data are published for the structured packings. This work is therefore focused on measurement of volumetric mass-transfer coefficient in the gas phase, k_Ga in the metal sheet structured packings under absorption conditions. Mass-transfer coefficient in the gas phase, k_G, is calculated from k_Ga using independent experiments for determination of effective interfacial area, a. Obtained data are compared with the predictions of the models.

Experiments have been performed with three Sulzer Mellapak packings differing in nominal geometric area (M250Y, M350Y, M500Y) and one high-capacity packing of the same vendor (M452Y). Therefore there was some base for study of the influence of the structure characteristic dimension on the mass-transfer characteristics. Experiments have been performed under various gas and liquid loads and therefore trends of the mass-transfer characteristics with phase velocities could be determined. All the experiments have been performed with the SO_2-air-lye system at 20°C and atmospheric pressure. The physical properties of the phases were therefore constant and study of their influence on the mass-transfer characteristics was not possible.

Gas-phase mass-transfer in structured packings

In the column packed with the metal sheet structured packing the gas phase flows upwards through the system of mutually perpendicular 45° (Y-type) inclined, one side open triangular channels - counter-currently to the flow of the liquid. There is usually assumed, that the gas follows

[†] corresponding author

the channel and doesn't change its direction until reaching edge of the element or the wall of the column.

Due to the inclination of the channels, true gas velocity through the channel is higher than the superficial velocity by factor $\sqrt{2}$ (for Y-type). Characteristic dimension of the packing is taken to be hydraulic diameter of the packing $d_P = \frac{4\varepsilon}{a}$ or length of the channel, *l*. Gas phase Reynolds number with hydraulic diameter of the packing taken as the characteristic dimension is relatively low; for superficial velocity 1m/s, air at ambient conditions and "ordinary 250 packing" *Re* value is 1455 only. Turbulent flow is usually tacitly assumed, however, although the flow falls perhaps more into transient regime.

As the mass-transfer occurs between gas phase and the liquid phase, which is also moving, determinative for hydrodynamic and mass-transfer calculations is the sum of their interstitial velocities, "effective velocity" - u_{GE}. Determination of the interstitial liquid velocity out of superficial velocity might be tricky and desires at least knowledge of liquid hold-up.

Dependence of the fundamental gas-phase mass-transfer quantity, mass-transfer coefficient k_G, on the hydrodynamic, geometrical and physical properties of the gas is usually provided in the generalized, dimensionless form

$$Sh = C \cdot Re^m \cdot Sc^n$$

It should be noted, that such formula binds together velocity and characteristic dimension dependence of k_G, as it can be rewritten as

$$k_G = C \cdot u_{GE}^m \cdot d_P^{(m-1)} \cdot D_G^{(n-1)} \cdot \nu_G^{(n-m)}$$

Examples of the models which can be in principle taken into account for description of the gas-phase mass-transport in structured packings follow:

Models

Fluid-fluid interface models

"Film model" proposed by Lewis and Whitman in 1923 assumes thin stagnant film of thickness δ adjacent to the interface, through which the transport is realized solely by the diffusion

$$k_G = \frac{D_G}{\delta}$$

The model doesn't predict dependence of k_G on gas velocity and doesn't provide any clue, how to relate film thickness, δ to geometrical and hydrodynamic parameters of the system. Its prediction of k_G dependence on the first power of diffusivity is in conflict with majority of the experimental data for gas-liquid systems.

The surface renewal models assume change of elements between turbulent, well mixed bulk of the phase and the interface. Throughout of retention of the elements at the interface, unsteady diffusion into the element takes place. The exchange of the elements is supposed to takes place after the period of time known as exposition time, τ_G, which is constant for all the elements in the "penetration model" variant proposed by Higbie 1935. No mass exchange is assumed among the neighbor elements.

$$k_G = \frac{2}{\sqrt{\pi}}\sqrt{\frac{D_G}{\tau_G}}$$

where

$$\tau_G = \frac{d_G}{u_G}$$

Model predicts dependence $k_G \sim u_G^{0.5}$, $k_G \sim D^{0.5}$ and $k_G \sim d_G^{-0.5}$. Regularity of the exchange process is perhaps justified for the regular architecture of the structured packings. Negligible mass-exchange between the elements at the interface desires flat velocity profile or stagnant fluid in the element, however. This is hardly imaginable for the gas phase, where velocity profile develops at the gas-liquid interface, where no-slip condition holds.

Boundary layer model

These models assume development of boundary layer along the surface. Bellow the critical Reynolds number this layer remains laminar. For laminar boundary layer developing along the flat plate, while bulk remains turbulent, following formula has been derived

$Sh = 0.664 \cdot Re^{1/2} Sc^{1/3}$

for $Re < 5 \times 10^5$

Turbulent flow models

These models assume dominant form of transport to the interface by turbulent eddies and analogy between momentum, heat and mass transfer. Well known "Chilton-Colburn analogy" postulates equality of "j-factors",

$$j_M \equiv \frac{f}{2} = j_H \equiv \frac{h}{C_P \varrho u} \cdot Pr^{2/3} = j_M \equiv \frac{k}{u} \cdot Sc^{2/3}$$

These factors themselves are functions of Reynolds criteria. This dependence is rather low for tubes

$j = 0.023 \cdot Re^{-0.2} \qquad Re > 10000$

resulting in

$Sh = 0.023 \cdot Re^{0.8} Sc^{1/3}$

and more pronounced for the flow past the single sphere, this dependence has been found to be

$j = 0.37 \cdot Re^{-0.4} \qquad 20 < Re < 100000$

$Sh = 0.37 \cdot Re^{0.6} Sc^{1/3}$

and for the flow through the packed bed

$j = 1.17 \cdot Re^{-0.415} \qquad 10 < Re < 2500$

$Sh = 0.37 \cdot Re^{0.585} Sc^{1/3}$

The obvious problem of utilization of these models for description of the gas flow in structured packings is in relatively low value of Reynolds number in structured packings of common sizes. The dependence of j-factors on Reynolds number must be determined experimentally.

Table 1: Summary of fundamental models

MODEL	$k_G = u_G^i$ i=	$k_G = D_G^i$ i=	$k_G = d_G^i$ i=	$k_G(nominal)$[m/s]
Film model	N/A	1	N/A	N/A
Penetration model	0.5	0.5	-0.5	0.0376m/s
Laminar boundary layer in turbulent flow in tube	0.5	2/3	-0.5	0.0214m/s ($d=d_P$) 0.0049m/s (d=l)
Chilton-Colburn: tube	0.8	2/3	-0.2	0.00659m/s
Chilton-Colburn: Sphere	0.6	2/3	-0.4	0.0247m/s
Chilton-Colburn : packed bed	0.6	2/3	-0.415	0.0221m/s

nominal conditions: u_G=1m/s, d_P=0.01552m, ρ=1.205kg/m^3, μ=18.18x10^{-6}Pas, D_G = 0.00001223604358 m^2/s (Re=1455, Sc=1.233)

Gas-phase mass transfer model packages developed for structured packings

Billet and Schultes (1993)[7] proposed gas-phase mass-transfer model which starts from penetration theory

$$k_G = \frac{2}{\sqrt{\pi}} \cdot \sqrt{\frac{D_G \cdot u_G}{(\epsilon - h_L) \cdot l_{\tau G}}}$$

where mixing length is equal to the hydraulic dimension of the packing

$$l_{\tau G} = d_P = \frac{4 \cdot \varepsilon}{a_g}$$

At this theoretical stage of their model development the dependence of the gas phase mass-transfer coefficient on gas velocity and on the packing dimension is as follows:

$$Sh_G = \frac{2}{\sqrt{\pi \cdot (\varepsilon - h_L)}} \cdot Re^{1/2} \cdot Sc^{1/2}$$

$$k_G \propto u_G^{1/2} d_P^{-1/2}$$

Model has been further adapted in order to fit results of experimental databank of Department of Thermal Separation Processes at Bochum University. Through this adaptation the dependence of k_G on u_G and d_P significantly changes to

$$k_G \propto u_G^{3/4} d_P^{-5/4}$$

where dependence on packing dimension is also included in variation of packing characteristic dimension, C_V, for different packings. In the model update from 1999 [6] the relation remains unchanged.

Model of Hanley (2012) [3] is based on the dimension analysis of the quantities describing gas-phase mass-transfer, which results in dimensionless equation

$$Sh_g = C \cdot Re^m \cdot Sc^n$$

Values of exponents *m*, *n* is specified by the assumption of validity of Chilton-Colburn analogy in the form

$$j_D = \frac{Sh_g}{Re_g^1 Sc_g^{1/3}} = \frac{f(Re_g)}{2}$$

For larger channels of metal sheet packings the Reynolds number dependence of the friction factor is assumed negligible and therefore

$$Sh_g = C \cdot Re_g^1 \cdot Sc_g^{1/3}$$

For mutual dependence of the quantities studied in the frame of this picture holds

$$k_G \propto u_G^1 d_P^0$$

For smaller channels assumed for gauze packings the Reynolds number dependence of the friction factor is assumed non-negligible and the resulting relation takes form of

$$Sh_g = C \cdot Re_g^{1/2} \cdot Sc_g^{1/3}$$

Constant C has been validated through fit to distillation data only, absorption experiments served for fit of liquid side mass-transfer coefficient solely.

Johnstone and Pigford (1942) [10] optimized the parameters *m* and *C* based on the experiments in the wetted wall column performed with four distillation and two absorption systems. They found the values $m = 0.77$ and $C = 0.0328$. The exponent n = 0.33 was taken over from Chilton and Colburn (1935) [13] because their experiments covered only a narrow range of Sc values and it was not possible to determine the effect of this group.

Bravo et al. (1985) [11] modeled mass-transfer in the gas-phase in the dimensionless form where characteristic dimension is calculated from geometry of corrugation. They used the HETP data of Sulzer BX packing measured on four distillation systems under various pressures. They correlated the HETP data with assumptions of fully wetted area of the packing and the mass transfer resistance limited to the gas phase and found the relation

$$Sh_g = 0.0338 \cdot Re^{0.8} \cdot Sc^{0.33}$$

Rocha et al. (1996) [18] rearranged this relation into

$$Sh_g = 0.054 \cdot Re^{0.8} \cdot Sc^{0.33}$$

Nawrocki et al. (1991) [12] and Brunazzi et al. (1995) [13] have took over the relation from Bravo et al. (1985) [11].

Xu et. al. (2000) [14] has took over the relation from Billet and Schultes (1993) [7]. They fitted the effective interfacial area to the experimental data.

As noted by Olujic' [13], for gas velocities prevailing in distillation columns hydrodynamic conditions correspond rather to transient than turbulent behavior. Therefore the author takes geometrical average of turbulent and laminar Sherwood number as a representative.

$$Sh_{g,turb} = \sqrt{Sh_{g,lam}^2 + Sh_{g,turb}^2}$$

Expressions for Sherwood numbers have been deduced from analogy with the heat transfer and the source relations are chosen that for short tubes with pronounced entrance effects for which the flow is not stabilized.

$$Sh_{g,lam} = 0.664 \cdot Re^{1/2} \cdot Sc^{1/3} \cdot \sqrt{\frac{d_P}{l_G}}$$

$$Sh_{g,turb} = \frac{Re_G \cdot Sc_G \frac{\xi_{GL}\varphi}{8}}{1+1.27\sqrt{\frac{\xi_{GL}\varphi}{8}}(Sc^{2/3}-1)}\left[1+\left(\frac{d_P}{l_G}\right)^{2/3}\right]$$

Nieuwoudt and Crause [2,13] considered earlier performed tests on wetted wall columns as inappropriate to be a basis for structured packings due to length to diameter ratio far greater than that occurring in structured packings. They performed experiments on the WWC with a length of four diameters using pure liquids. For low viscosity liquids they found

$$Sh_G = 0.00283 \cdot Re_G \cdot Sc_G^{0.5} \cdot Re_L^{0.08}$$

Table 2: Summary of model packages

Author	Dependences	$k_G(nominal)$ [m/s]
Bravo et. al.	$k_G \propto u_G^{0.8} d_P^{-0.2}$	9.7×10^{-3}
Hanley et. al. (2012)	$k_G \propto u_G^{1} d_P^{0}$	7.3×10^{-3}
Rocha et. al.	$k_G \propto u_G^{0.8} d_P^{-0.2}$	15.5×10^{-3}
Olujic' et al.		9.6×10^{-3}
Billet and Schultes	$k_G \propto u_G^{3/4} d_P^{-5/4}$	44.6×10^{-3}
Nieuwoudt and Crause	$k_G \propto u_G^{1} d_P^{0}$	4.73×10^{-3}
EXPERIMENTAL		32.1×10^{-3}

nominal conditions: u_G=1m/s, B=5m/h, d_P=0.01552m, ρ_G=1.205kg/m^3, μ_G=18.18x10^{-6}Pas, D_G = 0.00001224 m^2/s (Re_G=1455, Sc_G=1.233), ρ_L=1041kg/m^3, μ_L=1.236x10^{-3}Pas, σ=74.62mN/m.

Present time model packages for calculating hydraulic and transport characteristics of structured packings are developed from combination of basic principle thoughts and experimentally observed behavior. Gas-phase mass-transport is based on the imagination of the unidirectional and undisturbed flow through the triangular channel. Although Reynolds number for typical operations suggest laminar to transient regime of the gas-flow, mass-transfer relations are never based on solution of corresponding hydrodynamic and transport equations and normally assume turbulent character of the flow. In the Hanley model solution resulting from boundary layer is taken as a limit appropriate for small channel wire gauze packings.

Many mass-transfer models are based on the results which had been obtained through experiments with wetted wall columns. Recalculation for triangular cross-section of the structured packing channel is performed through usage of hydraulic diameter. As the typical length of the channel in the element of the structured packing is shorter than that used by early wetted-wall investigators, experiments in the equipment with lower length to diameter ratios were carried out. This arrangement also corrects somewhat improbable assumption of unidirectional flow of the gas in the frame of one element.

Another approach represented by DELFT model assumes analogy between heat and mass-transfer and therefore enables usage of more elaborated correlations for heat transfer. Chosen ones are able of description of laminar as well as turbulent flow in short channels, where the entrance effects can be significant.

Results of all model packages are finally confronted with pilot-plant and industrial scale efficiency results. Billet and Schultes use the large absorption and distillation database for determination of each packing specific constant. Other authors choose rearrangement of model equations to fit the experimental results and try to avoid usage of packing specific constants.

According to the authors tuning of the model on the basis of HETP data brings severe problems, however. While calculation of HETP from fundamental mass-transfer data (k_L, k_G, a) is straightforward, their backward deduction from overall HETP is uncertain. When values of Reynolds and Schmidt number exponents in gas-phase mass-transfer correlations are questioned, they must be evaluated together with those for liquid phase and with parameters of effective area equations. Determination of so many parameters is clearly difficult from dependence of single variable (HETP) on only several possible parameters (system, pressure, load).

Experimental

The experiments have been carried out in the column with a 0.29 m inner diameter. The packed bed consisted of four elements of the packing rotated mutually by 90° with a total bed height of 0.84 m. The packings used were Mellapak 250Y, 350Y, 452Y, and 500Y. Each element of the packing was horizontally drilled by electrical discharge machinery to enable installation of a sampling device for the withdrawal of the gas samples. This sampling device is a virtually tube of a diameter of 30mm with four downward conical openings collecting solely the gas and performing so uniformly along the packing diameter. Liquid distribution over the cross-section of the column was carried out by the pipe liquid distributor with drip-point density of 630 dp/m^2. Gas entered first a large drum where the flow was calmed and proceeded into the column through the grid which secured its uniform distribution.

Absorption liquid was the water solution of NaOH with concentration of 1 kmol/m^3. The solution was recirculated. The batch was superseded when the concentration of OH^- dropped under certain limit; 0.5 kmol/m^3 - in the case of interfacial area measurement, 0.1 kmol/m^3 - in the case of k_Ga measurement. The gas circuit was open. The experiments were performed with the temperature of the gas and liquid entering the column maintained at 20 °C. The gas was saturated with water vapor. The experiments were carried out up to flooding conditions with liquid velocities of 5, 10, 20, 40, 60, 80, 100 m^3/(m^2 h) and gas velocities of 0.5, 1.0, 1.5, 2.0, 2.5, 3.0 m/s.

The tracer gas was introduced into the gas phase long before it entered the column and was therefore well mixed. Tracer gas flow was set to maintain its concentration at the lowest sampling point at least 1000 ppm. The gas was sampled from each element at four different levels of the packed bed. The height of the packing between the sampling points corresponded to the height of the element and was 0.21 m. Mass-transfer characteristics were evaluated from concentrations measured in only two sampling points, however. Their choice has been done on the basis of an effort to cover as long as possible packed bed, but the top concentration must not drop under about 20ppm. Under this limit the error in concentration determination would be unbearable. For measurement of the interfacial area this politics meant most often utilization of 0.63m of the bed, while for k_Ga determination 0.21m or 0.42m of the bed was utilized.

The CO_2 and SO_2 concentration in the gaseous samples were measured continuously using IR analyzer S710 (Sick). Before entering the analyzer the gas was dried by a passage through the cartridge filled by crushed $CaCl_2$ (CO_2 analysis) or $MgClO_4$ (SO_2 analysis).

k_Ga determination

Measurement of the volumetric mass-transfer coefficients, k_Ga, is performed through experiments with systems, for which the mass-transfer resistance is limited to the gas phase. Typically, evaporation, or absorption accompanied by instantaneous interface reaction of the transferring gaseous solute is utilized [5]. In this work, absorption of SO_2 into the lye has been used as this is recommended and proven method used by the workplace routinely. Instantaneity of the interface reaction between SO_2 and OH^- ions has been experimentally verified by Linek 1995 [7].

Under the assumption of the plug flow of the gas phase, relation among process, geometrical, transport and hydrodynamic parameters takes shape

$$k_G a = \frac{u_G}{H} \cdot ln \frac{c_A^{IN}}{c_A^{OUT}}$$

and predicts arrangement of the experiment and necessary quantities to be measured.

Effective area determination

The effective area is usually measured using absorption of the solute *A* from gas into the liquid, in the system where the gas phase resistance is negligible and the rapid reaction of A with *B* consumes quantitatively the solute in the liquid film. In such case, the mass-transfer rate is determined by the reaction-transport phenomena in the liquid film and is predictable from physical quantities of the system. Simultaneously, mass-transfer rate of the solute is determinable from the solute balance over the packing height resulting in the relation

$$a = \frac{1}{He \cdot \sqrt{D_{A,L} \cdot k' \cdot c_B}} \cdot \frac{u_G}{H} \cdot ln \frac{c_A^{IN}}{c_A^{OUT}}$$

In this work, absorption of the CO_2 from air into the lye has been used. Value of the reaction-transport term $He \cdot \sqrt{D_{A,L} \cdot k' \cdot c_B} = 2.24 \times 10^{-3} m/s$ was determined by still level experiments as this procedure circumvents introduction of errors caused by accumulation of physical data uncertainties.

k_G determination

The absorption experiments related with $k_G a$ as well as *a* determination have been performed using lye as a liquid phase and air as the gas phase. Under the same liquid and gas flow rate, effective interfacial area for both experiments can be therefore considered equal, as usage of CO_2 or SO_2 cannot effectively change hydrodynamic conditions. Mass transfer coefficient was therefore determined as

$$k_G = \frac{k_G a}{a}$$

Results

Values of volumetric mass-transfer coefficient ($k_G a$) and effective interfacial area (a) resulting from the experiments have been successfully correlated in the power law form. The sum of squares of differences between measured and calculated quantity have been used as the objective function of correlation.

$$k_G a = C_1 \cdot u_G^{\alpha} \cdot B^{\beta} \quad (1)$$

$$a = C_2 \cdot u_G^{\gamma} \cdot B^{\delta} \quad (2)$$

Correlations have been performed for each of packing individually. Data whose contribution to the value of the objective function represented more than five times the value of the average contribution were excluded from set of correlated data and correlation procedure was repeated. Altogether 13 experimental values have been excluded from the overall number of 98 experiments performed.

Values of mass transfer coefficients (k_G) for each packing are presented in the power law form

$$k_G = C_3 \cdot u_G^{\varepsilon} \cdot B^{\zeta} \quad (3)$$

where constants for this correlation have been obtained by division of equation (1) by equation (2).Values of the constants can be found in the Table 1. These correlations fit experimental data well for phase velocities up to about 2/3 of those at flooding.

Table 3: Constants of correlations 1,2,3

	C_1	α	β	C_2	γ	δ	C_3	ε	ζ
M250Y	4.4	0.130	0.662	147	0.107	0.043	0.030	0.620	0.023
M350Y	5.6	0.164	0.643	202	0.068	0.048	0.028	0.595	0.096
M452Y	5.9	0.147	0.624	190	0.091	0.014	0.031	0.610	0.056
M500Y	6.0	0.279	0.762	297	0.118	0.175	0.020	0.587	0.161

M250Y packing provides lowest $k_G a$ values for given loads, but offers better capacity than M350Y and significantly better capacity than M500Y (Fig. 1). M350Y and M452Y packing provide similar mass-transfer behavior in the preloading region, but M452Y provides significantly higher capacity. In the range of loads above the maximum capacity of M350Y its high capacity analog M452Y retains very good efficiency (Fig. 2,3). The M500Y packing achieves highest absolute values of $k_G a$ for given loads, but provides just limited capacity (Fig.4).

Constants of the correlation (3) for mass-transfer coefficient k_G (C_3, ε,ζ) reveal, that dependence of k_G on the superficial gas phase velocity is for all packings close to
$k_G \sim u_G^{0,6}$
Dependence of k_G on liquid flow rate is rather low,
$k_G \sim B^{\leq 0.161}$
As can be seen from Figure 5, under the same phase velocities mass-transfer coefficient k_G exhibit highest values for Mellapak 350Y and Mellapak 452Y whereas its values are slightly lower for Mellapak 250Y and Mellapak 500Y. Generally can be the values considered similar, however.

Discussion of the results

Comparison of experimental data with results predicted by mass-transfer models provides several conclusions valid generally.

1) **Values of k_G predicted by fundamental mass-transfer models and by WWC experiments are far too low:** Experiments reveal values of k_G of about 0,035m/s for nominal conditions. These values are largely under-predicted by fundamental models, by WWC correlations and by analogies with heat transfer based on the flow through a channel, almost by an order of magnitude. Only BS model overpredicts actual k_G values. The BS data are obtained by readjustment of the raw model results to the large pilot-plant database.
2) **Dependence of k_G on gas superficial velocity predicted by WWC experiments is too high in comparison with the one found here for structured packings:** Experiments performed with WWC provided gas velocity dependence of k_G to be $k_G \sim u_G^{0.8\text{-}1.0}$. Experiments performed here on structured packings provide quite convincingly dependence $k_G \sim u_G^{0.6}$. Similar performance $k_G \sim u_G^{0.6\text{-}0.7}$ has been found earlier for random packings by [16] and is typical for turbulent flow around the sphere and cylinder. On the other hand, exponent of the dependence $k_G \sim u_G^{0.5}$ predicted by penetration theory and by laminar boundary layer model is bit too low.
3) **Mass-transfer coefficient k_G is similar for all studied packings of different characteristic dimension**: Mass-transfer coefficients k_G of Mellapak 250Y and of Mellapak 500Y differ in average by about 5%, those found on M350Y. Generally, the values of the k_G coefficients are found to be very similar for all the studied packings under identical conditions. However, correlation of the type $Sh_g = C \cdot Re^{0.6} \cdot Sc^n$ predicts increase of k_G for Mellapak 500Y by 32% with respect to Mellapak 250Y.

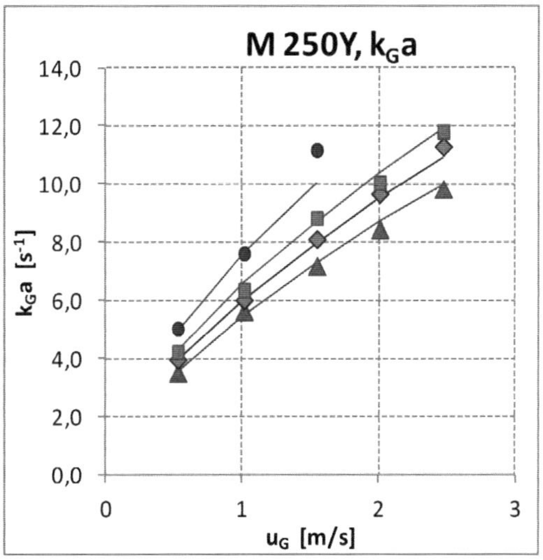

Figure 1

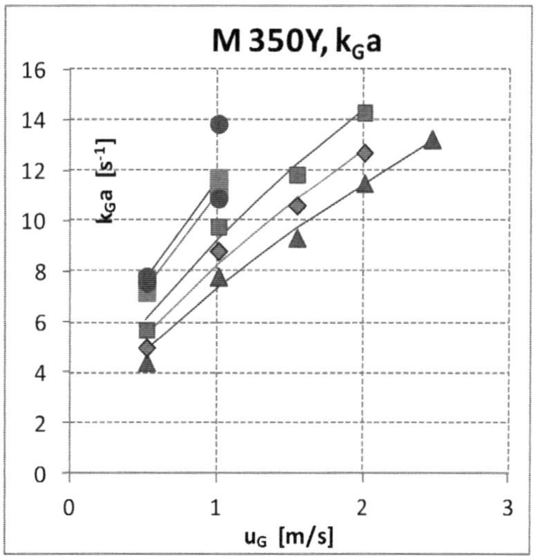

Figure 2

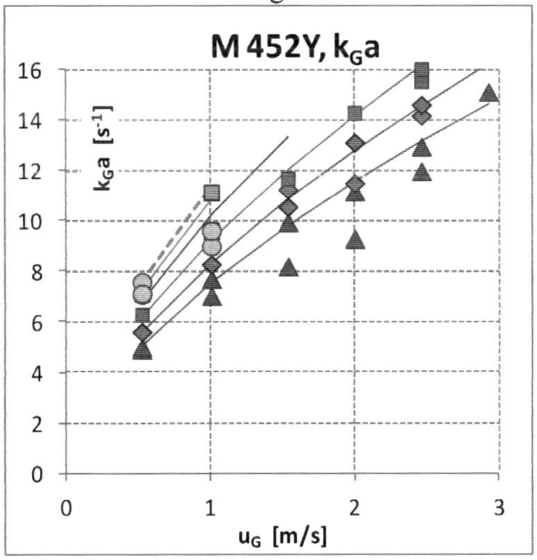

Figure 3

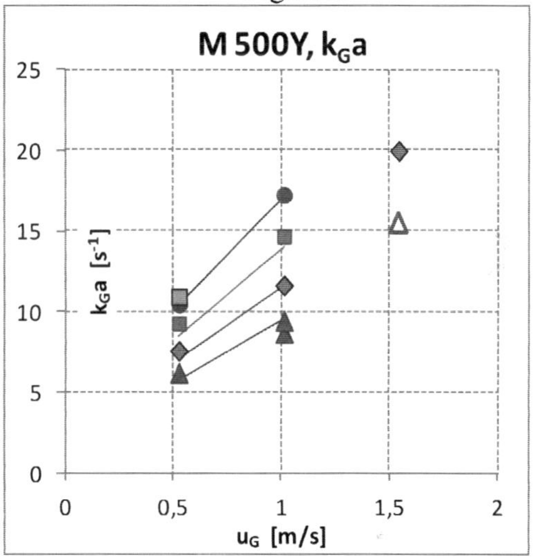

Figure 4

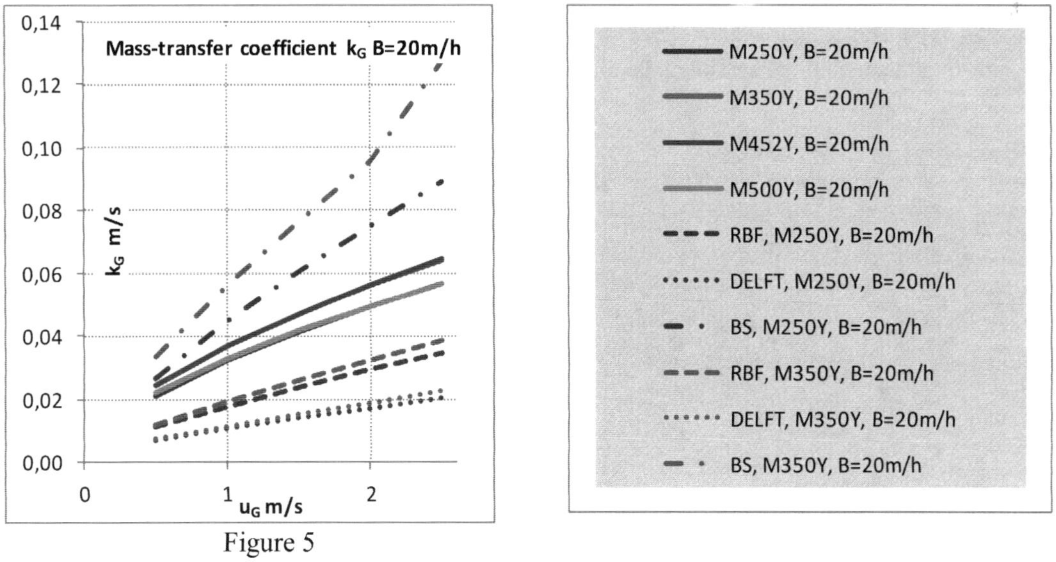

Figure 5

Conclusions

Gas flow through the structured packing is usually modeled on the basis of turbulent flow through the channel of the triangular cross-section. Long or short wetted-wall column is taken as a corresponding experimental set-up. Alternatively results of heat-transfer in the straight tube are utilized using analogy between heat and mass-transfer. Values of k_G acquired from this geometrical arrangement are substantially lower than those found experimentally in this work. The dependence of k_G on gas velocity is also different. It leads us to the belief, that modeling of the gas phase mass-transfer in the structured packings on the basis of the flow through the straight channel and on wetted-wall experiments is not (at least not universally) appropriate.

The value of exponent in velocity dependence of k_G dropping toward ½ is characteristic for flow around submerged objects, where abrupt change in gas flow direction and its velocity occurs. CFD simulations [15] indicate, that the gas is actually proceeding vertically through the packing in all directions from -45° to +45°, what suggests, that fair amount of the gas must perform "hops" around the crimps instead of flowing through the channels. Perpendicular and skewed arrangement of the channels also implies inevitable formation of rotational component of the velocity vector. There is also interesting, that absolute k_G value and its dependence on the gas-velocity are quite well described by venerable Onda's model [16] characterizing first generation packings. These packings (Raschig rings, spheres, saddles) certainly do not form straight channels, rather lead to the imagination of flow around an object.

The situation is not facilitated by the finding that usual dimensionless equation for Sherwood criteria in the form $Sh_g = C \cdot Re^m \cdot Sc^n$ does not enable simultaneous correct description of the k_G velocity and characteristic dimension dependence when hydraulic diameter of the channel is taken for the latter. Shape of the correlation and choice of appropriate characteristic dimension should be therefore also reconsidered.

As structured packings dominate the market due to their undoubted advantages, further progress in modeling of the structured packing mass-transfer behavior should be encouraged. There is still plenty problems to work on.

Acknowledgement:
Authors are grateful to Sulzer Co. for providing of the packings and to the Czech Science Foundation for supporting the project through Grant 13-01251S.

Symbols:

a	effective area	m^{-1}
a_g	geometrical area of the packing	m^{-1}
B	superficial liquid velocity	$m \cdot hr^{-1}$
c	molar concentration	$mol \cdot m^{-3}$
C_P	heat capacity	$J \cdot kg^{-1} \cdot K$
C_V	Billet and Schultes model constant	-
d	characteristic dimension, diameter of the tube	m
d_P	characteristic dimension of the packing	m
D	diffusion coefficient	$m^2 \cdot s^{-1}$
E	axial dispersion coefficient	$m^2 \cdot s^{-1}$
f	friction factor $f = \frac{1}{4}\frac{d}{L}\frac{P_0 - P_L}{\frac{1}{2}\rho u^2}$	-
h	heat transfer coefficient	$W \cdot m^{-2} \cdot K$
H	height of the packing	m
h_L	liquid hold-up	-
j_H	Chilton-Colburn's j-factor for the heat transport $j_H = \frac{h}{C_P \varrho u} \cdot Pr^{2/3}$	-

j_M	Chilton-Colburn's j-factor for the mass transport $j_M \equiv \frac{k}{u} \cdot Sc^{2/3}$	-
k	mass-transfer coefficient, concentration driving force	$m \cdot s^{-1}$
k'	second order reaction rate constant	$\frac{m^6}{mol \cdot s}$
$k_L a$	volumetric mass-transfer coefficient in the liquid phase	s^{-1}
$k_G a$	volumetric mass-transfer coefficient in the gas phase	s^{-1}
$k_i a^{\circ}$	volumetric mass-transfer coefficient in the ith phase, evaluated with the assumption of plug flow of the phase	s^{-1}
$k_i a^{\bullet}$	volumetric mass-transfer coefficient in the ith phase, evaluated with the assumption of plug flow with axial dispersion	s^{-1}
K	overall mass-transfer coefficient	$m \cdot s^{-1}$
$l_{G,pe}$	length of gas flow channel	m
$l_{\tau G}$	mixing length in the gas phase	m
M	molar mass	$kg \cdot mol^{-1}$
u	superficial velocity	$m \cdot s^{-1}$

<u>Subscripts:</u>

g	geometrical
G	gas phase
L	liquid phase

<u>Superscripts:</u>

IN	at the inlet
OUT	at the outlet

<u>Greek symbols:</u>

$\alpha, \beta, \gamma, \delta, \zeta$	constants of the relations 4,5,6	
ε	packing porosity	-
ρ	density	$kg \cdot m^{-3}$
μ	dynamic viscosity	$Pa \cdot s$
ξ_{GL}	gas-liquid friction factor	-
φ	fraction of the channel occupied by liquid	-

<u>Dimensinless criteria:</u>

Re_i	Reynolds number in the i-phase	$Re_i = \frac{u_i \cdot d \cdot \rho_i}{\mu_i}$
Sh_i	Sherwood number in the i-phase	$Sh_i = \frac{k_i \cdot d}{D_i}$
Sc_i	Schmitt number in the i-phase	$Sc_i = \frac{\mu_i}{\rho_i \cdot D_i}$

Literature cited:

[1] Valenz, L.; Rejl, F.,J.; Šíma, J.; Linek, V.: Absorption Mass-Transfer Characteristics of Mellapak Packing Series. *Ind. Eng. Chem. Res.* **2011**, *50*, 12134-12142.

[2] Crause, J., C.; Nieuwoudt, I.: Mass Transfer in a Short Wetted-Wall Column. 1. Pure Components. *Ind. Eng. Chem. Res.* **1999**, *38*, 4928-4932.

[3] Hanley, B.; Chen, Ch. New Mass-Transfer Correlations for Packed Towers. *AIChE Journal* **2012**, *58* (1), 132–152.
[4] Wang, G. Q.; Yuan, X. G.; Yu, K. T. Review of Mass-Transfer Correlations for Packed Columns. *Ind. Eng. Chem. Res.* **2005**, *44*, 8715–8729.
[5] Rejl, F. J.; Linek, V.; Moucha, T.; Valenz, L. Methods Standardization in the Measurement of Mass-Transfer Characteristics in Packed Absorption Columns. *Chemical Engineering Research and Design* **2009**, *87*, 695-704.
[6] Billet, R.; Schultes, M. Prediction of Mass Transfer Columns with Dumped and Arranged Packings. Updates Summary of the Calculation Method of Billet and Schultes. *Trans IChemE* **1999**, *77,* Part A, 498-504.
[7] Billet, R.; Schultes, M. Predicting Mass Transfer in Packed Columns. Chem. Eng. Technol. **1993**, *16*, 1-9.
[8] Rejl, F.,J.; Valenz, L.; Linek, V.: "Profile Method" for the Measurement of k_La and k_Va in Distillation Columns. Validation of Rate-Based Distillation Models Using Concentration Profiles Measured along the Column. *Ind. Eng. Chem. Res.* **2010**, *49*, 4383–4398.
[9] Bird, R.B.; Stewart, W.E., Lightfoot, E.N.: Transport Phenomena. *John Wiley & Sons*, **2002.**
[10] Henley, E.,J.; Seader J.,D.; Roper, D., K.: Separation Process Principles. Third Edition, International Student Version. *John Wiley & Sons*, **2011**.
[11] Olujic', Ž.: Development of a Complete Simulation Model for Predicting the Hydraulic and Separation Performance of Distillation Columns Equipped with Structured Packings. Chem. Biochem. Eng. Q. **1997**, *11 (1)* 31-46.
[12] Olujic', Ž., Behrens, M., Colli, L. and Paglianti, A.: Predicting the Efficiency of Corrugated Sheet Structured Packings with Large Specific Surface Area. *Chem. Biochem. Eng. Q.* **2004** *18 (2),* 89–96
[13] Olujic´, Ž., Kamerbeek, A.B., de Graauw, J.: A corrugation geometry based model for efficiency of structured distillation packing. *Chemical Engineering and Processing* **1999**, 38, 683–695.
[14] Nieuwoudt, I; Crause, J., C.: Mass Transfer in a Short Wetted-Wall Column. 2. Binary Systems. *Ind. Eng. Chem. Res.* **1999**, *38*, 4933-4937.
[15] Saleh, A., R.; Hosseini, S., H.; Shojaee, S.; Ahmadi, G.: CFD Studies of Pressure Drop and Increasing Capacity in MellapakPlus 752.Y Structured Packing. *Chem. Eng. Technol.* **2011**, *34*, No. 9, 1402-1412.
[16] Onda, K.; Takeuchi, H.; Okumoto, Y.: Mass Transfer Coefficients Between Gas and Liquid Phases in Packed Columns. *Journal of Chemical Engineering of Japan*. **1968**, 1, No. 1, 56-62.
[17] Schultes, M.: Research on Mass Transfer Columns "Old Hat or Still Relevant". *Lecture at AIChE 2013 Spring Meeting.*
[18] Rocha, J., A.; Bravo, J., L.; Fair, J. R.: Distillation Columns Containing Structured Packings: A Comprehensive Model for Their Performance. 2. Mass Transfer Model. *Ind. Eng. Chem. Res.* **1996**, *35*, 1660-1667.

(29b) Modeling of Mass Transfer of Random Packings for Gas-Liquid Systems

Jerzy Mackowiak, ENVIMAC Engineering GmbH, Oberhausen, Germany

This paper presents a new relationship for the prediction of the efficiency of random packings of various shapes and sizes from 15 to 90 mm for distillation and absorption systems, based on the droplet flow model in packed beds.

The new relationship in connection with the model of the instationary diffusion for short residence times [1] allows the prediction of the separation efficiency values HTUOV or nt/H up to the flooding point for classical packing types such as Raschig rings, Intalox saddles or Pall rings, as well as for modern lattice-type packings with partially perforated wall surface such as Hiflow rings, VSP rings, Mc-Pac, Tellerette, Envipac, Dtnpac, Ralu-Flow, RSR, CMR rings and others [2].

The presented model has been derived from an experimental data bank containing data from 30 different test systems under vacuum and normal pressure and for above mentioned packings [2] and with mass transfer systems with major gas-side resistance, such as NH3-air/water, NH3-air/1n H2SO4, SO2-air/water and SO2-air/1n NaOH and CO2-water/air in columns with diameters dS= 0.15-1.4 m.

The model was verified by means of more than 60 different packings and for a total number of more than 4000 experimental mass transfer absorption and distillation data points for above mentioned systems. The new model described the experimental data very well with a mean relative error of ±15-below the loading line Fv<0,65 FV,Fl and ± 23% below the flooding point. The satisfactory accordance of calculated data with experiments is confirmed by an additional comparison with the model of Billet and Schultes [3] for chosen packing types such as metallic and plastic Pallrings and Hiflowrings. This additionally confirms the general validity of the new relationship. In comparison to other known correlations from literature, the new model does not contain any characteristic, empirical packing specific constants.

The limitations of the new model are discussed and the calculation examples and comparisons with other models are presented.

With the presented model, it is possible to predict the volumetric mass transfer coefficient in the gas phase βV.a and finally, in combination with [1], the efficiency of any random packing type and size, nt/H- value or HTUOV-value, in the whole loading range up to the flooding point.

[1] MAÆKOWIAK, J., Model for the prediction of liquid phase mass transfer of random packed columns for gas–liquid systems, Chem. Eng. Res. Des. 2011, 89, 1308–1320.

[2] MAÆKOWIAK, J., Fluid Dynamics of Packed Columns - Principles of the fluid dynamic design of columns for gas/liquid and liquid/liquid systems, Springer, Berlin, 2009.

[3] BILLET, R., SCHULTES, M., Prediction of mass transfer columns with dumped and arranged packings, Trans IChemE 1999, 77, 498-504.

A RIVULET FLOW MODEL FOR WETTING EFFICIENCY IN A PACKED BED

Duvvuri Subbarao*, Fairuz Asyika Rosli, Faud Deen Azmi, Prakash Manogaran and Shuhaimi Mahadzir

Chemical Engineering Department, Univerisiti Teknologi PETRONAS, Tronoh, Malaysia.

E-mail: duvvuri_subbarao@petronas.com.my

ABSTRACT:

Wetting efficiency in a packed / trickle bed has direct impact on gas-liquid as well as liquid-solid mass transfer. Based on experiments on rivulet flow on an inclined plane surface, rivulet flow width is correlated to rivulet flow rate through the friction factor. This correlation is adopted to develop an equation for wetting efficiency in packed beds and evaluated with the literature data. Wetting efficiency is controlled mainly by gravitational and inertial forces. Effect of liquid surface interaction characterized by contact angle can influence the wetting efficiency.

Key words: Rivulet flow, Wetting efficiency, Liquid Holdup, Packed Bed Absorber, Trickle Bed Reactor

1. INTRODUCTION:

Packed towers are widely used in process industries in unit operations such as distillation, absorption, water cooling and catalytic gas-liquid reactions in which flowing gas and liquid need to be brought into intimate contact. The liquid is introduced through a distributor at the top of the packed bed and it trickles down over the particles in the form of rivulets or films. Wetting efficiency is defined as the ratio of gas-liquid contact area to the particle surface area. Wetting efficiency directly affects the performance of the operations in packed towers. Direct observation of flow and wetting using techniques such as dye adsorption [1] and computer assisted tomography [2-5] did provide qualitative understanding of the wetting phenomena. Quantitative measurement of wetting efficiency was made possible based on its impact over tracer response [6-13] or the rate processes [14-16]. It has been observed that wetting efficiency can be less than one depending on the liquid flow rate, type of liquid distributor, particle shape, size and materials of construction. Wang et al. [17] presented a review on the available correlations to estimate the effective interfacial area and mass transfer coefficients in packed beds.

Rivulet flow on inclined plane surfaces received attention to develop a scientific basis to explain wetting efficiency in packed beds. The width of rivulets increases with increase in liquid flow rate and eventually the liquid film spreads all over the surface of the particles.

Towell and Rothfeld [18] analyzed hydrodynamics of liquid rivulets flowing down an inclined surface in laminar flow to relate rivulet width to liquid flow rate and physical properties of the liquid. Extending this to turbulent flows and packing elements of various shapes can be a formidable exercise. In recent times, there is renewed interest on liquid rivulet flow on inclined surfaces due to recent advances in instrumentation and computational fluid dynamics [17-20]. Application of these results to understand the wetting efficiency in packed beds can still be formidable. An attempt is made in the present work to develop an engineering hydrodynamic model for rivulet flow which can be extended to explain wetting efficiency in packed beds.

2. EXPERIMENTAL STUDY:

A simple experimental facility to measure width of a rivulet flowing down an inclined plane glass surface as a function of liquid flow rate, as shown in Fig.1, is constructed. Water from a constant level over head tank is fed to the top of the inclined glass plate through a tube of 4 mm diameter to let the water flow in the form of rivulet. The rivulet flow is collected to measure water flow rate. An arithmetic graph sheet is pasted below the transparent glass plate to facilitate measurement of the width of rivulet by photography.

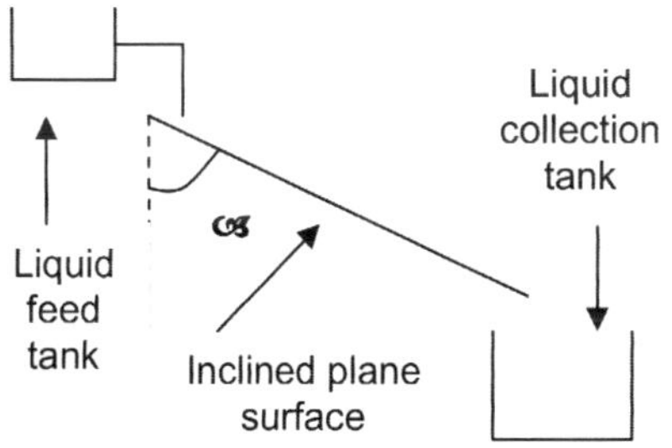

Fig.1a. Experimental set up

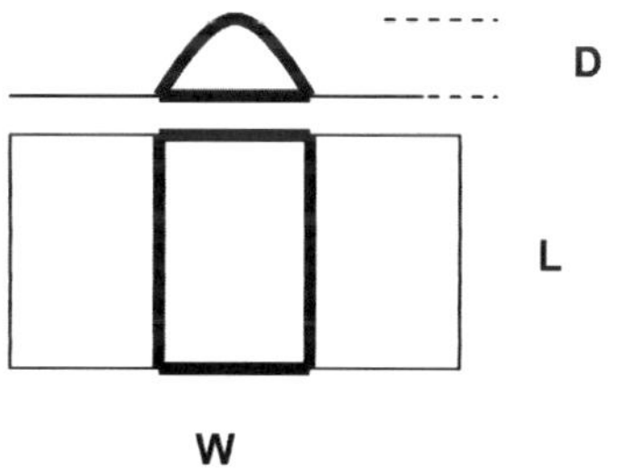

Fig.1b: Dimensions of a rivulet

For each liquid flow rate, rivulet width is measured at different locations along the length of flow of the rivulet. Average rivulet width was noted as a function of liquid flow rate for a plate inclination α of 63.6°.

3. A MODEL FOR RIVULET FLOW:

A rivulet flowing on a plane surface inclined at an angle of α to the vertical, as shown in Fig.1, is considered. Let the width, length, maximum thickness and cross sectional area of the rivulet be W, L, D and A_r respectively. Let the contact angle of the film be θ_r radians (θ degrees). Mass of the liquid in the rivulet on the inclined plane will be (A_r L ρ_l) and the gravitational force (A_r L ρ_l g cos α) acts on the fluid to make it flow downwards. Under steady state, the flow is resisted by frictional force at the wall which is proportional to wall area wetted by the rivulet (WL) and kinetic energy of the flowing liquid in the rivulet ($\rho_l u_l^2/2$). The proportionality constant is the Fanning friction factor f. This may be expressed in the form of an equation as

$$A_r L \rho_l g \cos(\alpha) = fWL\frac{\rho_l u_l^2}{2} = fWL\frac{\rho_l Q_l^2}{2A_r^2} \qquad (1)$$

Assuming the cross section of the rivulet to be a part of an arc with contact angle ϑ radians, cross sectional area of such rivulets can be obtained as

$$A_r = W^2 \frac{(\theta - \sin\theta\cos\theta)}{4\sin^2\theta} = W^2\psi(\theta) \quad \textbf{(2)}$$

From equations (1) and (2), friction factor f can be obtained as

$$f = \frac{2W^5 g\psi^3(\theta)\cos(\alpha)}{Q_l^2} \quad (3)$$

The friction factor is expected to be a function of Reynolds Number Re and it can be expressed as

$$Re = D_h u_l \frac{\rho_l}{\mu_l} = \frac{4A_r}{W}\frac{Q_l}{A_r}\frac{\rho_l}{\mu_l} = \frac{4Q_l\rho_l}{W\mu_l} \quad (4)$$

The friction factor is generally expected to be inversely proportional to Reynolds Number for laminar flow and to be a constant for highly turbulent flows.

$$f = \frac{2W^5 g\psi^3(\theta)\cos(\alpha)}{Q_l^2} = \frac{k_1}{Q_l\rho_l / W\mu_l} = k_2 \quad (5)$$

Numerical values of k_1 and k_2 can depend on surface curvature, roughness, texture and need to be experimentally evaluated. It can be seen that width of the rivulets can be expressed as:\

for laminar flow

$$W \propto \left[\frac{Q_l\mu_l}{g\rho_l}\right]^{0.25} \quad (6)$$

and for turbulent flow

$$W \propto \left[\frac{Q_l^2}{g}\right]^{0.2} \quad (7)$$

4. FRICTION FACTOR FOR RIVULET FLOW:

Experimental observations on rivulet width as a function of liquid flow rate are analyzed in terms of friction factor and Reynolds Number and presented in Fig.2 along with the data of Ataki and Bart [19] and Luo et al. [21]. These results indicate that friction factor for rivulet flow is independent of Reynolds Numbers greater than 3 to 4 hundred (Q_l®l/W/③$_l$) >100) indicating that rivulet flow could be in turbulent regime. The friction factor was around 0.001 for the data of Ataki and Bart [19] and Luo et al. [21] while it is around 0.0001 for the measurements in the present work. This could be due to sensitivity of friction factor to rivulet width measurement (f•W^5).

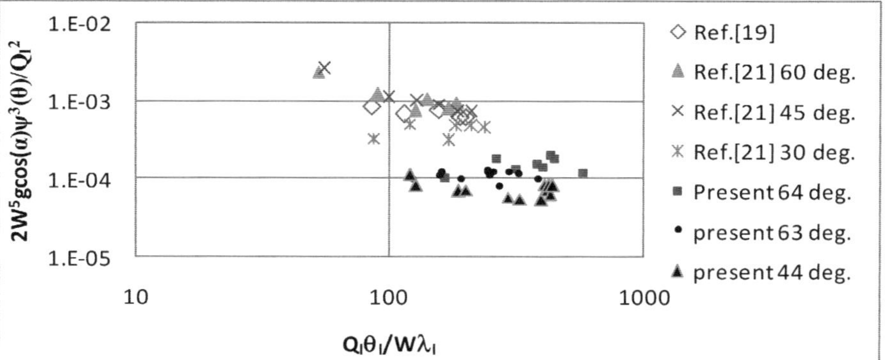

Fig.2. Friction Factor for Rivulet Flow on inclined surfaces as a function of Reynolds Number

5. A MODEL FOR WETTING EFFICIENCY IN A PACKED / TRICKLE BEDS :

Extending the information on rivulet widths on plane surfaces to packing elements is not an obvious exercise. A point source of liquid will yield one rivulet on a plane surface. However on a sphere it may lead to many rivulets in different directions. Still it is worth attempting to extend these results as a means to obtain model based equation for wetting efficiency in packed beds.

Liquid flowing on a single spherical particle in the form of a rivulet is considered. The wetted area can be expected to be proportional to (W Dp). The wetting efficiency is the ratio of wetted area to particle surface area and hence can be written as.

$$\eta \propto \frac{WD_p}{D_p^2} \propto \frac{W}{D_p} \qquad (8)$$

Assuming a perfect and uniform liquid distribution, liquid flow rate per particle Q_l can be expected to be total liquid flow rate divided by number of wetted particles in the cross sectional plane.

$$Q_l \propto \frac{Q}{\eta\left[A(1-\varepsilon)/D_p^2\right]} \propto \frac{u_{ol}D_p^2}{\eta(1-\varepsilon)} \qquad (9)$$

With this wetting efficiency can be obtained as

For laminar flow

$$\eta \propto \left[\frac{u_{ol}^2}{gD_p}\frac{1}{(1-\varepsilon)^2}\frac{\mu_l(1-\varepsilon)}{D_p u_{ol}\rho_l}\right]^{1/5} \qquad (10)$$

and for turbulent flow

$$\eta \propto \left[\frac{u_{ol}^2}{gD_p}\frac{1}{(1-\varepsilon)^2}\right]^{1/7} \qquad (11)$$

6. PACKED BED WETTING EFFICIENCY MODEL VALIDATION

Julcour-Lebique et al. [13] presented extensive measurements on the effect of viscosity (using heptane, (heptane+gasoil) and gasoil with viscosities of 0.0039, 0.012 and 0.034 cP rerspectively) and effect of bed packing size (using particles of size 1.8, 2.85, 3.65 and 7 mm) on wetting efficiency in trickle bed reactors. Their data are compared with laminar flow equation [10] in Fig. 3. The data of Julcour-Lebique et al. [13] are also compared with turbulent flow equation in Fig. 4.

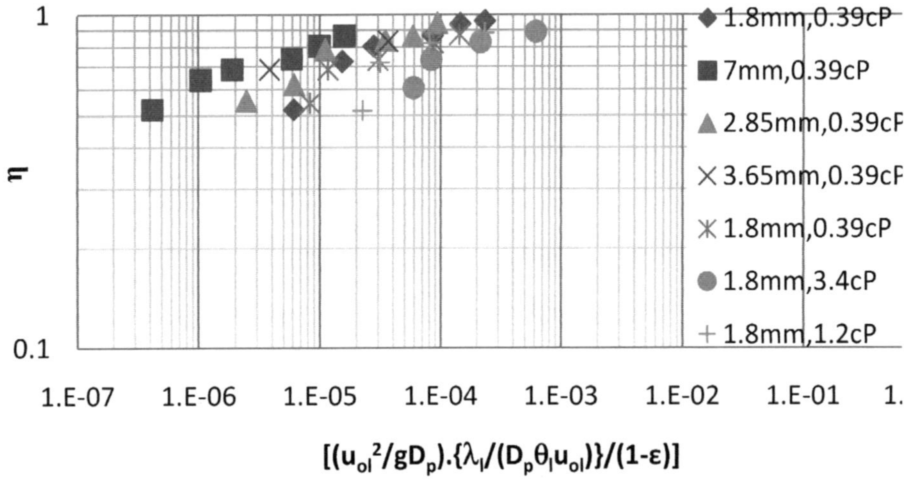

Fig.3:Evaluation of data of Julcor-Lebique et al.[13] with

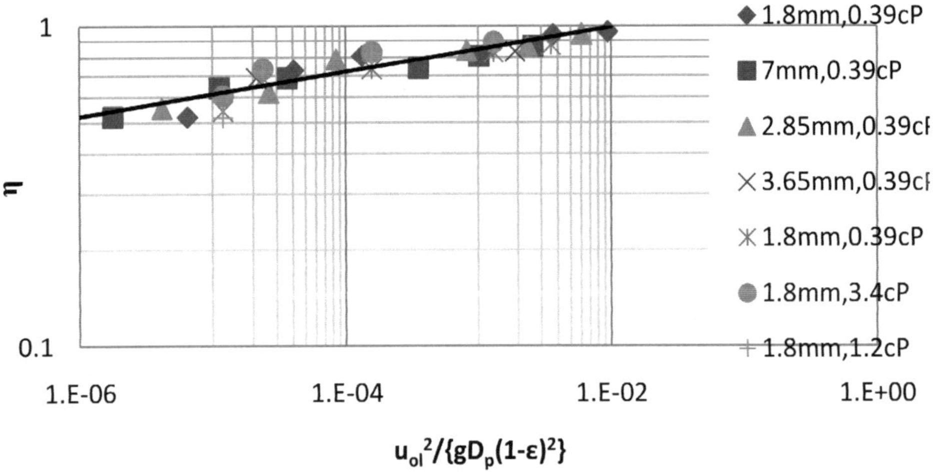

Fig.4 Evaluation of data of Julcor-Lebique et al.[13] with

It can be seen that the correlation with turbulent flow equation is better than the one for laminar flow.

$$\eta = 1.373\left[\frac{u_{ol}^2}{gD_p}\frac{1}{(1-\varepsilon)^2}\right]^{0.07} \qquad (12)$$

Data of Baussarona et al. [12] on the effect of liquid/solid interaction (wettability), investigated by changing liquid phase (heptanes, ethanol and water),, on wetting efficiency are compared in fig.5.

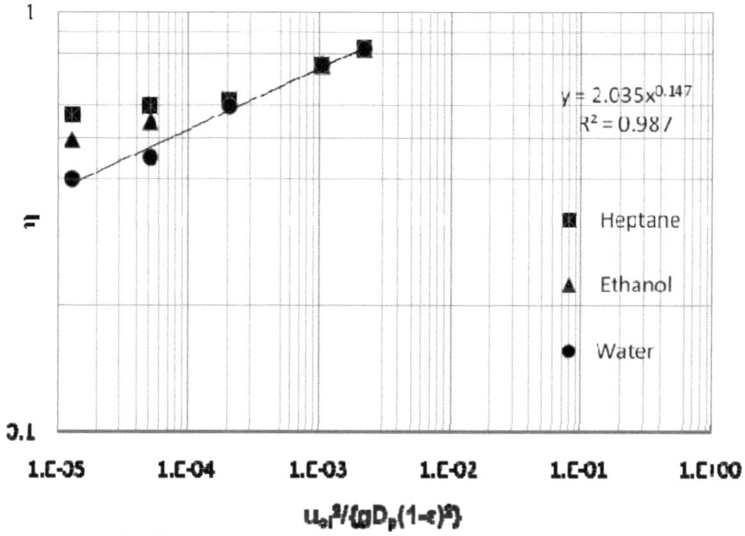

Fig.5. Effect of wettability of surface by liquid: Data of Baussarona et al.[12]

The data on water are correlated as

$$\eta_{\mathrm{water}} = 2.035\left[\frac{u_{ol}^2}{gD_p}\frac{1}{(1-\varepsilon)^2}\right]^{0.147} \qquad [13]$$

The exponent of 0.147 for water is closer to the value (1/7) expected by the model equation. At low flow rates, the wetting efficiency tapers off to a limiting value dependent on the surface-liquid wettability reflected by contact angle. Thus, liquid-surface interactions reflected in contact angle can lower the exponent for heptane and other hydrocarbons. It is interesting to note that wetting efficiency approaches 1 for $\left[\frac{u_{ol}^2}{gD_p}\frac{1}{(1-\varepsilon)^2}\right] \geq 0.01$.

6. CONCLUSIONS:

a). Width of rivulet flowing down an inclined plane surface is measured as a function of liquid flow rate and correlated through friction factor.

b). The correlation for rivulet width with rivulet flow rate on inclined plane surfaces is adopted to liquid flow down the packing elements in a packed bed to obtain

c). Literature data indicates that wetting efficiency in packed beds is dominated by gravity and inertial forces; liquid-surface interaction characterized by contact angle influences wetting efficiency.

d). Wetting efficiency approaches 1 for $\left[\frac{u_{ol}^2}{gD_p}\frac{1}{(1-\varepsilon)^2}\right] \geq 0.01$

NOMENCLATURE:

c_1, c_2:	Constants	
D:	Maximum thickness of the rivulet,	[cm]
D_p:	Diameter of packing element	[cm]
f:	Fanning Friction Factor	[-]
k_1,k_2:	Constants	
L:	Length of the rivulet,	[cm]
Q_l:	Liquid flow rate,	[cm^3/s]
Re:	Reynolds Number	[-]
u_l:	Average liquid velocity in the rivulet	[cm/s]
u_{ol}:	Superficial liquid velocity	[cm/s]
W:	Width of the rivulet,	[cm]

Greek Letters

[illegible]	Angle of inclination of the plate	
[illegible]	Wetting efficiency	
③$_l$	Liquid viscosity	[g/cm/s]
⑧$_l$	Density of liquid	[g/cm^3]
[illegible]	Contact angle	

ACKNOWLEDGEMENT

We thank Universiti Teknologi PETRONAS for the financial assistance and encouragement for carrying out this work.

REFERENCES:

[1]. Ravindra P.V., Rao D.P, Rao M.S: Liquid flow texture in trickle bed reactors: an experimental study, Ind Eng Chem Res., 1997, 36, 5133–5145.

[2]. Lutran P.G, Ng K.M, Delikat E.P. Liquid distribution in trickle beds: An experimental study using computer-assisted tomography., Ind Eng Chem Res., 1991, 30, 1270–1280.

[3]. Sederman A.J, Gladden L.F., Magnetic resonance imaging as a quantitative probe of gas–liquid distribution and wetting efficiency in trickle-bed reactors, Chem Eng Sci., 2001, 56, 2615–2628.

[4]. Toye D, Marchot P, Crine M, Pelsser A.M, L'Homme G., Local measurements of void fraction and liquid holdup in packed columns using X-ray computed tomography., Chem Eng Process. 1998, 37, 511–520.

[5]. Boyer C, Fanget B. ,Measurement of liquid flow distribution in trickle bed reactor of large diameter with a new gamma-ray tomographic system., Chem Eng Sci., 2002, 57, 1079–1089.

[6]. Schwartz J.G, Weger E, Dudukovic′ M.P., A new tracer method for determination of liquid–solid contacting efficiency in trickle bed reactors, AIChE J., 1976, 22, 894–904.

[7]. El-Hisnawi A.A, Dudukovic′ M.P, Mills P.L., Trickle-bed reactors: dynamic tracer tests, reaction studies, and modeling of reactor performance., ACS Symp Ser. 1981,196, 421–440.

[8]. Mills P.L, Dudukovic′ M.P., Evaluation of liquid–solid contacting in trickle-bed reactors by tracer methods., AIChE J., 1981, 27, 893–904.

[9]. Ring Z.E, Missen R.W., Trickle-bed reactors: tracer study of liquid holdup and wetting efficiency at high temperature and pressure., Can J Chem Eng., 1991, 69, 1016–1020.

[10]. Al-Dahhan M.H, Dudukovic′ M.P., Catalyst wetting efficiency in tricklebed reactors at high pressure., Chem Eng Sci., 1995, 50, 2377–2389.

[11]. Julcour-Lebigue C, Baussaron L, Delmas H, Wilhelm A.M., Theoretical analysis of tracer method for the measurement of wetting efficiency., Chem Eng Sci., 2007, 62, 5374–5379.

[12]. L. Baussarona, C. Julcour-Lebiguea, C. Boyerb,, A.M.Wilhelma, H. Delmas, Effect of partial wetting on liquid/solid mass transfer in trickle bed reactors, Chemical Engineering Science, 2007, 62, 7020 – 7025

[13] Julcour-Lebigue. C, Frederic Augier, Harold Maffre, Anne-Marie Wilhelm and Henry Delmas, Measurements and Modeling of Wetting Efficiency in Trickle Bed Reactors: Liquid Viscosity and Bed packing Effects, Ind.Eng.Chem.Res., 2009, 48, 6811-6819

[14]. Herskowitz M, Carbonell R.G, Smith J.M. Effectiveness factors and mass transfer in trickle bed reactors., AIChE J., 1979, 25, 272–283.

[15]. Llano J.J, Rosal R, Sastre H, Diez F.V., Determination of wetting efficiencyin trickle-bed reactors by a reaction method., Ind Eng Chem Res., 1997, 36, 2616–2625.

[16]. Pironti F, Mizrahi D, Acosta A, Gonzalez-Mendizabal D., Liquid– solid wetting factor in trickle bed reactors: its determination by a physical method., Chem Eng Sci., 1999, 54, 3793–3800.

[17]. Wang.G.Q., Yuan X.G. and Yu.K.T, Review of Mass transfer Correlations for packed columns, Ind.Eng.Chem.Res., 2005, 44, 8715-8729.

[18]. Towell, G. D. and Rothfeld, L. B., Hydrodynamics of rivulet flow. *A.I.Ch.E. Journal, 1966,* 972-980.

[19]. A. Ataki, H.-J. Bart,CFD-Simulation of Rivulet Flow on Inclined Plates, Proceed. Distillation & Absorption 02, 30.9./2.10.2002, Baden-Baden.

[20] A.Ataki, Wetting of Structured Packing Elements-CFD and Experiment, Thesis, 2006, TU Kaiserslautern, Germany.

[21]. Luo. S, Li. H, Fei. W and Wang .Y, Liquid Film Characteristics on Surface of Structured Packing, Chinese Journal of Chemical Engineering, 2009, 17(1), 47-52

[22]. Hoffmann. A_, I. Ausner, J.-U. Repke and G. Wozny, Detailed Investigation of Multiphase (Gas–Liquid and Gas–Liquid–Liquid) Flow Behavior on Inclined Plates, Chemical Engineering Research and Design, 2006, 84(A2), 147–154

(29d) Comparison of Distillation Packing Efficiency Through HETP or Mass-Transfer Coefficients Obtained by Profile-Method

Lukas Valenz, Chemical Engineering, Institute of Chemical Technology, Prague (ICT), Prague, Czech Republic, Frantisek J. Rejl, Department of Chemical Engineering, Institute of Chemical Technology, Prague (ICT), Prague, Czech Republic, Vaclav Linek, Department of Chemical Engineering, Prague Institute of Chemical Technology (VSCHT), Prague, Czech Republic and Jan Haidl, Chemical Engineering, ICT Prague, Pargue, Czech Republic

Present paper deals with two effects which seriously limit HETP to provide the objective comparison of different packings. HETP values depend on the concentration range in which the measurement was performed. Using HETP it is not possible to identify or to distinguish whether the resistance against the mass transfer is limited to the side of liquid or vapor phase. Therefore the standardization of the measuring method for the determination of the packing separation efficiency is desirable and it should start by a deep understanding of all conditions influencing the efficiency in order to keep them on the same level during the measurements on different packings. The analysis of HETP dependence on the measured concentration range and the comparison of the shortcut and the rigorous HETP evaluation method is shown. The profile method is offered to provide the fundamental data for the objective comparison of the separation efficiency of the packings and for the rate-based calculations of the columns.

(29e) Gas-Liquid Flow Characteristics On Large-Scale Float Valve Trays Operating At Low Loads

Jun Wang, Chemical Engineering, China University of Petroleum-Beijing, Beijing, China

Abstract:this paper has established a model for calculating the total energy of gas-liquid layer on the float valve trays operating at low loads. In the model the fractions of bubbling area and weeping area are defined, through the analysis of liquid flow characteristics of the different areas, the expressions of kinetic energy, potential energy and surface energy are obtained, which are related with the fractions of bubbling area and weeping area. If bubbling and weeping coexists, this definition will eliminate the inconvenient of dealing with singular point mathematically, which comes from the model. When the energy of system is at minimum, the system is at the most stable state, in this condition the fractions of bubbling area and weeping area change with the time being according to the optimized model. And because of the change of the weeping holes, the liquid heights on the trays vary and the gas pressures under the trays fluctuate. With the restriction of the model, the wave equations of liquid and gas cannot be obtained, and this unstable flow pattern will cause the pressures of the liquid on the trays and under the trays fluctuating, thus induce the gas-liquid layers on the trays to vibrate. Such vibration may cause a series of safety issues when valve trays are operated at low loads(start-up or shut-down).

The system of air-water in two-phase flow conditions are studied on the 6400×800 mm^2 two-pass rectangle Glisch-V1 trays which are near-industrial scale, and the valve weight is 32g, the fractional hole area is from 5% to 14%, the weir height is from 25mm to 65mm, the liquid flow rate is from 5.6 m3/h to 32 m3/h, and the superficial gas velocity is from 0.1 to 0.65m/s. Through the experiments, in the range of gas velocity at or below the seal point, all the trays bubble and weep intermittently (the fraction of bubbling area and the fraction of weeping area changes in the range of 0 to 1 alternately), meanwhile the gas-liquid layers vibrate up and down periodically, and the whole tower also shakes, which are in accordance with the flow pattern theoretically predicted in this paper. In addition, the clear liquid heights of different positions on the trays are very uniform, and the weeping rates of every valve hole are almost equal. The uniformity of these hydraulics parameters can prove the hypothesis made in the setup procedure of physical model to be reasonable.

Dry tray pressure drop of double weight float valve trays

Le-huan Wu, Jun Wang, Yan-sheng Liu, Xu-qiang Guo , and Tian-xin Chen, State Key Laboratory of Heavy Oil Processing Research, China University of Petroleum, Beijing, China

Abstract

This paper calculated the dry tray pressure drop of double weight float valve trays based on a new dry tray pressure drop model proposed by Wang[1]. Firstly, pressure drop is related with the positions of valves on the trays, and the opening process of double weight valves can be divided into 7 sections by 6 special points (ignoring the influence of light valve inclined state on pressure drop). Secondly, gas velocities at every apparent balance point can be calculated as well as the fractions of float valve state in different gas velocity sections. Finally, dry tray pressure drop can be obtained via the physical model. A series of hydraulics experiments have been done on a large-scale experimental tower, and two types of float valve combination (32&58g and 32&80g) were studied. As the influence of light valve inclined state on pressure drop was ignored, the deviation caused by this neglect was acceptable, however, the effect of heavy valve inclined state cannot be ignored and it will cause a relatively large positive deviation, which have been proved. The model calculated values fit the experiment data very well.

Keywords: double weight valve, dry tray pressure drop, large-scale experimental tower , balance point

1. Introduction

Distillation is the most widely applied technique in petroleum and chemical engineering field so far, it's also relatively mature, with little improved or developed, it will cause great economic benefit. As distillation tower internals, compared with sieve trays, valve trays have some advantages: high tray efficiency, low pressure drop and a large operation range. The dry tray pressure drop is a significant parameter which not only reflects the tray operating condition but also can calculate other hydraulic parameters. Thus to study the dry tray pressure drop is of much importance. The most authoritative works on the dry tray pressure drop were done by Wood[2], Bolles[3], Klein[4], Lockett[5] etc.. However, their models all adopted the segmentation method, that means the equations for calculating pressure drop in each section were not the same, the corresponding force analysis on float valve was simple, which decreased the prediction accuracy on gas velocity through the valve orifice at each balance point. What's more, the influence of the intermediate state in the valve opening process on pressure drop was ignored, which was unallowed in the case of a greater valve weight. Recently, a new dry tray pressure drop model for valve tray was introduced by Wang, which may solve the above questions effectively, and a good prediction result was obtained compared with the experimental data.

Double weight valve trays have been used in industry since 1960s, which are defined that two weights of valves are placed in alternate rows on the tray. This design will reduce the pressure drop as well as avoid weeping (Bolles 1976a). At low gas velocity, the gas-liquid contact on the tray is more sufficient and reasonable, thus the tray will have a larger operation range and a better operating performance. Bolles[3] analyzed the opening process of double weight valves based on single valve pressure drop model and considered four balance points were involved including: 1) light valves just starting to open, 2) light valves just completely open, 3) heavy valves just starting to open, and 4) heavy valves just starting to open. And Bolles used his model to calculate the dry tray pressure drop at different balance points. Lockett[5] also discussed the opening process of double weight valves, and the intermediate orifice coefficient for calculating pressure drop between 2) and 3) was also given, which was related with the numbers of different valves. Through theoretical analysis of both the float valve tray with Norton Company patent and the F-1 type float valve tray generally used in China, a theoretical model was established for dry tray pressure drop of the double weight float valve tray by Li[6]. Based on Klein's model[4] on single valve pressure drop, Qi[7] studied the characters of the double weight float valve trays. However, the above models are extended from the single valve pressure drop models which also exist some disadvantages.

Therefore, this paper focused on the double weight float valve opening process based on the Wang's model for single valve pressure drop. According to Wang' idea, the influence of the inclined state of valve on pressure drop should be considered, which has been observed and identified in large-scale experimental equipment. Thus the opening process of double weight valve can be divided into 7 sections by 6 balance points (ignoring the influence of light valve inclined state on pressure drop), pressure drop in each opening section was obtained by calculating the orifice gas velocity at balance points, finally the accuracy of the model for predicting the double weight valve pressure drop was testified by the experiments.

2. Dry tray pressure drop calculations

Pressure drop, which generated from gas through the float valve deck, is related to the positions of valves on the tray. And it can be obtained from the following equation proposed by Wang[1]:

$$\Delta P = \frac{1}{2}\rho_g \left(\frac{u}{\phi_C C_{DC} + \phi_I C_{DIn} + \phi_O C_{DO}} \right)^2 \quad (1)$$

Where ΔP is dry tray pressure drop, ϕ_C, ϕ_I and ϕ_O are the opening fractions which represent valves in the state of closed, inclined and open respectively, C_{DC}, C_{DIn} and C_{DO} are the orifice coefficients when valves are in the state of closed, inclined and open respectively. Obviously the opening fractions should satisfy the following equation:

$$\phi_C + \phi_I + \phi_O = 1 \quad (2)$$

2.1 Gas velocity calculations at apparent balance points

According to the dry tray pressure drop model presented by Wang[1], the opening process of double weight float valves can be divided into 9 stages: ①double weight float valves all closed; ②light valves from closed to full inclined; ③heavy valves closed, light valves full inclined; ④ light valves from inclined to full open; ⑤heavy valves closed, light valves full open; ⑥heavy valves from closed to full inclined; ⑦heavy valves full inclined, light valves full open; ⑧heavy valves from inclined to full open; ⑨double weight float valves full open. When the weight of light valve is rather small, the influence of the light valve inclined state to pressure drop can be ignored, thus the opening process of valves can be divided into 7 stages shown in Figure 1.

At the point A, all valves are closed, thus the point A is called apparent light valve closed balance point. The Euler number E_u is only related with the positions of valves when valves are at the balance point, therefore the hole F factor of point A is equal to that of light valve closed balance point, the detailed calculation method can be referred to literature 1.

$$u_A = u_{lc} \tag{3}$$

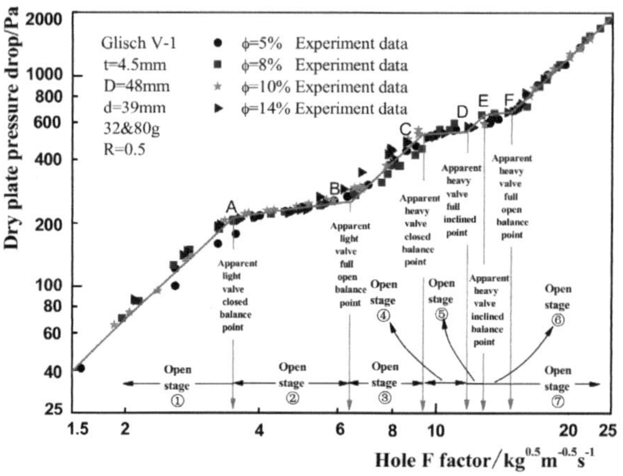

Figure 1. The typical distribution of dry tray pressure drop for double weight float valves

The point B is apparent light valve open balance point, at which point light valves are just all open, heavy valves all closed. The influence of light valve inclined state to pressure drop is supposed to be ignored, for this reason, average valve hole gas velocity can be obtained via the equation of float valve opening fraction,

$$\phi_1 = \frac{u - u_{lc}}{u_{lo} - u_{lc}} \tag{4}$$

u_{lo} is the gas velocity for light valves at full open balance point, which can be calculated from the model. When the ratio of the number of light valves to the number of total valves is a constant, R, the point B can be represented as

$$u_B = R\left(u_{lo} - u_{lc}\right) + u_c \tag{5}$$

The point C is apparent heavy valve closed balance point, heavy valves are just at closed balance point. Due to the great influence of inclined valves to pressure drop when valves are relatively heavy, the influence cannot be neglected. Here, the gas velocity at point C cannot be predicted by the opening fraction equation of float valve, which must be calculated by the following method:

When float valves are at point C, heavy valves become unstable and tend to incline, however, not inclined yet. According to the equilibrium condition of pressure drop, we have

$$\Delta P = \frac{1}{2}\rho_g\left(\frac{u_{wc}}{C_{DC}}\right)^2 = \frac{1}{2}\rho_g\left(\frac{u_l}{C_{DO}}\right)^2 \tag{6}$$

Where u_{wc} denotes the valve hole gas velocity for heavy valves at closed balance point, calculated from Wang' model; ul denotes the velocity of gas through the light valve holes.

From equation 6 the average valve hole gas velocity at point C can be obtained:

$$u_C = \left(1 - R\right) \times u_{wc} + R \times \frac{C_{DO} u_{wc}}{C_{DC}} \tag{7}$$

The point D is apparent heavy valve full inclined point, and heavy valves are just in the inclined state while all light valves are open. The average valve hole gas velocity at point D can also be obtained:

$$u_D = (1-R)\times u_{wI} + R\times\frac{C_{DO}u_{wI}}{C_{DI}} \tag{8}$$

Where u_{wI} denotes the valve hole gas velocity for heavy valves at full inclined point, calculated from literature 1.

The point E is apparent heavy valve inclined balance point, and heavy valves are just at inclined balance point while all light valves are open. The average valve hole gas velocity at point E can also be obtained:

$$u_E = (1-R)\times u_{wIB} + R\times\frac{C_{DO}u_{wIB}}{C_{DI}} \tag{9}$$

Where u_{wIB} is the valve hole gas velocity for heavy valves at inclined balance point, and it can also be calculated from the model.

The point F is apparent heavy valve full open balance point, at which point heavy valves are just in the open state, because the states of light valves and heavy valves are the same, the velocity of gas through light valve holes is equal to the gas velocity at heavy valve all open balance point, and can be calculated from literature 1:

$$u_F = u_{wO} \tag{10}$$

2.2 Calculations of float valve state fraction

There exists two different weight float valves on the tray, thus the condition number for valves is much larger and the states are much more complicated. Based on the analysis above, we can know: the entire open process of valves can be divided into 7 sections by 6 special balance points. In these sections, all the states are not the same, therefore should be calculated stage by stage.

When all the valves are closed (the average float hole gas velocity u in the range from 0 to u_A), there exists only one state, thus the fraction of float valve state on the tray is:

$$\phi_C \equiv 1 \tag{11}$$

A-B stage: when average valve hole gas velocity is in the range from u_A to u_B, a part of light valves are open, other light valves and all heavy valves are closed, thus the opening fraction of valves is:

$$\phi_o = \frac{u - u_{lc}}{u_{lo} - u_{lc}} \tag{12}$$

If the average valve hole gas velocity verges to the gas velocity at the apparent light valve balance point, the opening fraction of light valves verges to R infinitely.

B-C stage: when the average valve hole gas velocity is in the range from u_B to u_C, all light valves are open, while all heavy valves are closed, so there are two states: closed and open, the fractions the different states occupied respectively are:

$$\begin{aligned}\phi_C &= 1-R\\ \phi_O &= R\end{aligned} \tag{13}$$

C-D stage: when the average valve hole gas velocity is in the range from u_C to u_D, all light valves are open, while a part of heavy valves are closed, others are inclined, so there are three states: closed, inclined and open, the fractions occupied respectively are:

$$\phi_C = 1 - R - \phi_I$$
$$\phi_I = \frac{u - u_C}{u_{wI} - u_{wc}} \tag{14}$$
$$\phi_O = R$$

D-E stage: when the average valve hole gas velocity is in the range from u_D to u_E, all light valves are open, while all heavy valves are inclined, so there are two states: inclined and open, the fractions occupied respectively are:

$$\phi_I = 1 - R$$
$$\phi_O = R \tag{15}$$

E-F stage: when average valve hole gas velocity is in the range from u_E to u_F, all light valves are open, while a part of heavy valves are open, others are inclined, so there are two states: inclined and open, the fractions occupied respectively are:

$$\phi_I = 1 - \phi_O$$
$$\phi_O = \frac{u - u_{wIB}}{u_{wO} - u_{wIB}} \tag{16}$$

When all valves are open, there exists only one state: open, thus the fraction of valve state is:

$$\phi_O \equiv 1 \tag{17}$$

2.3 Dry tray pressure drop calculation procedure

For double weight float valve plates, the dry tray pressure drop can be calculated as following steps:

(1) according to literature 1, the gas velocities at different points can be calculated, the results are shown in Table 1;

(2) different orifice coefficients in different states of float valves are shown in Table 2, combined with equation (3) to (10), the gas velocities at every apparent balance point for double weight float valves can be calculated and the calculation results are shown in Table 3;

(3) the fractions of float valve state in different gas velocity sections can be calculated through equation (11) to (17), then on the base of orifice coefficients of Glistch V-1 float valves in different states listed in Table 2, dry tray pressure drop can be obtained from equation (1).

Table 1. Gas velocities at each balance point for single valve [1]

Orifice Velocity/ms^{-1} \ Balance Point; Valve mass /g	Closed balance Point	Full inclined point	Inclined balance point	Open balance point
80	5.09	9.49	9.97	13.69
58	4.33	8.08	8.49	11.66
32	3.22	-	-	8.66

Table 2. Orifice coefficients of Glistch V-1 float valves in different states [1]

	Valve Closed state	Valve Inclined state	Valve Open state
C_D	0.169	0.306	0.402

Table 3. Gas velocities at apparent balance points for double weight float valves

	u_A	u_B	u_C	u_D	u_E	u_F
32g&58g	3.22	5.94	7.31	9.19	10.08	11.66
32g&80g	3.22	5.94	8.66	10.8	11.83	13.69

3. Experiments

A series of hydraulics experiments have been done on a 6400×800mm2 experimental equipment, which is a rectangular column of double overflow, the same as the equipment used by Wang. There are two types of float valve combination, the weights are 32&58g and 32&80g respectively, besides the ratio of light valves to heavy valves is 1(R=0.5), and the valves are distributed as the Figure 2 shows. The objective is to study the relationship between dry tray pressure drop and different parameters for double weight valves, the parameters include operating gas velocity and the weight of valves. The experiment conditions and valve characteristics are shown in Table 4 and 5, and the experimental results are shown in Figure 6.

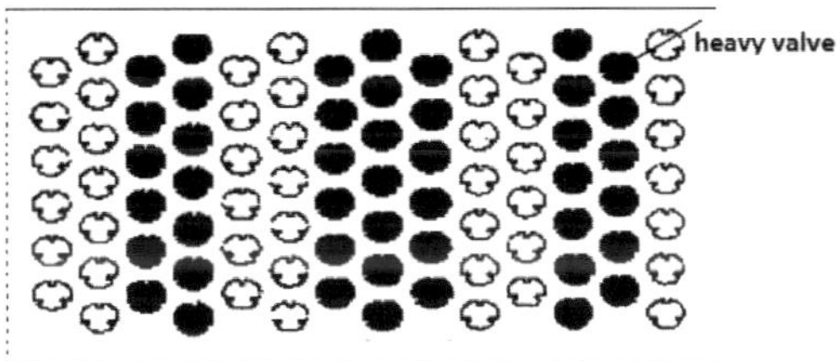

Figure 2. Configuration of double weight float valves

Table 4. Experiment conditions

	Fractional hole area，%	**Experiment medium**	**Superficial gas velocity, m/s**	**Temperature，℃**	**Pressure**
Parameters	14	Air-water	0~2.3	25	Atmosphere

Table 5. Valve characteristics

	Valve diameter(m)	**Hole diameter(m)**	**Valve weight(g)**	**h_{max}(m)**	**h_{min}(m)**
Glistch V-1	0.048	0.039	32&58;32&80	0.0025	0.009

h_{max}: maximum valve height above the tray.
h_{min}: minimum valve height above the tray

4. Results and discussions

The experimental results of dry tray pressure drop for two different combinations of double weight valves are shown in Figure 3. There exists an obvious pressure drop gradient in the middle part of the pressure drop profile, it's due to that when light valves are full open, with the gas velocity increasing, heavy valves remains stable, thus the open area and apparent perforation coefficient of the plate doesn't change, therefore a sudden change of pressure drop at apparent light valve full open point occurs. Comparing the results of 32&80g with 32&58g, before the apparent heavy valve closed balance point for 32&58g, the pressure drop profiles for two combinations are mainly coincident, with the gas velocity increasing , the pressure drops begin to be different. The pressure drop of 32&80g keeps on increasing, while for 32&58g the pressure drop changes suddenly, the main reason is that when heavy valves start to incline, the open area on the plate increases as well as perforation coefficient, although the gas velocity increases, the final pressure drop varies slowly. This dynamic procedure can be described and expressed very well by the results calculated through the model. Not only the predicted values of pressure drop fit the experiment values very well, but also do the predicted gas velocities at each apparent balance points fit the experiment values (the points where pressure drop change suddenly). From the model described in this article and the experiment results we can know, the gas velocities at apparent light valve closed balance point and apparent heavy full open balance point are only related with the weights of valves, independence with the ratio of light valves to heavy valves; the gas velocities at apparent heavy valve closed balance point, apparent heavy valve full inclined point and apparent heavy valve inclined balance point are dependence with the weights of heavy valves and the ratio of light valves to heavy valves, but independence with the weights of light valves; relatively speaking, the gas velocity at apparent light valve full open balance point is related with the weights of light valves and the ratio of light valves to heavy valves, independence with the weights of heavy valves.

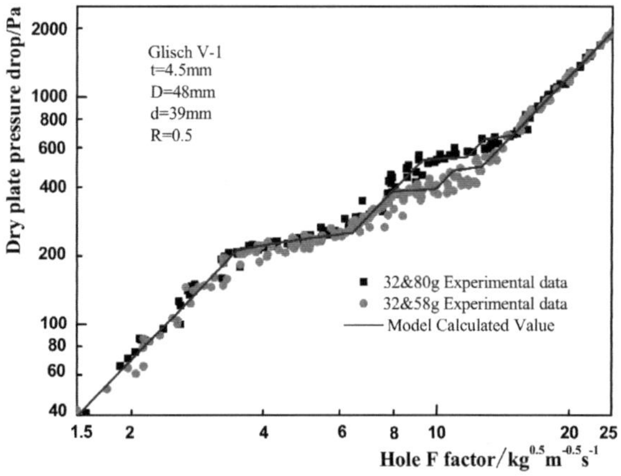

Figure 3. Comparison between experiment values and calculation values

The deviations between the model predicted values and experiment values are shown in Figure 4, and the average error of model prediction is 6.11%. Because of the neglection of the influence of the light valve inclined state, there will be a certain deviation for predicting the pressure drop between apparent light valve closed balance point and apparent light valve full open balance point as well as the gas velocity at apparent light valve full open balance point. When predicting the gas velocity at apparent light valve full open point, the inclined state is ignored and the gas velocity is calculated by Equation (5), the result is shown in Table 3. However, if using the pressure drop equilibrium condition, the calculated value of true u_B is 6.15m/s(32&80g), so the absolute deviation is -0.21m/s and the relative error is -3.41%, which is acceptable in engineering. If the heavy valve inclined state is ignored, according to the equation of valve opening fraction the gas velocity at apparent heavy valve closed balance point is 9.39m/s(32&80g), while the actual value is 8.66m/s shown in Table 3, so the absolute deviation is +0.73m/s, the relative error is +8.42%. It also seems to be acceptable in engineering application, but the dry tray pressure drop is proportional to the square of the gas velocity, the deviation of gas velocity will be amplified gradually in the calculation process and cause an unacceptable result. For instance, with the valve inclined state being ignored, the dry tray pressure drops at apparent light valve full open balance point and apparent heavy valve closed balance point will increase to -6.94% and +17.5% respectively because of the deviation of predicted gas velocity value, that's to say, ignoring the light valve inclined state will cause a certain negative deviation to the gas velocity and the pressure drop calculation value at apparent light valve balance point, which can be proved in Figure 6, the pressure drop predicted value is much lower than the experiment value at apparent light valve full open balance point; while ignoring the heavy valve inclined state will cause a major positive deviation to the calculation values between apparent heavy valve closed balance point and apparent heavy valve inclined balance point, which can also be proved clearly.

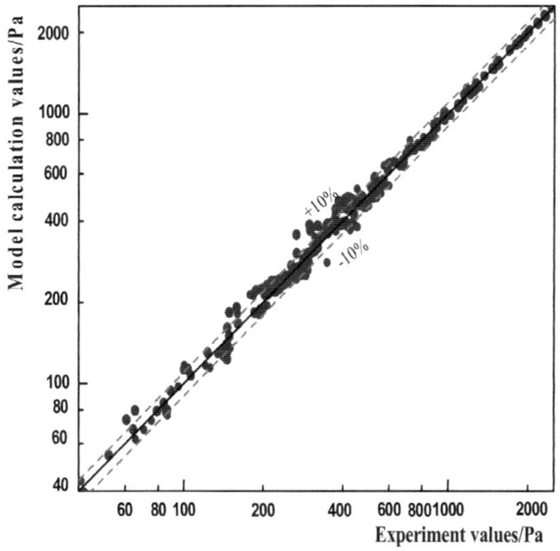

Figure 4. The deviations between model calculation values and experiment values

5. Conclusions

On the large-scale cold mold experimental tower, the influence of two combination types of double weight valves(32&58g and 32&80g) on dry tray pressure drop were studied. With the gas velocity increasing, the float valves on the plates undergo three conditions: closed, inclined and open. Because of two different weight valves on the tray, the states of float valve change more complicatedly than those of single valve. In the opening process of float valves, there are 6 balance points.

According to dry tray pressure drop model for single valve, a model for calculating dry tray pressure drop for double weight valve was derived, and the effect of light valve inclined state on the pressure drop was ignored, furthermore the deviation caused by this neglection was acceptable, which has been proved, however, the effect of heavy valve cannot be ignored. The model calculated values fit the experiment data very well.

References

1. Jun Wang. A new dry tray pressure drop model of float valve trays. AIChE J. 2013
2. Wood RM. The stability of the valves of floating valve plates. Trans Inst Chem Eng. 1961;39:313–317.
3. Bolles WL. Estimating valve tray performance. Chem Eng Prog. 1976;72:43–47.
4. Klein GF. Simplified model calculates valve–tray pressure drop. Chem Eng (N.Y.). 1982;89:81–85.
5. Locket MJ. Distillation tray fundamentals. London: Cambridge University Press, 1986:82–86.
6. Li Yun-peng. Correlation of pressure drop with double valve weighed float valve tray. Petro-Chemical Equipment.1999;28:31-34.
7. Qi Fu-lai. Dry tray pressure drop of two-weight valve tray. Chemical Engineering (China). 2011;39:19-22.

Study on Gas Short-cut through Dynamic Seal of Rotating Packed Bed

Yumin Li, Xiaohua Li, Yunliang Yu, Zhichao Xu, Xuejun Liu, Jianbing Ji
Zhejiang Province Key Laboratory of Biofuel, College of Chemical Engineering & Materials Science, Zhejiang University of Technology, 310014, Hangzhou, Zhejiang, China

Abstract: By rotating a rigid rotor in the stationary casing, rotating packed beds adopt higher centrifugal acceleration in mass and heat transfer operations to achieve process intensification. The inside dynamic seal between the rotating rotor and the stationary casing usually uses packing or centrifugal liquid ring dynamic seal to ensure gas passage through the rotor. The gas in the casing may bypass the rotor and make a short-cut through the inside dynamic seal so that the purity of the gas product is reduced, if the dynamic seal system is not perfect. A model of gas short-cut through inside seal was established to predict the effect of short-cut gas on gas concentration at the inner periphery of the rotor. Both the theoretical plate number of total-reflux distillation with ideal system and the total number of gas transfer unit of low concentration gas absorption decrease with the increases in the ratio of short-cut gas rate to total gas rate, which indicates that the reliability of dynamic seal has a direct impact on mass-transfer in rotating packed bed.
Keywords: dynamic seal, rotating packed bed, inside dynamic seal, gas short-cut, model

Introduction

By rotating a rigid rotor in the stationary casing, rotating packed beds adopt higher centrifugal acceleration in mass and heat transfer operations to achieve process intensification. So the sizes of rotating packed beds are greatly reduced, compared to the conventional trayed and packed column. However, two types of dynamic seals between the rotating rotor and stationary casing are needed. The shaft seal at the bottom is used to prevent leakage from the casing, and inside seal at the top is used to ensure gas passage through the rotor. And the inside seals are most likely to suffer from wearing and are the most common trouble spots[1]. Usually, packing dynamic seals[2] and centrifugal liquid ring dynamic seals[3] are employed as the inside seals in the rotating packed beds. The gas in the casing may bypass the rotor and make a short-cut through the inside seal, so that the gas purity at the inner periphery of the rotor is reduced. Therefore, a model of gas short-cut through inside seal was established to predict the effect of short-cut gas on gas concentration at the inner periphery of the rotor.

1. Model of gas short-cut through inside seal

The packing dynamic seal and centrifugal liquid ring dynamic seal as the inside seals are shown in Fig.1. For the packing inside seal, long-term rotation of the rotor and direct contact between packing and rotor make the packing worn, so that the gas at the outer periphery of the rotor in the casing bypasses the rotor and makes a short-cut through the worn packing to enter directly the inner periphery of the rotor without the gas-liquid contact and mass-transfer. For the centrifugal liquid ring inside seal, the liquid secondary flow in the liquid rings will lead to the occurrence of gas short-cut. Therefore, a model of gas short-cut through inside seal was established, as shown in Fig.2.

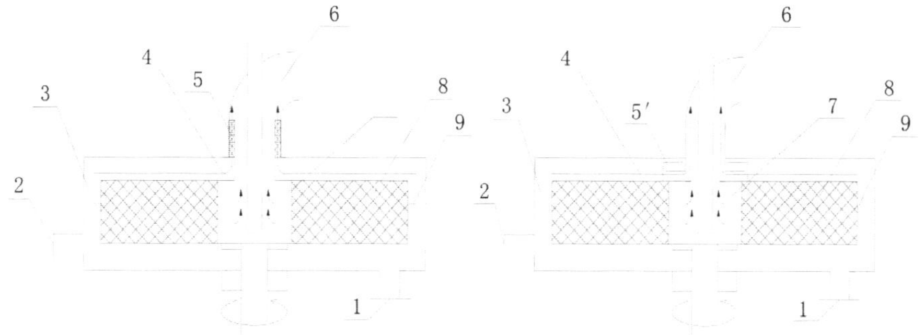

(a) packing inside seal (b) centrifugal liquid ring inside seal

1. liquid outlet; 2. gas inlet; 3. outer periphery of the rotor; 4. gas of short-cut through the inside seal; 5. packing inside seal; 5′. centrifugal liquid ring inside seal; 6. gas outlet; 7. inner periphery of the rotor; 8. rotor; 9. gas passing through the rotor

Fig.1 Structure of packing and centrifugal liquid ring inside seals

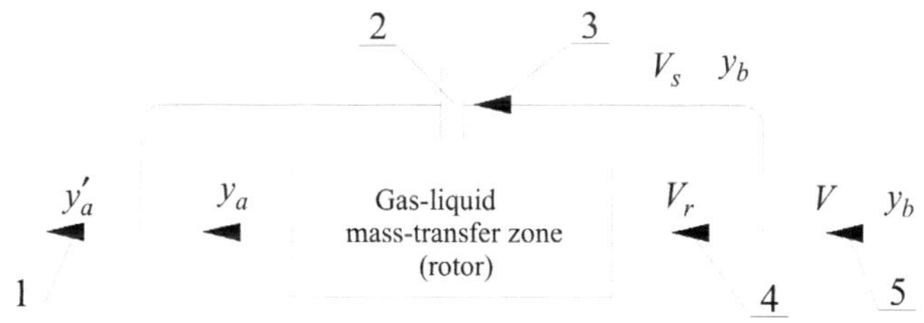

1. gas at the inner periphery of the rotor; 2. inside seal; 3. gas of short-cut through the inside seal; 4. gas passing through the rotor; 5. gas at the outer periphery of the rotor

Fig.2 Model of gas short-cut through inside seal

The rotor of rotating packed bed is regarded as the gas-liquid mass-transfer zone. It is assumed that the mass-transfer efficiency of the gas-liquid mass-transfer zone is invariable with the decrease in the molar flow rate of the gas passing through the rotor.

The molar flow rate of the gas is V with mole fraction y_b of light component for distillation or solute component for gas absorption. If the gas short-cut occurs, the molar flow rate of the short-cut gas is V_s with mole fraction y_b due to no contact of gas-liquid; and the molar flow rate of the gas passing through the rotor is V_r with mole fraction y_a for the gas leaving the gas-liquid mass-transfer zone. The gas passing through the rotor is mixed with the gas of short-cut through the inside seal. The molar flow rate of the mixed gas is still V with mole fraction y'_a.

Total material balance is given by

$$V = V_s + V_r \tag{1}$$

Light or solute component material balance is given by

$$V \cdot y'_a = V_s y_b + V_r y_a \tag{2}$$

The ratio Φ of short-cut gas rate V_s to total gas rate V is

$$\phi = \frac{V_s}{V} \tag{3}$$

Substituting eq(3) into eq(1) gives

$$1-\phi = \frac{V_r}{V} \tag{4}$$

Substituting eq(3) and eq(4) into eq(2) gives

$$y'_a = y_b\phi + y_a(1-\phi) \tag{5}$$

2. Total reflux distillation

For total reflux distillation in a single-rotor rotating packed bed with binary ideal mixture system, the number of theoretical plates N plus a reboiler is given using the Fenske equation, that is,

$$N+1 = \frac{\lg\left[\left(\frac{x_D}{1-x_D}\right)\left(\frac{1-x_W}{x_W}\right)\right]}{\alpha} \tag{6}$$

where x_D and x_W are mole fractions of liquid phase at the top and bottom, and α is geometric mean of the relative volatility at the top and bottom.

Substituting y_a and y_b into eq(6) gives

$$N = \frac{\lg\left[\left(\frac{y_a}{1-y_a}\right)\left(\frac{1-y_b}{y_b}\right)\right]}{\alpha} \tag{7}$$

If the gas makes a short-cut through the inside seal with the ratio Φ less than 5%, the operating line is basically unchanged, and the number of theoretical plates with gas short-cut N' is given by

$$N' = \frac{\lg\left[\left(\frac{y'_a}{1-y'_a}\right)\left(\frac{1-y_b}{y_b}\right)\right]}{\alpha} \tag{8}$$

Substituting eq(5) into eq(8) gives

$$N' = \frac{\lg\left[\left(\frac{y_b\phi + y_a(1-\phi)}{1-y_b\phi - y_a(1-\phi)}\right)\left(\frac{1-y_b}{y_b}\right)\right]}{\alpha} \tag{9}$$

The drop percentage of the number of theoretical plates β_d is

$$\beta_d = \frac{N-N'}{N} \tag{10}$$

Substituting eq(7) and eq(9) into eq(10) gives

$$\beta_d = 1 - \frac{\lg\left[\left(\frac{y_b\phi + y_a(1-\phi)}{1 - y_b\phi - y_a(1-\phi)}\right)\left(\frac{1-y_b}{y_b}\right)\right]}{\lg\left[\left(\frac{y_a}{1-y_a}\right)\left(\frac{1-y_b}{y_b}\right)\right]} \quad (11)$$

As a result, the effect of the ratio Φ on the drop percentage β_d is shown in Fig.3. The drop percentage β_d increases with increasing Φ, because the short-cut gas reduces the concentration of light component at the inner periphery of the rotor, i.e. the concentration of overhead product. And the drop percentage β_d increases with increasing y_a at the same Φ, which indicates the following fact. The higher the concentration of overhead product from separation requirements is, the more the short-cut gas reduces the concentration of overhead product.

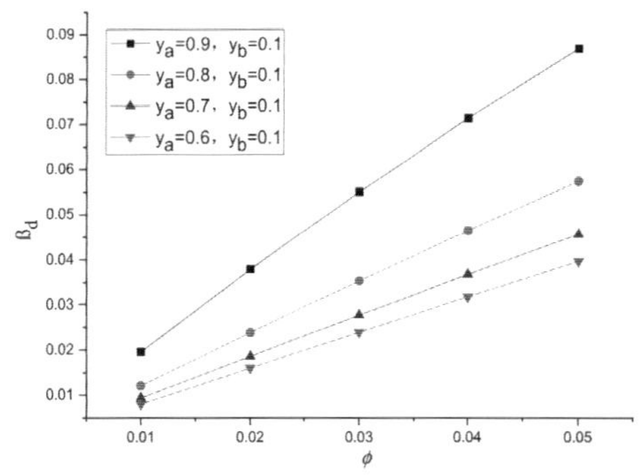

Fig.3 Effect of the ratio Φ on the drop percentage β_d

3. Low concentration gas absorption

For low concentration gas absorption in a single-rotor rotating packed bed with acetone in the air absorbed by pure water, the total number of gas phase transfer units N_{OG} is given by

$$N_{OG} = \frac{1}{1-S}\ln\left[(1-S)\frac{y_b}{y_a} + S\right] \quad (12)$$

where S is stripping factor, expressed as $S=mG/L$, m is phase equilibrium constant, L/G is liquid-gas ratio, y_b and y_a are the mole fraction of acetone of gas phase at the outer and inner periphery of the rotor, respectively.

If the gas makes a short-cut through the inside seal with the ratio Φ less than 5%, the ratio L/G is basically unchanged, and the total number of gas phase transfer units with gas short-cut N'_{OG} is given by

$$N'_{OG} = \frac{1}{1-S}\ln\left[(1-S)\frac{y_b}{y'_a} + S\right] \quad (13)$$

Substituting eq(5) into eq(13) gives

$$N'_{OG} = \frac{1}{1-S}\ln\left[(1-S)\frac{y_b}{y_b\phi + y_a(1-\phi)} + S\right] \quad (14)$$

The drop percentage of the total number of gas phase transfer units β_a is

$$\beta_d = \frac{N_{OG} - N'_{OG}}{N_{OG}} \quad (15)$$

Substituting eq(12) and eq(14) into eq(15) gives

$$\beta_a = 1 - \frac{\ln\left[(1-S)\dfrac{y_b}{y_b\phi + y_a(1-\phi)} + S\right]}{\ln\left[(1-S)\dfrac{y_b}{y_a} + S\right]} \qquad (16)$$

As a result, the effect of the ratio Φ on the drop percentage β_a is shown in Fig.4, where m is 1.68 for the acetone-water system at normal temperature and L/G is taken as 1.93, leading to $S=0.87$. The drop percentage β_a increases with increasing Φ, because the short-cut gas increases the acetone concentration at the inner periphery of the rotor, i.e. the concentration of overhead product. And the drop percentage β_a increases with decreasing y_a at the same Φ, which indicates the following fact. The lower the concentration of overhead product from separation requirements is, the more the short-cut gas increases the concentration of overhead product.

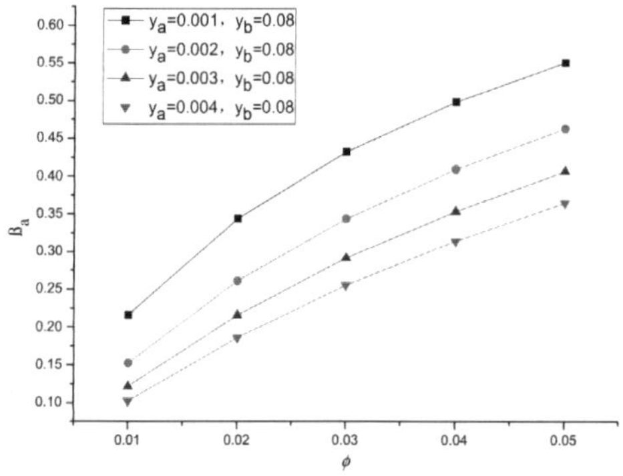

Fig.4 Effect of the ratio Φ on the drop percentage β_a with $S=0.87$

4. Conclusion

(1) A model of gas short-cut through inside seal was established to predict the effect of short-cut gas on gas concentration at the inner periphery of the rotor. For the two cases of total reflux distillation and low concentration gas absorption, the greater the short-cut gas rate is , the more the number of theoretical plates and the total number of gas phase transfer units reduce.

(2) The inside seals in rotating packed beds directly influence the performance of mass-transfer, so the reliable inside seals are necessary.

REFERENCES

(1) Wang, G. Q.; Xu, Z.C.; Ji, J.B. Progress on Higee distillation—Introduction to a new device and its industrial applications. Chemical Engineering Search and Design, 2011, 89 (8), 1434−1442.

(2) Zheng, C.; Guo, K.; Feng, Y.D.; Yang, C. Pressure drop of centripetal gas flow through rotating beds. Ind. Eng. Chem. Res., 2000, 39, 829−834.

(3) Kelleher, T.; Fair, J.R. Distillation studies in a high-gravity contactor. Ind. Eng. Chem. Res., 1996, 35, 4646−4655.

(45a) Pilot and Full-Scale Application of Technologies for Mercury Control

Noah Meeks, Research & Environmental Affairs, Southern Company Services, Inc., Birmingham, AL

With the final MATS rule going into effect in 2015, electric utilities must now consider mercury control as a needed technology in their emissions train. This talk wiill cover recent full-scale injections of activated carbon and alkali injection, as well as pilot-scale studies of non-halogenated carbons or non-carbon sorbents. We will also discuss recent mixed matrix membrane pilot studies. **(Most recent data will be included in the abstract closer to the meeting)

(45b) Acid Tunneling Vs Acidizing : A Case Study of Rudatain Oilfield-Kuwait

Adel Elsharkawy, Petroleum Engineering, Kuwait University, Kuwait, Kuwait and Mohamed Al-Hadher, Pet Eng, Kuwait Oil Company, Safat, Kuwait

Acid Tunneling Vs Acidizing : A Case Study of Rudatain Oilfield-Kuwait

By Mohammed Al-Hadher and Adel Elsharkawy

Kuwait University, College of Engineering & Petroleum

Po Box 5969 Safat 13060 Kuwait

Abstract

Acidizing is one of the most important techniques used in oil industry in order to enhance the reservoirs and increase oil production. There are three types of acidizing techniques which are matrix acid stimulation, acid fracturing, and acid tunneling. In this paper a comparison these types of acidizing is discussed. The case study of RA-0007 acid tunneling completed in Kuwait Oil Company is presented. It represents the first dendritic with extensive worm holes created by acid tunneling for maximum reservoir contact. The whole effect is aptly demonstrated by the change in PI of the well before and after the acid tunneling as mentioned before. Analysis of the well designed and implemented job showed that two tunnels were created successfully in MA-C formation through a two days rig-less operation. The results indicated an improvement in the well productivity index (PI) which had been increased from 1.3 to 3.87 bpd/Psi.

(45e) Sustained Casing Pressure Modeling and Well Integrity

Tony Rocha-Valadez, Mary Kay O'Connor Process Safety Center, Artie McFerrin Department of Chemical Engineering, Texas A&M University, College Station, TX, A. Rashid Hasan, Harold Vance Department of Petroleum Engineering, Texas A&M University, College Station, TX, C. Shah Kabir, EPT, Hess Corporation, Houston, TX and Sam M. Mannan, Artie McFerrin Department of Chemical Engineering, Mary Kay O'Connor Process Safety Center, Texas A&M University, College Station, TX

SUSTAINED CASING PRESSURE MODELING AND WELL INTEGRITY

T. Rocha-Valadez[1*], A.R. Hasan[2], C.S. Kabir [3] and S.M. Mannan[1]

[1]*Mary Kay O'Connor Process Safety Center, Artie McFerrin Department of Chemical Engineering, Texas A&M University, College Station, Texas 77843-3122, US., email: *tonyrochav@tamu.edu, mannan@tamu.edu.*

[2] *Harold Vance Department of Petroleum Engineering, Texas A&M University, College Station, Texas 77843-3122, US, email: rashid.hasan@pe.tamu.edu.*

[3]*Hess Corporation, 1501 McKinney Street, Houston, TX 77010, US, email: skabir@hess.com.*

With an increasing demand for hydrocarbons and other sources of energy, the methods for delivering such energy needs have become complex. In the case of the oil and gas industry, exploration and production have gone to deeper waters and deeper horizons than ever before, resulting in drilling several miles into the earth's core. Deeper well penetration means encountering increasingly high formation pressures and temperatures. For example, in deepwater drilling, well construction and production management deal with significant heat transfer between the produced fluids and its surroundings. The high-rate production and the attendant heat transfer demand close attention in terms of operational safety and potential environmental impact. One phenomenon that occurs in producing wells is sustained casing pressure (SCP), which presents a risk to both process safety and the environment.

In oilfield terminology, the casing is a steel pipe cemented to the surrounding formation. Multiple concentric casings are involved in a typical well construction. The casing immediately surrounding the production tubing is known as the production casing, which is most vulnerable to thermal and mechanical stresses and fluid leakage. The cement sheaths provide zonal isolation, support the casing, and protect it from corrosive fluids. Ideally, pressures monitored at the wellhead for multiple casings should change minimally after being bled off through a needle valve at the surface. When the pressure rebuilds after being bled off, that casing is said to exhibit SCP [1].

SCP occurs because of stresses that the system experiences owing to fluids production up the tubing. Besides the normal source of SCP, a poor cement job may allow hydrocarbon migration from the producing or other horizons and accumulation into one or more of the casings. The risk of not attaining well integrity can range from the activation of rupture discs to the blowout of hydrocarbons. Because of the safety and environmental dangers of SCP, regulations require remedial treatments for a casing string that exhibits significant SCP problems. Unfortunately, available methodologies to test for SCP are often qualitative and empirical in nature [2, 3], rendering them inapplicable in many situations. No practical systematic testing procedure has emerged to quantify SCP.

In this paper, we present a model for SCP that is rooted in the transport processes of the system. The model assumes gas leaking through the cement barrier, and then migrating up the annular-liquid column. It accounts for the fluid-temperature profile along the wellbore and liquid compressibility. Some robust assumptions about the transport processes allow us to write the governing equation as a first-order, linear-differential equation that is solved with appropriate boundary conditions. The resulting algebraic expression for the casing pressure rise as a function of time paves the way for all applications. A comparative study of the model's performance with field data suggests excellent agreement for wells where the casing pressure increase is owes gas influx into the annulus. The proposed model could form the basis for standardization of SCP testing. This study also offers some test procedures to assess SCP-related safety concerns.

References

1. Bourgoyne Jr., A.T.J., S.L. Scott, and W. Manowski, ***A Review of Sustained Casing Pressure(SCP) Ocurring on the OCS***, 2000, Final Report Submiteted to MMS.

2. Xu, R. and A.K. Wojtanowicz, ***Diagnosis of Sustained Casing Pressure from Bleed-off/Buildup Testing Patterns.*** SPE, 2001: p. Productions and Operations Symposium.

3. Wojtanowicz, A.K., S. Nishikawa, and R. Xu, ***Diagnosis and Remediation of Sustained Casing Pressure in Wells***, July 2001, MMS.

(45f) Evaluation of Particle Growth in an Electrochemical Fluidised Bed Reactor

Emmanuel Ehirim, Rivers State University of Science and Technology, Port Harcourt, Nigeria

The process of electrodeposition within electrochemical reactor results in the increase of particle sizes, which increases the volume of the bed. This work entails a mathematical model to describe the effect of this phenomenon on the behaviour of a three-dimensional fluidised bed electrode operating under the condition of electron transfer. The model incorporates the pseudo-stationary behaviour in the reactor since the bed increases in volume due to the particle size growth. The Orthogonal Collocation method was used to solve the developed differential equations. The overpotential distribution and the reaction rate within the reactor have been calculated. The principal results are the effects of the electron transfer operation on the superficial displacement with time. The effects of total applied current, electrode geometry and the bed porosity on the behaviour of the system have been analysed. The result shows a very gradual increase in the bed height.

(45g) Debottlenecking Gas Dryers and Desiccant Adsorbers

Benjamin A. Schmitt, Hydrocarbon Processing Industry (HPI) - Business Unit, Johnson Screens, Inc., New Brighton, MN and Michael Ekholm, Engineering, Johnson Screens, Inc., New Brighton, MN

Debottlenecking Gas Dryers and Desiccant Adsorbers

Though Maximization of the Useable Bed Volume with Johnson Screens' SSG^{TM}

Benjamin Schmitt, PE

Senior Mechanical Engineer, Technology

Michael Ekholm, PE

Applications and Product Design Manager

Johnson Screens Inc. St. Paul, MN

AIChE Spring Meeting, April 2013

13th Topical Conference

San Antonio, TX

Introduction:

The following presentation of the Shaped Support Grid (SSG^{TM}) releases additional supporting data and updates to the initial presentation given by Robert G. Norell at the 2010 AIChE Spring Conference.

Limitations on ethylene dryers and desiccant adsorbers can create bottlenecks in the production process which can lead to a need for increased dryer and adsorber capacity. Capital intensive investments may be necessary in order to eliminate the bottleneck which may include the addition of an additional vessel. With limits on the plot plan, this may not be a workable solution. Adding to the problem is the amount of time consumed in engineering, planning and construction, which can have negative effects on the operation of the ethylene unit. Increased desiccant volume and operational efficiency can increase the unit's production with minimal capital expenditures. Johnson Screens', patent pending, SSG^{TM} can add the increased process efficiency and cycle life necessary to extend the operating capability of the unit.

Extending the operation, though debottlenecking, can be achieved by maximizing the time before breakthrough of the media bed. There are three primary functions that maximize the breakthrough time:

Proper gas distribution in both normal operation and regeneration conditions which maximizes the bed utilization.

Increased molecular sieve or desiccant bed volume are achieved through the elimination of traditional bed supports comprised of support grids and beams. The support interface for the screen configuration also allows for thermal expansion typically present in vessels with a thermal cycle as part of the process. The design allows for thermal cycling without the risks of seal or structural failure inherent in other solutions.

Minimizing the bed volume consumed by the distribution and support system increases the net desiccant volume lost when utilizing alternative bed supports.

Johnson Screens ***SSG*** is designed to provide the industry with an alternative, cost effective, approach to increasing capacity. Supporting data and preliminary case studies will be depicted to quantify the benefits that are achievable through the implementation of the ***SSG***.

Conclusions:

The use of a Johnson Screens ***SSG*** provides a reliable immediate solution to reduce capacity bottlenecks. The ***SSG*** increases the bed volume, when compared to traditional bed support arrangements comprised of grids and beams. The increase in desiccant volume will allow for increases in feed flow rate and/or adsorbtion cycle time. The ***SSG*** maximizes the flow distribution during desiccant regeneration, maximizing the utilization of the complete cross section of the vessel. This solution can be implemented in both new and existing vessels, providing consistency in design between planned increases in performance of current vessels and future expansions.

The modular design of the ***SSG*** allows for easy installation and maintenance. Connection of the ***SSG*** to the vessel is performed within the outlet nozzle, eliminating all welding on the pressure vessel and eliminating any need for recertification of the pressure vessel after installation. The design allows for ease of dismantling for inspection of the vessel head during routine maintenance and inspections. The low profile will also reduce interference with any required media exchanges. The design package of the ***SSG*** provides the end user with a holistic solution that fulfills debottlenecking needs with limited capital expenditures.

Biography of Speaker:

Benjamin Schmitt is a Senior Mechanical Engineer in Johnson Screens' Hydrocarbon Processing Industry Technology Group. Ben graduated from North Dakota State University in 2004 with a Bachelor of Science degree in Mechanical Engineering. He currently working toward his Master's in Business Administration from Hamline University. Ben is member of AIChE and is a registered professional engineer in the state of Minnesota.

AMACS
PROCESS TOWER INTERNALS

AIChE 2013 Spring Meeting Poster board

Design Considerations for effective liquid removal In Suction Drums

Presented By:
Kanti Patel, PE

Compressor down again?

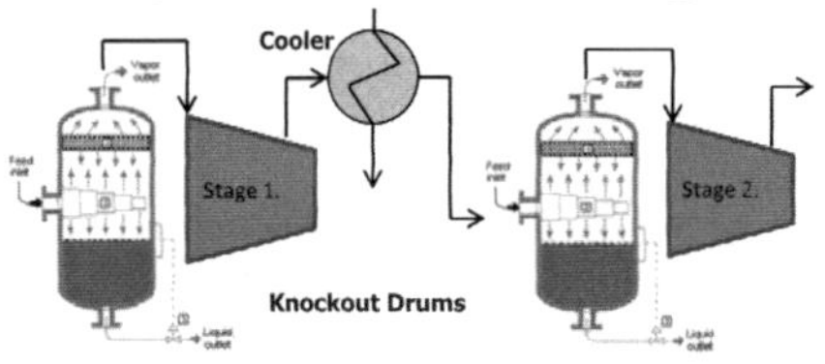

Is liquid carryover the culprit?

<u>We have heard it all before</u>:

-Premature compressor shutdown.

-Production time lost.

-High yearly compressor maintenance costs.

-Impeller damage due to pitting corrosion.

-Corrosion due to salt deposits.

AMACS
PROCESS TOWER INTERNALS

Common Design Problems with Suction Drums which promotes liquid carryover

1. Choosing the correct mist eliminator (M.E. equipment)
2. Knowing where to install the M.E. in a suction drum.
3. Consider a high mist loading in the gas when selecting your M.E.
4. Fouling considerations.
5. M.E. disengaging space, bad gas distribution & liquid slug in the feed.
6. One micron and great droplets, versus submicron mist removal.
7. High viscosity liquids.
8. Dislocated M.E. due to pressure surges.
9. Through put exceeds original design capacity.
10. Capacity range too large for suction drum.

AMACS
PROCESS TOWER INTERNALS

Choosing the correct Mist eliminator equipment

- Proper sizing

\+ Use Souders-Brown equation to find optimum velocity. $\mathbf{Vd = K \cdot \sqrt{\frac{Pl}{Pg} - 1}}$

\+ Select the correct vapor velocity factor K. based on the operating pressure, and type of equipment.

- Lower K-factor 0.01 for every 100 psi
- K factor to select when using:

Mesh Pad	K= 0.1 - 0.55
Chevron Vane	K= 0.45 - 0.65
MultiPocket® Vane	K= 0.75 – 1.0

AMACS
PROCESS TOWER INTERNALS

Choosing the correct Mist eliminator equipment cont.

- Selection based on droplet size.

Table 2. Droplet sizes (water in air) typically captured with 99.9% efficiency by mist eliminator elements of various types

Element type	Size range microns
Fiber candles or panels	0.1 and larger
Mesh with co-knit yarn	2.0 μm and larger
0.006-inch knitted mesh	5.0 μm and larger
0.011-inch knitted mesh	10 μm and larger
AMACS MultiPocket ® Vane	10 μm and larger
Conventional vane arrays	15 μm and larger

AMACS
PROCESS TOWER INTERNALS

Where to install the M.E. in the suction drum

- Improperly installed M.E. can cause severe liquid carryover problems.
- General accepted spacing guidelines.

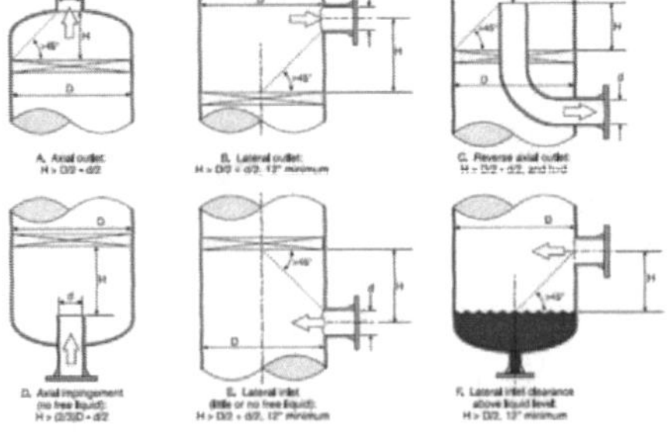

AMACS
PROCESS TOWER INTERNALS

Consider high mist loading when selecting your M.E. equipment

+ High mist loading in the gas stream can flood your M.E. equipment and cause liquid carryover.

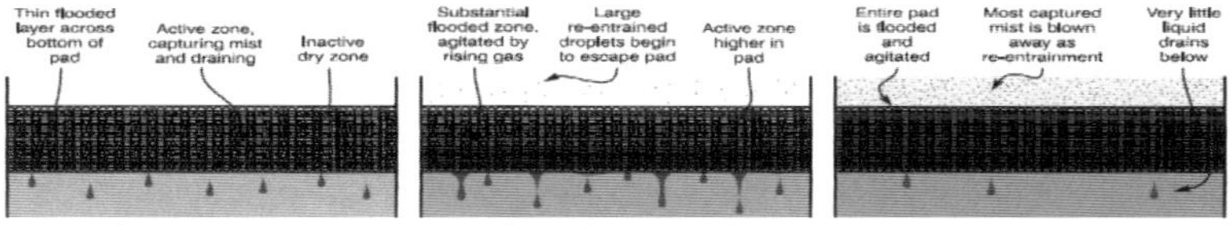

Figure 4. Envisioning mesh pad performance degradation as liquid and gas loads increase (vertical cross-sections)

AMACS
PROCESS TOWER INTERNALS

Consider high mist loading when selecting your M.E. equipment cont.

+ Limited drainage capabilities
    - mesh M.E. <0.5 gal/FT2/min
    - Vane type M.E. <2 gal/FT2/min

+ Select a combination of vane & mesh mist eliminators

AMACS
PROCESS TOWER INTERNALS

Fouling Applications

+ Fouling of mesh mist eliminator will cause liquid carryover.

+ For mild fouling install a water spray system below the mist eliminator for occasional inline cleaning (based on Delta-P)

+ For consistent fouling problems, install a Vane mist eliminator upstream of the mesh mist eliminator.
  - Vanes remove solids and bulk liquids therefore, protecting the downstream mesh mist eliminator

AMACS
PROCESS TOWER INTERNALS

Fouling Applications cont.

+ For consistent fouling problems where the solids are "sticky" install a water spray system below the vane mist eliminator.

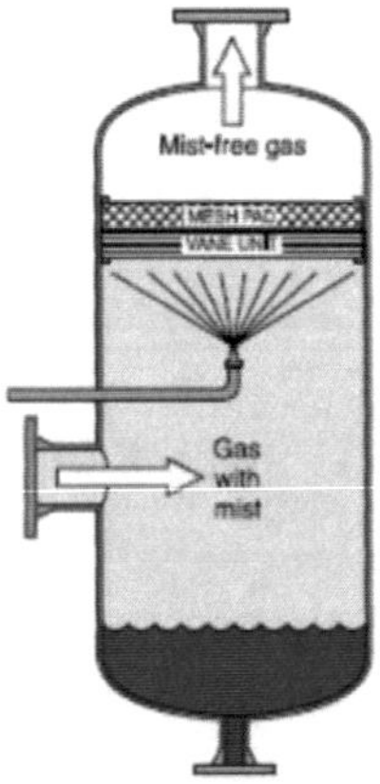

AMACS
PROCESS TOWER INTERNALS

Bad Gas distribution, limited disengaging space and liquid slugs.

+ Bad gas distribution, limited disengaging space in a suction drum can cause severe liquid carryover.
+ CFD analysis can assist in locating high velocity areas.

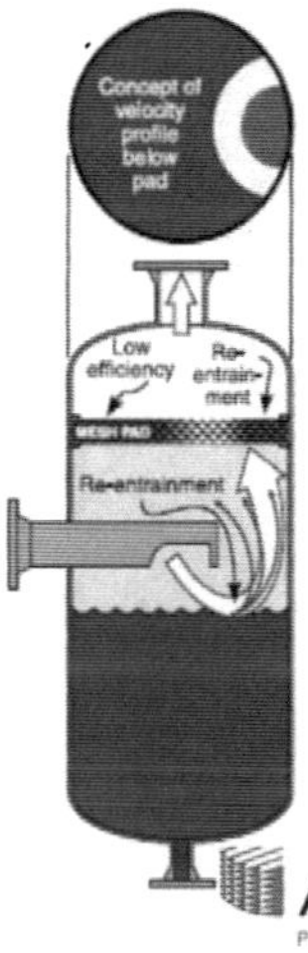

AMACS
PROCESS TOWER INTERNALS

Bad Gas distribution, limited disengaging space and liquid slugs cont.

+ Use inlet diffuser to force uniform gas distribution i suction drum.

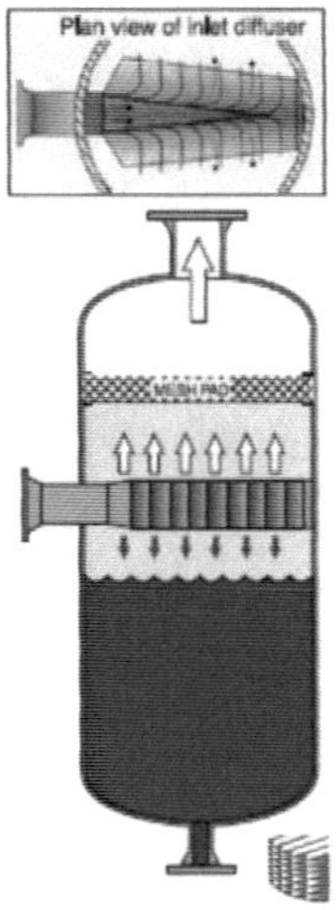

AMACS
PROCESS TOWER INTERNALS

Inlet Feed Device

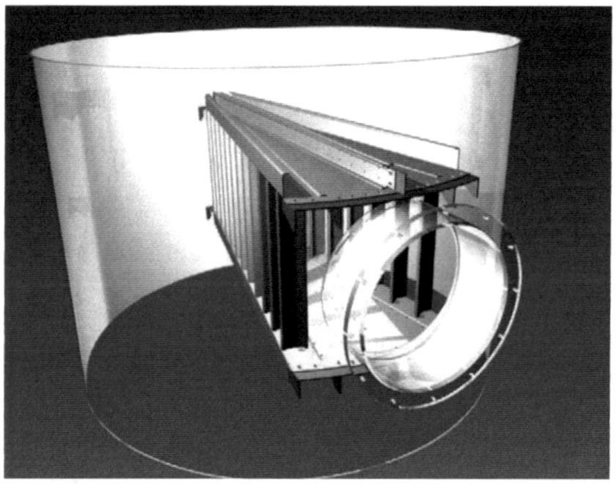

Inlet Feed Device

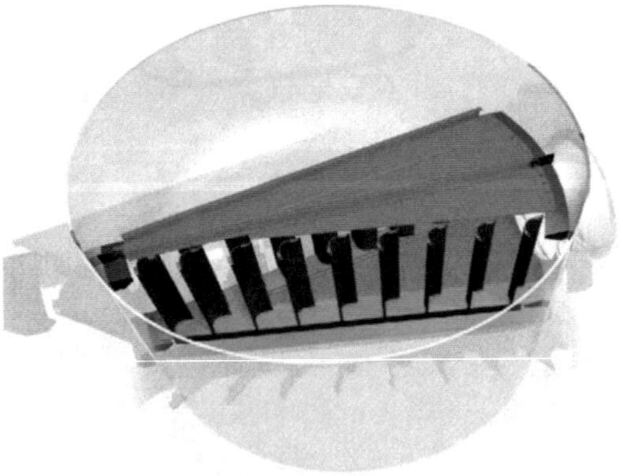

Bad Gas distribution, limited disengaging space and liquid slugs cont.

+ Use a perforated distribution plate downstream of the M.E. for additional distribution assistance.

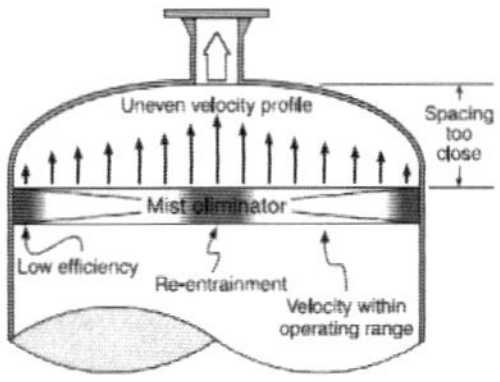

Figure 7. Example of mist eliminator performance degradation due to uneven velocity profile

+ Inlet diffusers also handle incoming liquid slug which otherwise can cause flooding of the Mist eliminator.

AMACS
PROCESS TOWER INTERNALS

Know what mist size droplets You are dealing with

+ Know if your mist is > 1 micron or submicron.
+ When gas travels from A to B, condensation in the pipeline will cause a fine mist (fog or submicron droplets)

  Examples:

  - Incoming gas from an offsite location
  - Fuel gas to furnaces/heaters with low NOx burners

AMACS
PROCESS TOWER INTERNALS

Know what mist size droplets You are dealing with cont.

+ Submicron mist is not captured by conventional M.E.

+ Submicron mist requires a coalescing filter, typically
installed in a two stage filter separator.

AMACS
PROCESS TOWER INTERNALS

Know what mist size droplets You are dealing with cont.

1c stage; submicron droplets are coalesced to > 10 micron droplets.

2c stage; greater than > 10 micron droplets are ejected from the coalescing filter and captured by a vane, which captures droplets > 8 micron.

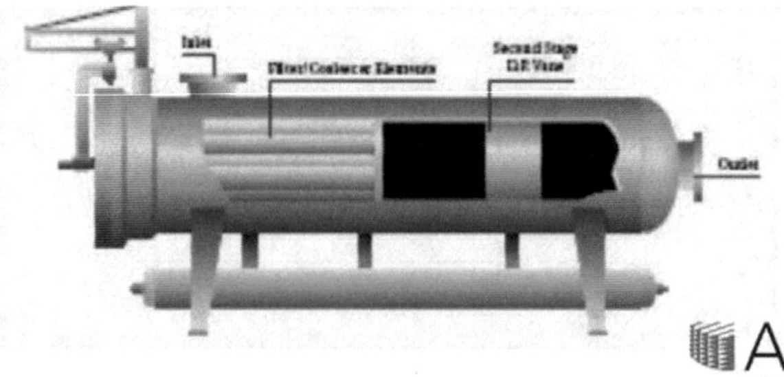

AMACS
PROCESS TOWER INTERNALS

High viscosity liquid as mist

+ Viscous liquids drain slower out of the M.E. and could cause localized flooding.
+ Use a mesh mist eliminator with an open drainage layer on the bottom to improve liquid drainage.
+ Use a lower K- value (lower face viscosity) for the M.E. To allow additional drainage time.
+ Use a vane type M.E. upstream of the mesh mist eliminator, to remove the bulk of the viscous liquid, so the M.E. can effectively handle the residual liquids.

Dislocated M.E. due to System pressure surges

+ M.E. sections can be dislodged into the downstream equipment, which then creates massive carryover.
+ Standard M.E. equipment often gets installed and secured to the support ring with tie wire.

Dislocated M.E. due to System pressure surges cont.

+ Where system pressure surges are common, there are more secure ways to install the M.E.
  - dual support rings
  - through bolts
  - install a pressure relief door

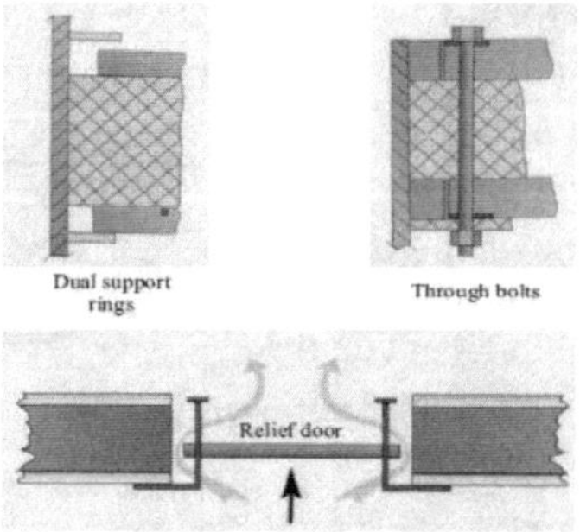

AMACS
PROCESS TOWER INTERNALS

When throughput exceeds original capacity

+ There are various solutions to increase capacity in an existing vessel.
  - Patented High capacit
    mist eliminator 0% - 1

AMACS
PROCESS TOWER INTERNALS

When throughput exceeds original capacity cont.

-Patented MISTFIX™
cylindrical mist eliminator 0% - 25%

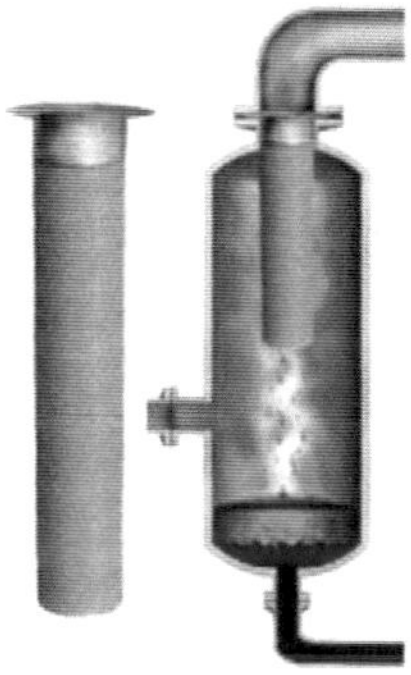

When throughput exceeds original capacity cont.

- High capacity Patented MultiPocket® vane. 0%-50%

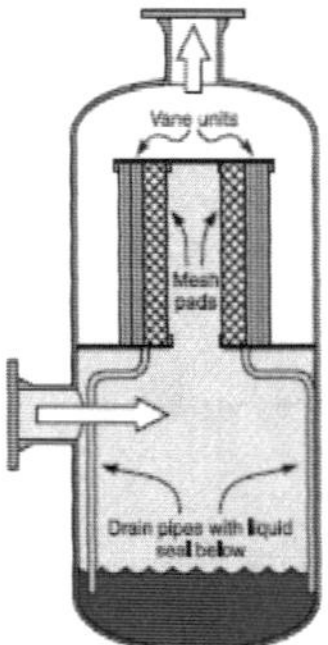

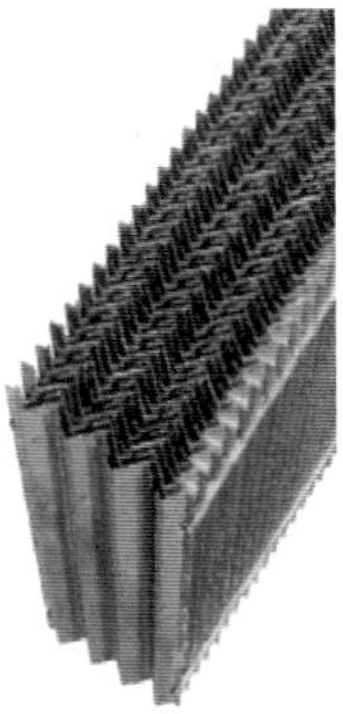

AMACS
PROCESS TOWER INTERNALS

When throughput exceeds original capacity cont.

+ MistFix™ is installed through the gas outlet, all other M.E. equipment comes in sections, installed through the manway.
+ With increased capacity comes increased bad gas distribution.
  - Use an inlet diffuser.
+ To avoid welding new support rings and housing for the high capacity new equipment, use double expansion rings, a compression fit.

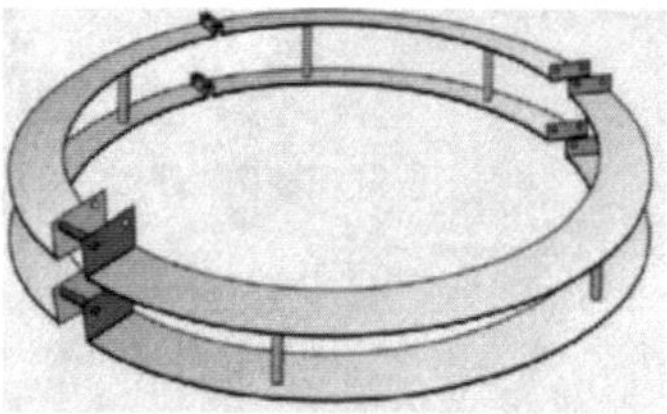

AMACS
PROCESS TOWER INTERNALS

Support Rings

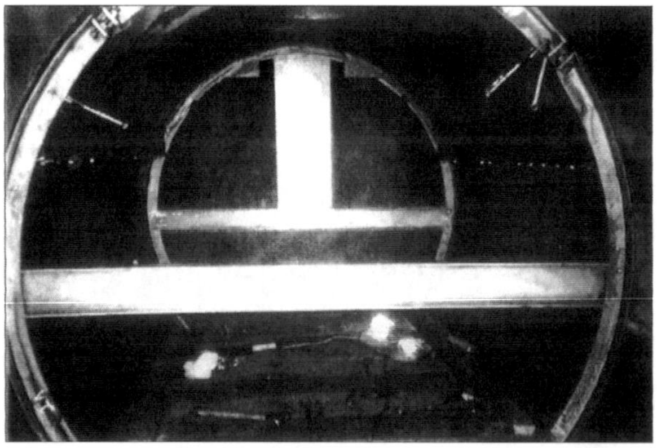

AMACS
PROCESS TOWER INTERNALS

Support Rings

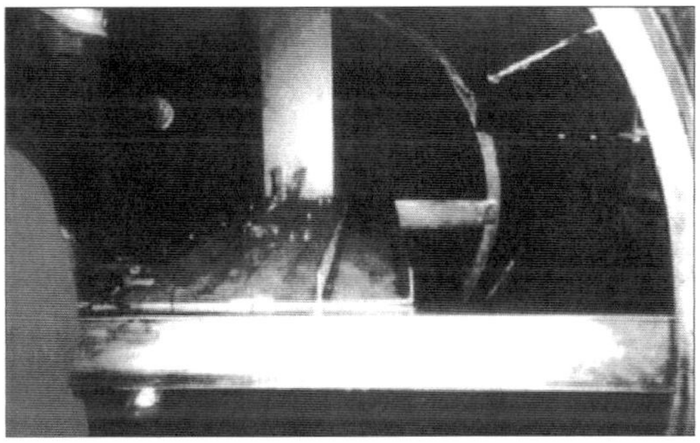

When the capacity range is too large for a suction drum

+ Typical M.E. equipment will operate between
  30% – 110% capacity.

+ When a suction drum often operates below 30% liquid carryover may occur.

When the capacity range is too large for a suction drum Cont.

To facilitate a capacity range of 10% - 110% consider a 2 phase M.E. in the same drum.

- one that operates 30% - 1
- second that operates between 10% - 30%

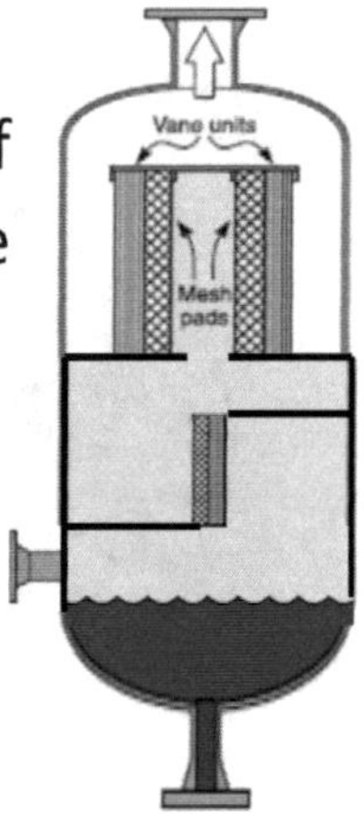

(45i) Renewable Feedstock for Steam Crackers: Catalytic Upgrading of Crude Tall Oil (CTO) Into Bio-Naphtha

Jinto Anthonykutty, Juha Linnekoski, Antero Laitinen and Ali Harlin, VTT Technical Research Centre of Finland, Espoo, Finland

Renewable feedstocks from bio resources are better alternatives to petroleum derived naphtha for the production of olefins and aromatics by steam cracking or reforming.[1, 2] Mostly, these renewable feedstocks (bio-naphtha) rich in n-alkanes are produced by catalytic upgrading methods via hydrotreating/hydroprocessing; in which oxygenates are removed from bio-oil by various deoxygenation processes. At present, hydrotreated vegetable (HVO) oil is produced in many countries on industrial scale using existing petroleum refinery methods[3]. However, employing higher priced vegetable oils could be a major concern in these processes and the main challenge is therefore finding an inexpensive and non-food chain affecting feedstock. Tall oil, obtained as a by-product from Kraft pulping process, is abundant and cheap;and hence, meets the criteria of a low cost starting material for upgrading[4]. Moreover, the utilization of tall oil in conventional oil refineries could serve as an eco-friendly method to convert bio-wastes into bio-naphtha[1].

In the present research approach, catalytic upgrading (hydrotreating) of crude tall oil (CTO) into bio-naphtha suitable for a steam cracker is discussed. CTO, which contains mainly fatty acids (42.9 wt. %) and resin acids (25.7 wt. %) was obtained from Stora-Enso Kaukopää (Finland) mill and used as received. A commercial NiMo catalyst was employed and the experiments were conducted in a continuous, down flow, fixed bed reactor at different process conditions (WHSV 1-3h^{-1}, Pressure 5 MPa and Temperature 350 - 450 ^{o}C). After the stabilization period (6 h.), the liquid sample was collected and separated into organic and aqueous phases.The product distribution in organic phase was analyzed with a Shimadzu GC/MS equipped with a Zebron MS 30m × 0.25 mm × 0.25µm column.

This current approach reports that CTO produces mainly paraffins, cycloalkanes and aromatics during upgrading process. Paraffinic hydrocarbons mainly comprise n-octadecane (20-25 wt. %) and n-heptadecane (15-20 wt. %), are obtained in maximum yield (55- 60 wt. %) in low temperature reactions. The yield of paraffinic hydrocarbons increased as a function of increasing space time in low temperature reactions. 18-Norabietane and other mono-, di- and tri- cycloalkanes are obtained from resin acids in CTO, and these cyclic structures are converted to aromatics such as naphthalene and phenanthrene in high temperature reactions (>400 ^{o}C). Stigmastane, apparently produced from sterols in CTO are also present as a major three ring component in the product stream. Additionally, gaseous products obtained and the yield increased as a function of temperature. The trend observed with the distribution of gaseous products reveal that decarboxylation and cracking reactions are favored at higher temperatures, while hydrogenation and hydrodeoxygenation (HDO) are suppressed.

The elemental composition of CTO shows that it contains relatively high sulfur content (1800 mg/kg), could be useful to keep the catalyst's activity with time on stream (TOS). However, the considerable amount (87 mg/kg) of alkali metals (Na, K, Ca and P) in CTO might be critical for the catalyst activity with TOS, will be discussed in this contribution.

The product composition obtained with CTO is assessed in comparison with compositions and characteristics of refinery naphtha feeds, and will be discussed in detail. Furthermore, this contribution will discuss the achieved deoxygenation rate and success criteria for the product composition of upgraded CTO.

References

1. Gornay, J.; Coniglio, L.; Billaud, F.; Wild, G. Steam Cracking and Steam Reforming of Waste Cooking Oil in a Tubular Stainless Steel Reactor with Wall Effects. ***energy&fuels***, **2009**, 23(11), 5663-5676.
2. Pyl, S.P.; Schietekat, C.M.; Reyniers, M.F.; Abhari, R.; Marin, G.B.; Van Geem, K.M. Biomass to olefins: Cracking of renewable naphtha. ***Chemical Engineering Journal***, **2011**, 176-177, 178-187.
3. Sunde, K.; Brekke, A.; Solberg, B. Environmental Impacts and Costs of Hydrotreated Vegetable Oils, Transesterified Lipids and Woody BTL- A Review. ***Energies***, **2011**, 4(6), 845-877.
4. Coll, R.; Udas, S.; Jacoby, W.A. Conversion of the Rosin Acid Fraction of Crude Tall Oil into Fuels and Chemicals. ***energy&fuels***, **2001**, 15, 1166-1172.

AIChE Paper Number 45j
High Efficiency Ethylene Cracking Furnaces

Rajaram Ramesh

Product Development Manager - Ethylene
Technip
Netherlands

Marco W. M. van Goethem, PhD

Product Development Manager – Modeling and Testing
Technip
Netherlands

25th Ethylene Producers Conference
AIChE 2013 Spring National Meeting
San Antonio, Texas, April 29 – May 3, 2013

Abstract

Environmental and economic pressures generate a driving force to produce products more efficiently. A direct effect of more efficient processes is that the emission of green house gases, particularly carbon dioxide decreases. In households nowadays "high efficiency furnaces" are available with efficiencies in the range of 94-97%. This paper describes the outcome of a feasibility study on applying the high efficiency furnace concept to steam cracking furnaces. The high efficiency is achieved by enhanced heat recovery from the flue gases before they are released to the atmosphere at ca. 65°C.

Technip studied the application of this concept for cracking furnaces based on gas as well as liquid feedstocks. Natural gas was used as basis for the energy price. Since energy prices have a strong impact on project economics, the evaluations were done for an EU average natural gas price and a US natural gas price.

Two alternative heat sinks have been considered: preheating of combustion air and preheating of cold hydrocarbon feed. Both are viable for liquid cracking furnaces. In the case of gas cracking furnaces feed preheating results in a pinch and this makes it not viable.

Capital costs are based on Technip's cost database and quotations from air preheater vendors. Special attention has been paid to the material selection of the affected convection section banks.

This study shows that high efficiency ethylene cracking furnaces are economically viable. For Europe the study indicates an attractive payback period. In the US the payback period is longer due to the lower energy prices. If energy prices increase in the future, the concept's economic viability will become better. Inclusion of CO2 credits will make the payback period more attractive.

Furnace Efficiency

In this paper furnace efficiency has been calculated as net thermal efficiency as interpreted by API560[1]. Furnace efficiency is determined by drawing an envelope around the furnace and including in it the radiant section, convection section and any heat exchanger equipment that recover heat from the flue gas (air preheater, economizer, steam superheater, etc.). The efficiency is the ratio of the useful heat absorbed to the fired heat of the fuel and any other external heat inputs. Ambient temperature is a reference temperature for efficiency calculations and any heat inflow or outflow above this reference is taken into consideration as a heat input or heat loss.
Furnace efficiency can be determined as follows:

$$\frac{heat\ absorption\ in\ radiant\ section\ +\ convection\ section}{(firing\ rate\ \times\ fuel\ heating\ value) + sensible\ heat\ of\ fuel\ and\ air\ above\ ambient\ temp. + other\ external\ heat\ inputs}$$

Alternately, $\text{Furnace efficiency} = 100\% - \text{Total heat loss, where:}$

$$\text{Total heat loss} = \frac{stack\ loss\ +\ radiation\ heat\ loss}{(firing\ rate\ \times\ fuel\ heating\ value) + sensible\ heat\ of\ fuel\ and\ air\ above\ ambient\ temp. + any\ external\ heat\ inputs}$$

In a furnace where fuel and combustion air enter the furnace at ambient temperature and there are no other external heat inputs, the first expression can be simplified to the following:

$$Furnace\ efficiency = \frac{heat\ absorption\ in\ radiant\ section\ +\ convection\ section}{fuel\ firing\ rate\ \times\ fuel\ heating\ value}$$

Please note that net thermal efficiency, as defined by API 560, is based on the lower heating value of the fuel.

Study objectives

Technip undertook this study with the goal to reduce fuel firing in cracking furnaces by at least two percent. Although this figure does not look very ambitious, it must be remembered that since efficiencies in current designs are already in the mid nineties (upper part of the technology S-curve), further improvements will not be easy and can only be incremental.

Since stack loss is the main source of inefficiency (radiation heat loss is comparatively smaller and more difficult / expensive to reduce) it was decided to make stack loss reduction as the main focus of the study.

By a relatively straight forward calculation it can be shown that the two percent efficiency rise target would require the flue gas to be cooled to approximately sixty or seventy degrees Celsius.

The main challenge in reducing stack temperature to this level is finding an appropriate heat sink for the recovered heat. In a cracking furnace there are two suitable heat sinks, combustion air and hydrocarbon feed, and both were studied. The following sections discuss the results for liquid furnaces and gas furnaces.

Basis of Study

We have done the study on a naphtha furnace and an ethane furnace. We have chosen these two feedstocks since a large number of furnaces operate on these two feeds and we can draw conclusions that can be applied approximately to other liquid and gas feeds. The study has been performed on the following bases:

- The reference liquid and gas furnace designs used in this study are based on actual furnaces from Technip's past projects. The furnaces were simulated using Technip's proprietary tool SPYRO which is used in the design and simulation of cracking furnaces.
- The budgetary capital cost need to achieve the target furnace efficiency was calculated as an incremental cost. It was estimated from vendor quotes and Technip's internal cost database.
- The cost estimates are for a new project, not a revamp, and are for a West European location.
- Operating costs are based on cost of base utilities and derived utilities. Base utilities are natural gas and demineralized water. These are taken from market data.
- Since natural gas prices are very different in different parts of the world, the operating costs and economic analysis have been done for European and United States gas prices.
- EU natural gas price was fixed at €25.1 /MWh of higher heating value (HHV) which is equivalent to €28.0 /MWh of lower heating value (LHV). The prices are based on EUROSTAT data for 1st half of 2010 and were obtained from the publications of the Department of Energy & Climate Change in the UK. They correspond to the EU-15 average for Large Industrial Consumers (i.e. 2.13 pence/kWh HHV).
- US Natural gas prices were obtained from the US Department of Energy. The average value was 6 \$/MM BTU (HHV) or 17 €/MWh (LHV) for 1st half of 2010. The average HHV used is 1029 BTU per cubic foot or 1.029 MM BTU per thousand cubic feet.
- Cost of demineralized water was assumed at € 2.00 /ton.
- The prices of other utilities (boiler feed water, steam and electricity) have been calculated from those of the two base utilities. Electricity was priced assuming 35% overall efficiency of power generation.
- The economic calculations do not include any CO2 credit for the fuel saved.
- Assumed exchange rate: 1 US dollar = 0.75 euros
- The plant is assumed to operate 8400 hours per year

The resultant utility prices used in the study are summarized below.

UTILITY COSTS

		EU basis	**US basis**	
FUEL as NATURAL GAS	Base	**28.0**	**17.0**	€ / MWh
DEMINERALISED WATER	Base	2.0	1.5	€ / ton
VHP STEAM (net value)	Derived	23.7	14.4	€ / ton
MP STEAM (net value)	Derived	13.9	8.5	€ / ton
LP STEAM	Derived	9.7	5.9	€ / ton
ELECTRICAL POWER	Derived	80.1	48.6	€ / MWh

Study Results

Naphtha Furnace using Air Preheat

Combustion air can be preheated in a plate type air preheater (APH). These are cross flow type exchangers and have low pressure drops. Combustion air is at a lower temperature than any of the process or utilities streams entering the convection section; therefore, from the standpoint of efficiency, APHs are ideally suited to be the last heat recovery section before the stack.

Fig 1 shows the Base Case Naphtha furnace while Fig 3 shows the same furnace with an APH. Please also refer to Table 1 for a performance comparison of the APH design with the Base Case naphtha furnace. The following points are of interest:

1) The overall efficiency of the furnace will improve from 94.4 to 97.0% for the APH option. This is mainly due to the lower stack temperature but also partly due to the smaller excess air, which is explained below.

2) This option will require a forced draft combustion air fan to overcome the pressure drop in the APH. As a result the furnace operation will change from induced draft mode to balanced draft. The availability of air in forced draft will enable more accurate control of the excess air and it will be possible to operate the furnace with lower excess air (5% instead of 8%). This leads to a lower flue gas flow, a lower stack loss and therefore better efficiency.

3) There will be a drop of 4.8% in fuel firing which is greater than the efficiency increase. This is because of the lower absorbed heat. The furnace is designed and controlled in such a way as to supply all the required process heat and to absorb the surplus heat into utilities to the extent possible. When the furnace operates more efficiently and the heat demanded by the process media remains unchanged, as in the APH case, there will be less surplus heat available for VHP steam. As a result the VHP steam generation will go down from 79.7 to 76.0 t/h.

4) The heat transfer surface in some parts of the APH will be below the water dew point of the flue gas (around 60°C). This can lead to water condensation and create two possible issues: 1) accumulation of water in the APH and 2) corrosion due to the presence of liquid water, oxygen and carbon dioxide.

   Two measures have been included in the design to counter these problems:
   - Placement of unit such that flue gas flows downward to aid easy draining
   - Use of corten steel to resist corrosion

5) We have evaluated the impact of the air preheat option on NOx emissions using API 535[2)]. The rise in air temperature and the drop in excess air will have opposing effects on the NOx production. The net effect is that the NOx concentration will be close to the Base Case.

   If a DeNOx unit is present, as is often the case in the US, the APH option may help to slightly extend the life of the DeNOx catalyst. This is because the total NOx load will be smaller than the Base Case due to the smaller flue gas flow.

Table 2 summarizes the results of the economic analysis of this option. We have used payback period to compare different options. Payback period is the ratio of the incremental capital cost above the Base Case CAPEX divided by the savings in yearly operating cost.

1) The capital cost (CAPEX) of the APH option includes the cost of the APH, the combustion air fan, extra ducting and structural steel. The operating cost (OPEX) savings is the savings in furnace fuel but this must be offset against the value of the lost VHP steam production (which notionally must be made up in a utility boiler elsewhere) and the additional electricity for the two fans.

2) It can be seen that the payback period is three years based on European gas prices and five years based on American gas prices.

Naphtha Furnace using Feed preheat

Naphtha is normally stored at ambient temperature and is preheated to 60°C using quench water before it is routed to the furnaces. Most cracking furnaces are equipped with a Feed Preheater (FPH) bank in the convection section. If flue gas is utilized to preheat naphtha from ambient conditions the stack temperature can be lowered to the desired range of sixty to seventy degrees Celsius. Till now designers and operators have not taken full advantage of this mainly because of the higher CAPEX and fear of dew point corrosion.

One of the consequences of this feature will be that the quench water duty previously absorbed by the naphtha will now shift to the air coolers in the quench water circuit. The additional electrical power that the air cooler fans will require will be about 10 KW per furnace.

Please refer to Table 3 for an overview of the performance of a naphtha furnace with and without an extended feed preheater. Please note that the Base Case naphtha furnace is the same as the one considered in the APH option.

1) The overall efficiency of the furnace can be improved from 94.4 to 96.0% for this option.

2) Unlike the APH option a combustion air fan will **not** be required. The furnace will continue to operate in induced draft mode.

3) The fuel firing and the VHP steam generation will drop marginally.

4) The upper tube rows of the FPH will be below the water dew point of the flue gas and some condensation will occur. As a precaution we have considered that the top four rows will be of stainless steel.

5) The flue gas NOx concentration will be unaffected by this option. However the total hourly NOx emissions and load on the DeNOx unit, if any, will decrease to the same extent as the fuel firing, which is around 0.6%.

Table 4 summarizes the results of the economic analysis of this option.

1) The capital cost (CAPEX) of this option covers the incremental cost of the extended FPH, extra ducting and structural steel. The OPEX savings is the reduction in furnace fuel after adjusting for the reduction in VHP steam production and the additional electricity for the ID fan and the Quench Water Air Cooler fans.

2) In the Base Case the naphtha feed piping between the external feed preheater and the convection section is insulated for heat conservation. In the FPH option naphtha will be fed at ambient temperature to the furnace and so these lines do not have to be insulated. This will yield some capital cost saving but this has not been taken into account in our analysis.

3) The payback period excluding CO2 credit will be three and a half years with European natural gas and five and a half years with US natural gas.

Ethane Furnace using Air Preheat

Fig 2 shows the Base Case Ethane furnace that was used in the study. The APH arrangement on an ethane furnace will be quite similar to the one on the naphtha furnace. Table 5 shows the impact that the APH would have on the furnace performance. The main points are as follows:

1) The overall efficiency of the furnace will improve from 91.7 to 96.5% for the APH option. This is mainly due to the lower stack temperature but also partly due to the smaller excess air.

2) In this furnace also the APH will require a forced draft combustion air fan. This will allow the excess air to be reduced from 8% to 5% which will result in lower stack losses and a better efficiency.

3) The fuel firing will go down by 11.0%, from 8.2 to 7.3 t/h. The absorbed heat will drop from 116.8 to 110.6 MW. As a result of these the VHP steam generation will go down from 97.0 to 88.8 t/h.

4) A portion of the APH will operate below the flue gas water dew point. Consequently the unit must be designed to prevent water accumulation and corrosion.

5) The impact on NOx was evaluated here also. The rise in air temperature will have a greater impact on NOx than the fall in excess air. We estimate that as a result the flue gas NOx concentration will increase slightly (around two percent). This can be a cause for concern if the furnace does not have a DeNOx unit and is operating close to the limits imposed by authorities. However since the total firing will go down, the hourly NOx production will drop. Plants can consider discussing with authorities to obtain a higher allowable NOx limit,

   If the furnace has a DeNOx unit, the chemical load on the unit will reduce, which should help to obtain a longer catalyst life.

The operating cost savings and payback for this option have been summarized in Table 6.

1) The capital cost includes the cost of the APH, the combustion air fan, extra ducting and structural steel. The operating cost savings is the savings in fuel firing with an adjustment for the value of the reduced VHP steam generation and the extra electricity for the ID fan and FD fan.

2) The payback period with natural gas priced at European rates will be one and a half years. The corresponding figure for US gas prices will be three years.

Ethane Furnace using Feed Preheat

Using feed at ambient temperature does not significantly reduce the stack temperature in a gas furnace. This is because the feed preheat curve has a steeper slope than the flue gas cooling curve, causing a pinch on the hot side of the Feed Preheater. Fig 4 shows the heating and cooling curves of ethane, boiler feed water and flue gas in the convection section. The hatched shape on the plot shows the maximum additional heat duty that can be recovered into feed, over and above the Base Case. The plot clearly shows that when the ethane feed temperature is lowered from the base value of 55°C to 30°C, the stack temperature will drop from 170°C to about 150°C. The flue gas cannot be cooled any more than this modest amount.

For this reason the feed preheat option was not investigated any further.

Conclusions and Summary

The key results from the different study options are listed together below.

Cracking furnace feed	Naphtha		Naphtha		Ethane	
Cold sink	Combustion air		Hydrocarbon feed		Combustion air	
	EU nat. gas	US nat. gas	EU nat. gas	US nat. gas	EU nat. gas	US nat. gas
OPEX savings	€ 0.60 mill	€ 0.36 mill	€ 0.17 mill	€ 0.10 mill	€ 1.27 mill	€ 0.77 mill
Payback time	3 years	5 years	3.5 years	5.5 years	1.5 years	3 years
Salient features of each option	Small number of additional equipment		No additional equipment		Small number of additional equipment	
	Higher CAPEX		Lower CAPEX		Higher CAPEX	
	More OPEX savings		Less OPEX savings		Best OPEX savings	
	Slight effect on NOx concentration		No effect on NOx		Around 2% higher NOx concentration	

All three scenarios have attractive payback periods and will be worth considering in high energy cost countries such as Europe and Asia. In US and other energy rich nations, the economic incentive will be much lower. In the US, where ethane cracking is common and increasing even more, improvement of furnace efficiency with air preheat can be worth considering as the payback time is less than 3 years.

If the options are compared against each other, in our opinion the Air preheating option has more potential, irrespective of feedstock.

The Feed Preheating option in liquid furnaces is cheaper but it also gives proportionally modest savings. Moreover, as earlier mentioned, the feed preheat duty is being shifted from one waste heat source (quench water) to another (flue gas). This option will not lead to a net improvement or cash savings in the plant as a whole unless there is a shortage of quench water and there are other, better uses for the saved quench water duty

Acknowledgements

The authors wish to express their thanks to their colleagues Mr Ashok Laghate, Ms Susan Cooper, Mr Sachin Singare and others in Technip Benelux who contributed to this study.

References

1) API 560: Fired Heaters for General Refinery Service, 4th edition

2) API 535: Burners for Fired Heaters in General Refinery Service, 2nd edition

Table 1
Naphtha Furnace with Combustion Air Preheating – Performance Summary

		Base Case	Air Preheat Case
Naphtha feed	t/h	60.8	60.8
Ambient conditions	-	15°C, 85% R.H.	15°C, 85% R.H.
Air temp. to burners	°C	15	77
Stack temperature	°C	108	61
Excess air	%	8	5
Absorbed heat	MW	113.2	110.4
Overall efficiency	%	94.4	97.0
Fuel flow	t/h	8.4	8.0
Fired heat	MW	119.9	113.9
VHP steam production	t/h	79.66	76.01
NOx concentration		Base	Close to Base

Table 2
Economics of Naphtha Furnace with Combustion Air Preheating

		Air Preheat Case (EU location)	Air Preheat Case (US location)
Savings in fuel firing	million €/yr	1.42	0.86
Cost of lost VHP steam capacity	million €/yr	0.73	0.44
Cost of additional electricity	million €/yr	0.10	0.06
Net OPEX saving	million €/yr	0.60	0.36
Payback	Years	3	5

Table 3

Naphtha Furnace with Feed Preheating – Performance Summary

		Base Case	Feed Preheat Case
Naphtha feed	t/h	60.8	60.8
Feed temperature	-	60°C	30°C
Combustion air temp.	°C	15	15
Stack temperature	°C	108	71
Excess air	%	8	8
Absorbed heat	MW	113.2	114.3
Overall efficiency	%	94.4	96.0
Fuel flow	t/h	8.4	8.35
Fired heat	MW	119.9	119.1
VHP steam production	t/h	79.66	79.63
NOx concentration		Base	Base

Table 4

Economics of Naphtha Furnace with Feed Preheating

		Feed Preheat Case (EU location)	Feed Preheat Case (US location)
Savings in fuel firing	million €/yr	0.19	0.12
Cost of lost VHP steam capacity	million €/yr	< 0.01	< 0.01
Cost of additional electricity	million €/yr	0.01	< 0.01
Net OPEX saving	million €/yr	0.17	0.10
Payback	Years	3.5	5.5

Table 5
Ethane Furnace with Air Preheating – Performance Summary

		Base Case	Air Preheat Case
Ethane feed	t/h	51.9	51.9
Ambient conditions	-	15°C, 85% R.H.	15°C, 85% R.H.
Air temp. to burners	°C	15	129
Stack temperature	°C	169	64
Excess air	%	10	5
Absorbed heat	MW	116.8	110.6
Overall efficiency	%	91.7	96.5
Fuel flow	t/h	8.2	7.3
Fired heat	MW	127.3	114.5
VHP steam production	t/h	97.0	88.75
NOx concentration		Base	Base + 2%

Table 6
Economics of Ethane Furnace with Air Preheating

		Air Preheat Case (EU location)	Air Preheat Case (US location)
Savings in fuel firing	million €/yr	3.01	1.83
Cost of lost VHP steam capacity	million €/yr	1,64	1.00
Cost of additional electricity	million €/yr	0.10	0.06
Net OPEX saving	million €/yr	1.27	0.77
Payback	Years	1.5	3

Fig 1
Naphtha cracking furnace

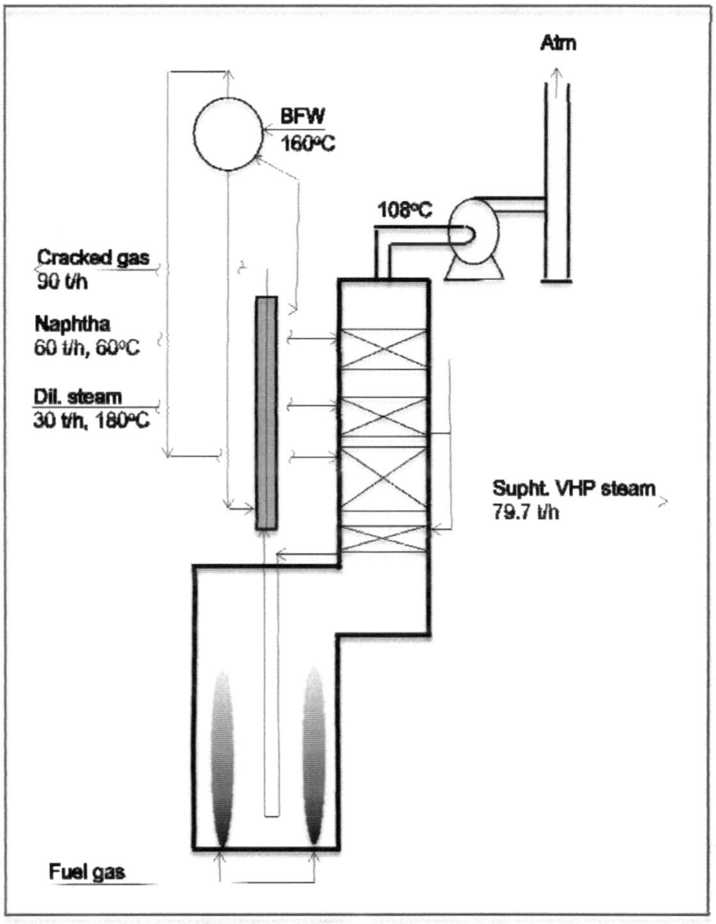

Fig 2
Ethane cracking furnace

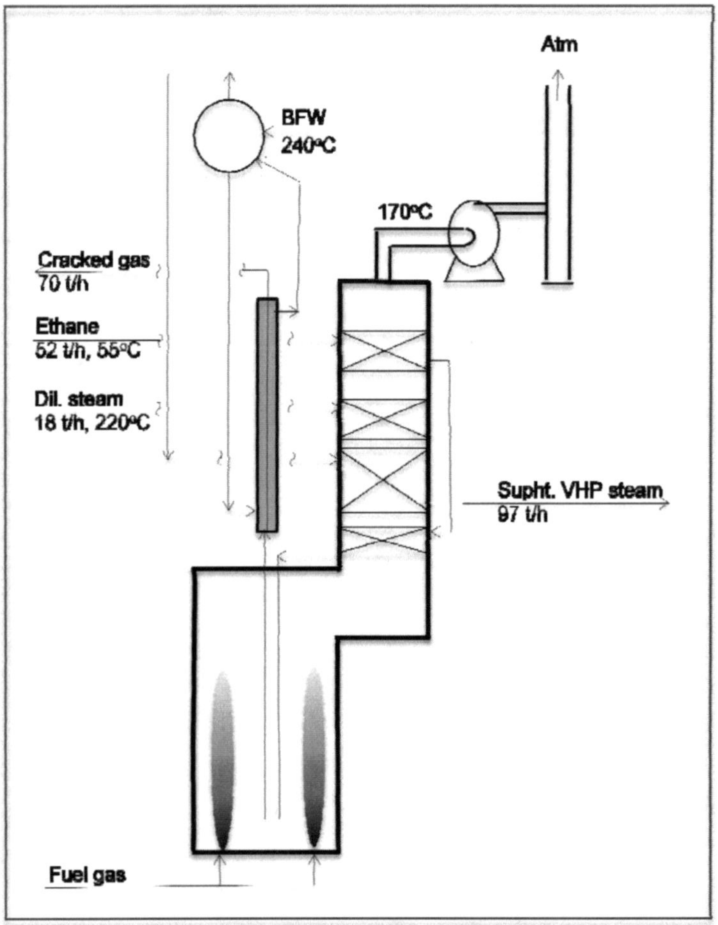

Fig 3

Naphtha cracking furnace – Air Preheater option

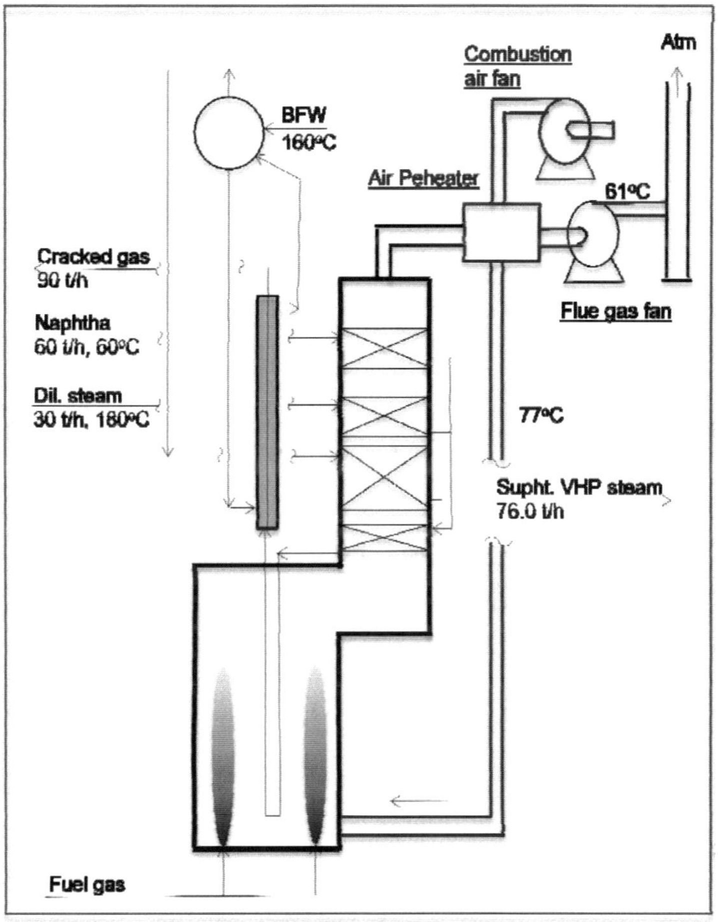

Fig 4
Ethane cracking furnace – Feed Preheater option

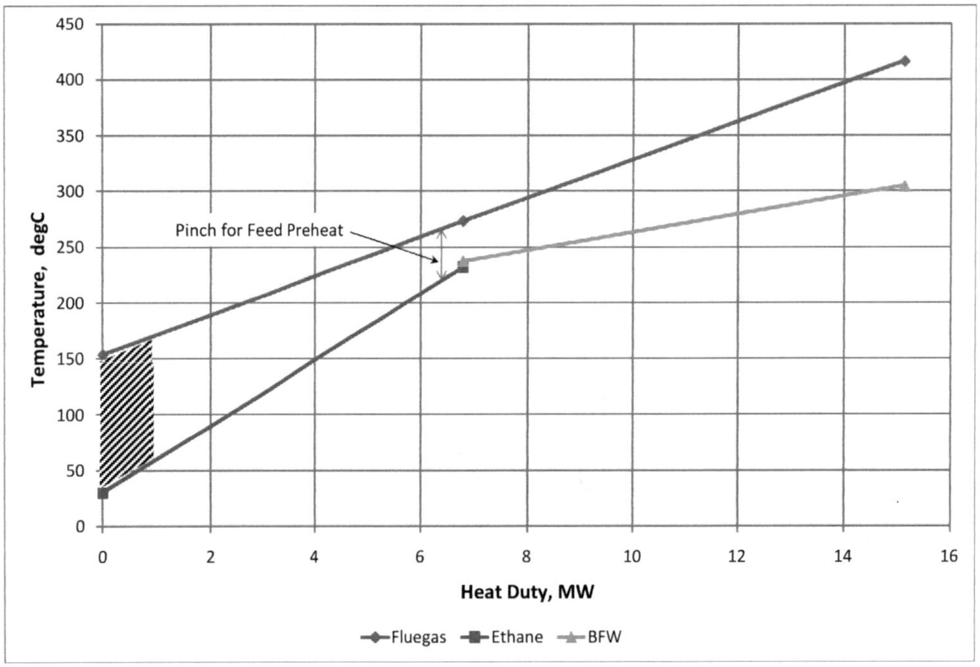

Modernize Your Control Functionality
Easy Steps to Start Process Optimization

13 AIChE
Spring Meeting, San Antonio, TX

Presenters

- Laurie Ben
- James Beall

EMERSON
Process Management

Agenda

- Introduction
- Examples – Best Practices & Lessons Learned
  - Split Range Duty
  - Smith Predictor
  - Gain Scheduling/Adaptive Control
  - Signal Characterization
- Summary

Introduction

- Modernizing Control System – Decision Done Check!
- 'Do I' or 'Don't I' take advantage of new technology
  - Implemented older cumbersome techniques?
  - Complex custom strategies that handle automation challenges?
  - Hard to implement & maintain?
- Standard features that are field proven, practical, and easy to implement
  - Easier to implement & maintain
  - More efficient and effective control strategies
  - Achieve improvements in process performance
  - Lays the foundation for optimization

Agenda

- Introduction
- Examples – Best Practices & Lessons Learned
  - Split Range Duty
    - Control Challenge
    - Legacy Solution
    - Newer Technology Solution
    - Summary
  - Smith Predictor
  - Gain Scheduling/Adaptive Control
  - Signal Characterization
- Summary

Split Range Duty – Control Challenge

- Use steam and hot oil as energy source for distillation column for temperature controller.
- Hot oil is "free energy" and should be maximized
- Control Scenarios
  - Maximize hot oil and supplement with steam
  - Base load steam (if meter bad) and supplement with hot oil
  - Base load hot oil flow (if bad meter) and supplement with steam
- Constraints
  - Minimum and maximum hot oil valve position and flow rate
  - Minimum steam flow rate

Split Range Duty – Legacy Solution

– Old legacy system implemented this feature using a velocity PID with incremental output and a "micro expert rules system"

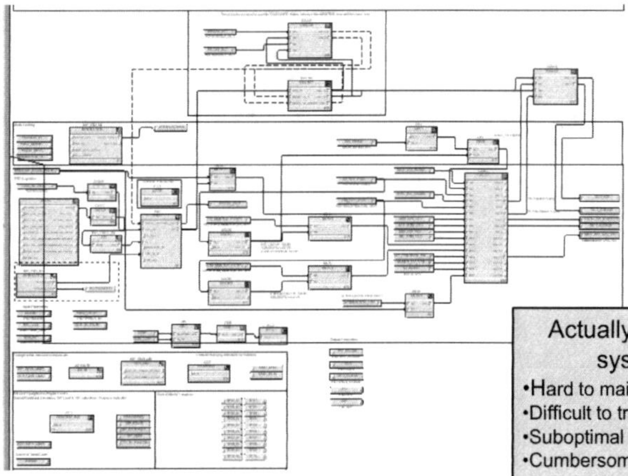

Actually Duplicated in New system, but still...
- Hard to maintain and modify
- Difficult to troubleshoot
- Suboptimal performance
- Cumbersome/Kludge/Ugly

Split Range Duty – New Technology Solution

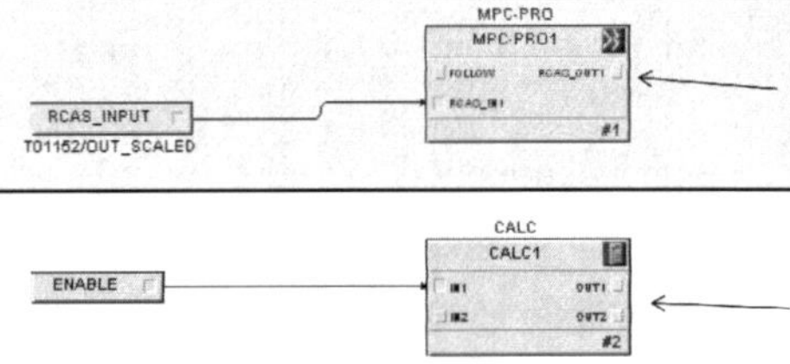

Embedded MPC Function Block provides all control functionality! EASY!!!

Simple calculations to calculate limits

Split Range Duty – New Technology Solution

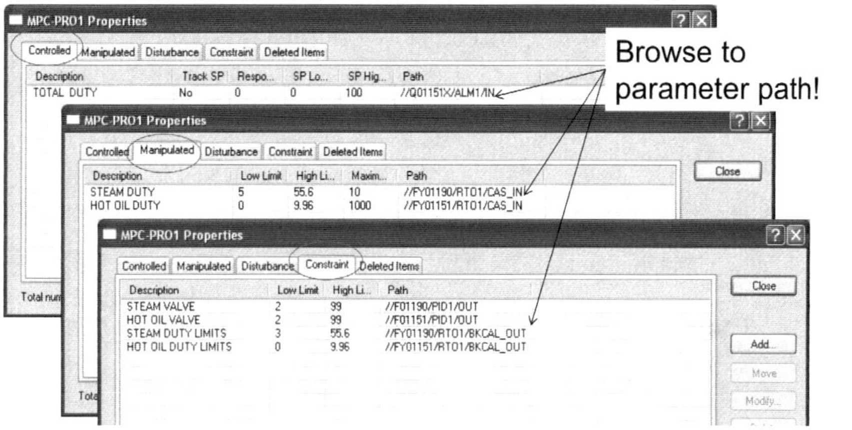

Split Range Duty – New Technology Solution

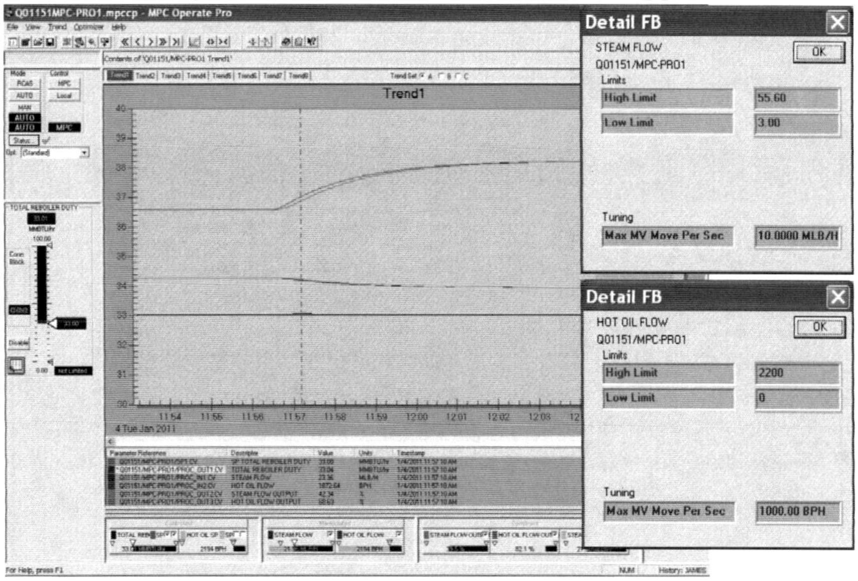

Summary Split Range Duty Control

Embedded MPC (New Technology) vs Legacy Solution

- Better performing solution
  - Bumpless transfers and constraint control
- Easier to Maintain
  - Change limits and settings
  - Simpler implementation
  - Easy to understand, train others, document
- Easily expands to handle future constraints
- Reduce energy costs!

Agenda

- Introduction
- Examples – Best Practices & Lessons Learned
  - Split Range Duty
  - Smith Predictor
    - Control Challenge
    - Legacy Solution
    - Newer Technology Solution
    - Summary
  - Gain Scheduling/Adaptive Control
  - Signal Characterization
- Summary

Smith-Predictor PID – Control Challenge

- Control of processes that are "dead time dominant" (dead time > time constant)
- Typically custom configured
  - some legacy control systems have preconfigured options, e.g. PROVOX
- Algorithm uses a process model
  - (gain, dead time and time constant)
- Challenge: as process dynamics change from model used, performance degrades rapidly

Smith-Predictor PID – Legacy Solution

- Traditional Smith-Predictor

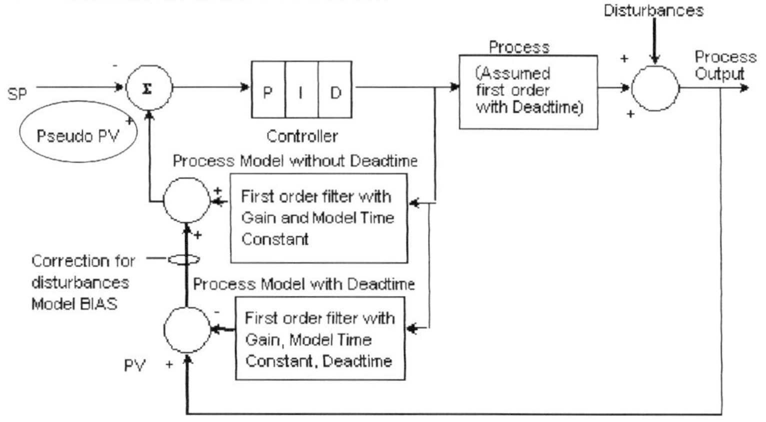

•High maintenance when process dynamics change
•Performance degrades rapidly as process dynamics change from configured model

Smith-Predictor PID – New Technology Solution #1

Easy to implement the Smith-Predictor but…

- High Maintenance when process dynamics change
- Performance degrades rapidly as process dynamics change from configured model

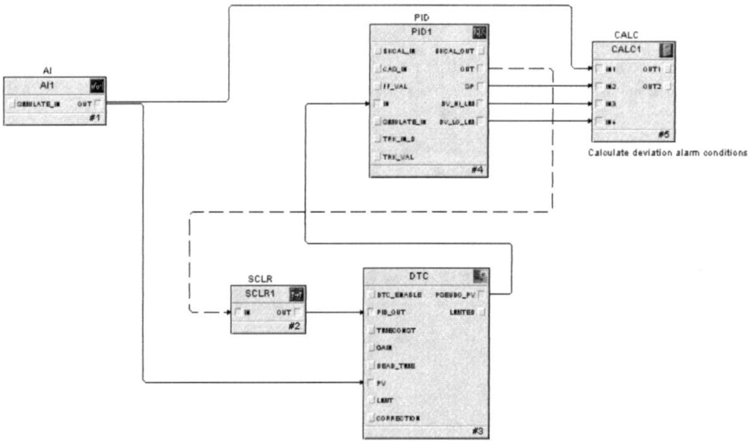

Smith-Predictor PID – New Technology Solution #2

- Embedded MPC Solution

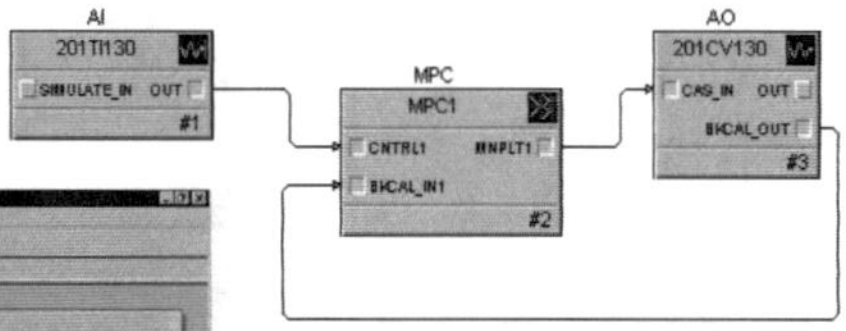

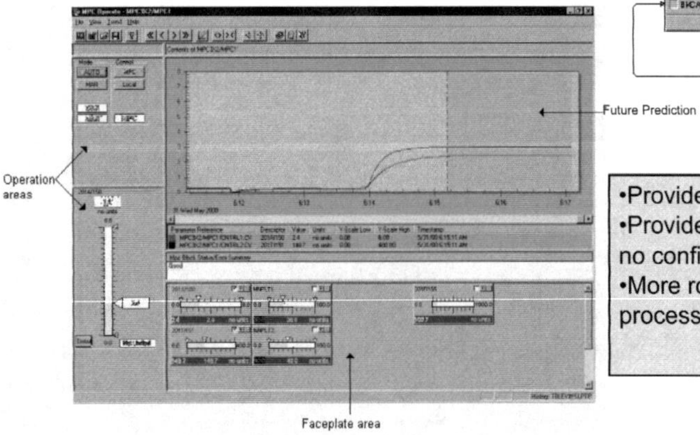

- Provides Future Prediction
- Provides MPC Faceplate – no configuration needed
- More robust under changing process dynamics

Smith-Predictor PID – New Technology Solution #2

- Embedded MPC Solution

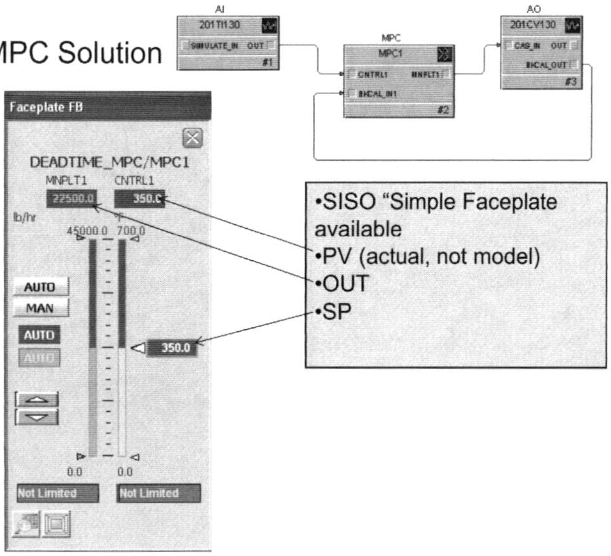

Summary Smith-Predictor PID

New Technology Solutions

- Standard Smith Predictor (#1) and Embedded MPC (#2)
- Both easy to implement and understand!

New Technology Solution #2 (Embedded MPC)

- More robust solution under a changing process model
- Provides visual display of the "future" PV
- Save money through better control performance

Agenda

- Introduction
- Examples – Best Practices & Lessons Learned
  - Split Range Duty
  - Smith Predictor
  - Gain Scheduling/Adaptive Control
    - Control Challenge
    - Legacy Solution
    - New Technology Solution
    - Summary
  - Signal Characterization
- Summary

Gain Scheduling – Control Challenge

- Legacy systems have limited capability to modify tuning based on process conditions
- Others had add-on application software, e.g. PROVOX "Gain Scheduler" system provided 3 sets of Gain, Reset, and Rate applied in 3 separate regions of a "state variable"
  - State variable = PV, OUT or "other"
  - Transition between tuning sets interpolated over "dead band"
- Beware: some suspend loop execution when tuning parameters change, so don't change the tuning every scan

Gain Scheduling – Legacy Solution

- Custom code
- Multiple points, registers, displays
- Restricted (slower) update rate

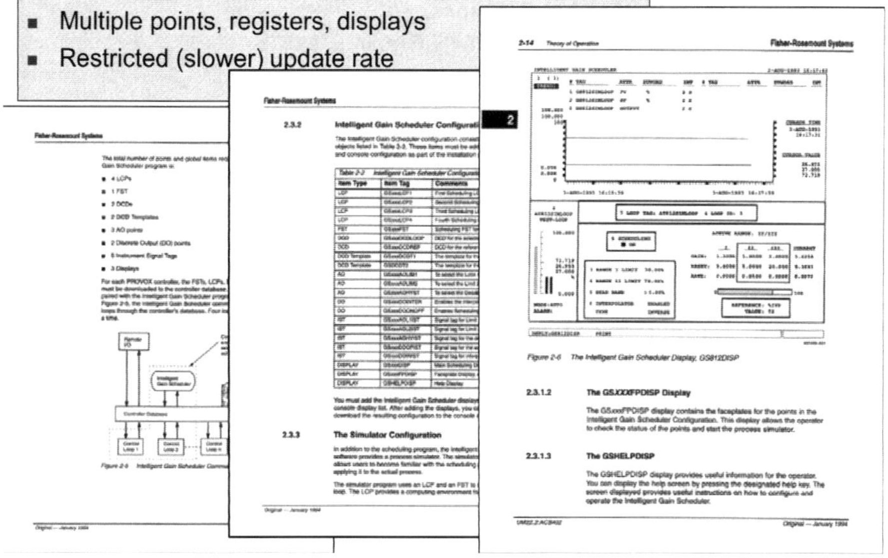

Gain Scheduling – New Technology Solution #1

Standard Modules

- Standard configuration
- Preconfigured faceplate & detailed display
- Easy to implement
- Tuning updates at loop execution

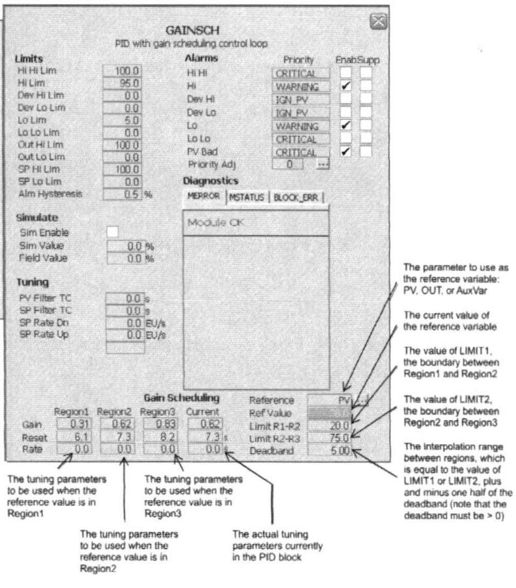

Gain Scheduling – New Technology Solution #2

New Technology "Adaptive Control"

- "Learns" process models from normal operating moves
- Allows up to 5 tuning regions of the "state variable"
- Partial Adaptive Control
  - Use "approved model" only & selected tuning rule in each region
  - Lambda PI or IMC PID; "Closed loop time constant" or "Lambda Factor".
- Optional Full Adaptive Control
  - continuous "Learning" of process model in each region
  - updates tuning while in each region per selected tuning rule
  - set allowable deviation (%) from approved model in each region

New Technology – Adaptive Control

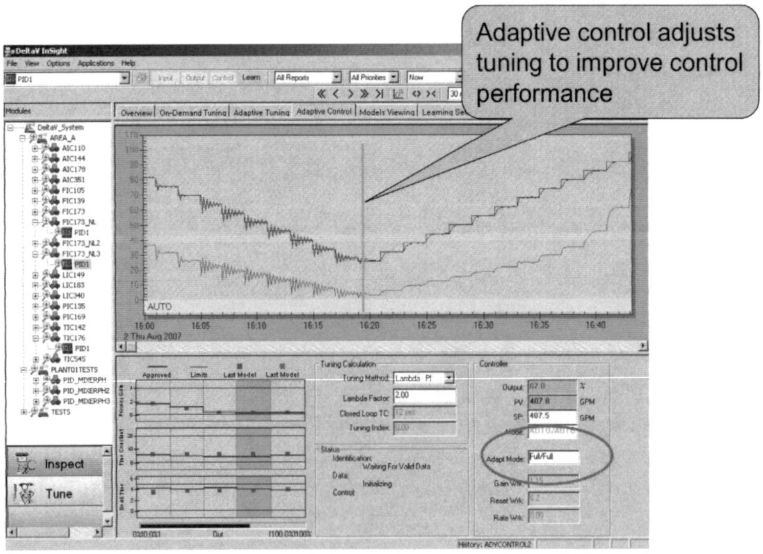

Summary Gain Scheduling

New Technology Solution(s) vs. Legacy Solution

- Gain Scheduling easy to implement and understand!
  - Standard configuration
  - Preconfigured faceplate & detailed display
  - Tuning updates at loop execution
- Adaptive Control provides additional benefit and ease of implementation
  - Learns models during normal operation – NO STEP TESTs!
  - More Regions – Up to 5
  - Automatic tuning updates when process model changes
- Improved control performance saves money!

Summary Gain Scheduling

New Technology Solution(s) vs. Legacy Solution

- Gain Scheduling easy to implement and understand!
  - Standard configuration
  - Preconfigured faceplate & detailed display
  - Tuning updates at loop execution
- Adaptive Control provides additional benefit and ease of implementation
  - Learns models during normal operation – NO STEP TESTs!
  - More Regions – Up to 5
  - Automatic tuning updates when process model changes
- Improved control performance saves money!

Agenda

- Introduction
- Examples – Best Practices & Lessons Learned
  - Split Range Duty
  - Smith Predictor
  - Gain Scheduling/Adaptive Control
  - Signal Characterization
    - Control Challenge
    - Legacy Solution
    - New Technology Solution
    - Summary
- Summary

Signal Characterization - Control Challenge

- Input Characterization (AI)
  - Combustion control – characterize fuel and air
  - Level – characterize pv (e.g. level to volume)
- Output characterization (AO)
  - modify installed flow characteristics of valves, dampers or other final control elements
  - Use to modify loop gain, e.g. output path to slave loop
- Legacy System Challenges
  - Lack of resolution (few points)
  - External black box characterizers may shift, lose documentation
  - Potential operator/maintenance confusion on signal going to final control element

Signal Characterization – Legacy Solution

- Legacy system using external analog characterizer

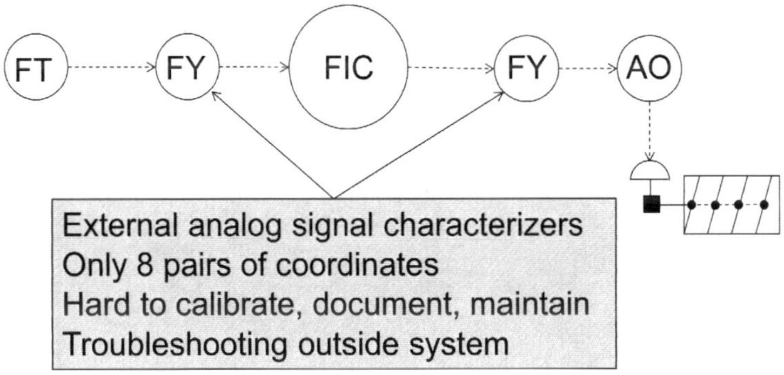

Signal Characterization – New Technology Solution #1

- Standard Module implementation provides 21 point signal characterizer function block (SGCR)

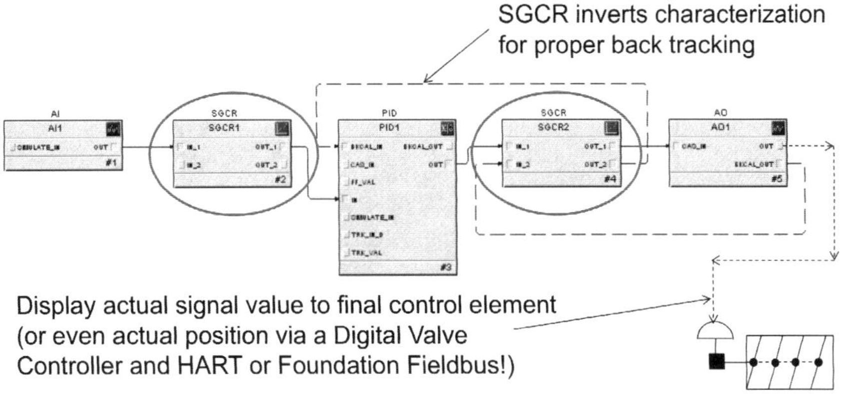

Signal Characterization – New Technology Solution #2

- Adaptive Control
  - Adjust tuning to compensate for final control element characteristics

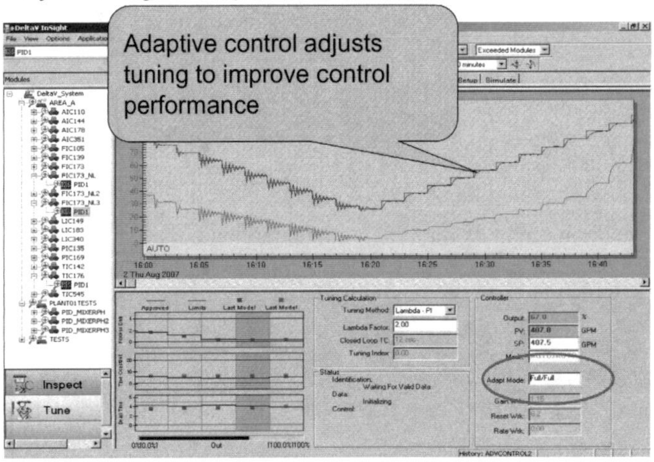

Summary Signal Characterization

- Newer Technology Signal Characterization function block
  - Better resolution (21 pairs of coordinates)
  - Self documenting (exists in control system)
  - No calibration required vs. 'external black box'
  - Provides "inverse" characterization for analog output characterizations
  - Reduce confusion: display signal being sent to final control element
- Applying Adaptive Tuning may provide a superior solution for AO characterization of final control element

Agenda

- Introduction
- Examples – Best Practices & Lessons Learned
  - Split Range Duty
  - Smith Predictor
  - Gain Scheduling/Adaptive Control
  - Controller Characterization
- Summary

Summary

- Practical application of control features that deal with common challenges
- Highlighted best practices & lessons learned
  - Split Range Duty
  - Smith Predictor
  - Gain Scheduling/Adaptive Control
  - Signal Characterization
- Do look at new functionality available when modernizing
  - Allows jump start to process optimization
  - Leads to improved process performance
  - Easier to implement and maintain over the life investment

Contacts & Questions?

James.Beall@Emerson.com
903-235-7935

Laurie.Ben@Emerson.com
512-834-7237

Paralleled Double-Effect Distillation: Simulative Case Studies

*Pei Wang, Leilei Dai, Ming Shi , Fan Wang, Hong Gao, Jinsheng Sun**
School of Chemical Engineering and Technology, Tianjin University, No. 92 Wei Jin Road, Tianjin 300072, P. R. China

Abstract

The increasing scarcity and growing demand for energy have led to an increasingly attention to the energy-efficient methods. Double-effect distillation is a common method to achieve remarkable energy conservation, which has three kinds of typical configurations including the feed splitting (FS), the light split/forward (LSF) and the light split/reverse (LSR). In spite of a good pathway for energy saving, better options are expected to cut down energy expenditure by a bigger margin than double-effect distillation does.

In this paper, case studies of paralleled or combined double-effect distillation schemes are proposed on the basis of current double-effect distillation. In order to verify the feasibility of this method, the improved methanol distillation scheme and organosilicon distillation scheme are optimized under this principle as examples. Exergy losses and the energy consumptions are measured as a tool to compare the two prevalent schemes with their corresponding optimized counterparts, respectively. The results of the total exergy loss and the energy consumption of these schemes show that the optimized schemes can cut the corner of the exergy loss by 22.8% and 40.41% and the energy consumption by 15.61% and 44.28%, respectively, in comparison to their popular cousins.

Keywords: double-effect distillation, paralleled double-effect distillation, methanol distillation, organosilicon monomer distillation, distillation process simulation, energy saving, exergy analysis

1. Introduction

Distillation, in chemical process and related industries, is by far the most widely applied for the separation of liquid mixture. In spite of its well-known benefit and widespread application, the major drawback is its high energy consumption. In classical distillation process, heat is transferred to the reboiler of distillation column. Along with the distillation column to separate the mixture, a large number of heat are directly discharge into the surrounding environment and cooling water. Araujo AB [1] has done some research to achieve that distillation has a relatively low thermodynamic efficiency and it requires the input of high quality energy in the reboiler to push the separation of liquid mixture. Therefore, the focus of energy conservation is to consider how to recover or reuse the discharged heat, or how to reduce the quantity of transferred heat to the column. In the early 20th century, many scholars have devoted to the study of saving energy in distillation, and now most of correlative techniques are widely used in distillation process [2,3]. These energy saving methods involve multi-effect distillation, heat pump distillation, heat integrated distillation and etc [3-8]. But without extra rotating machines by electricity or other power producers, multi-effect distillation has been more favored in research and application [9].

Many researches confirmed multi-effect distillation to be one of the effective methods to reduce the energy consumption [10-12]. The energy saving effect of those schemes used multi-effect distillation is associated with the number of the effects and the characteristics of separation system. Generally speaking, those schemes can offer energy savings around 30% with respect to the original or classical multicomponent distillation scheme [23, 27, 28]. Considering the interaction between investment of plants and energy saving effect, the double-effect distillation has been the most commonly used in not only research but also field application.

The use of the double-effect distillation technology for optimizing the conventional distillation schemes can cut a large amount of the energy requirement in comparison with the correspondingly prevalent schemes [13]. Although the system is very difficult to control in the sense that it is nonlinear, multivariable and interacting [14]. More and more published studies [15-19] have been focused on further questing the way of energy saving on the basis of the double-effect distillation schemes. However, the overhead vapor of one high pressured column only can integrate with the reboiler of one low pressured column in those published studies. In view of this, in this paper, the paralleled or combined double-effect distillation is proposed to broaden the thought for the application of the double-effect distillation, in which the overhead vapor from one of the high pressured column is divided into two or more streams to successively integrate with the reboiler of the other columns [9].

2. *Theory*

2.1 Double-effect or heat integrated distillation system

A double-effect distillation system is made up of two adjacent distillation columns running in different pressure, which can lead to a temperature difference between the top of the high pressured column and the bottom of the low pressured column. Through the thermodynamic connection, the reboiler of the low pressure column plays a role of the condenser for the high pressure column. According to the direction of the mass flow and the heat integration, the double-effect distillation can fall into three kinds of configuration: the feed splitting (FS), the light split/forward (LSF) and the light split/reverse (LSR), as showed in Figure. 1.

One of the configurations with a high thermodynamic efficiency is the feed splitting (FS), in which the feed is divided roughly equal between the two columns running in different pressure, and the products can be obtained from both of the top and bottom of the two columns, as illustrated in figure. 1 (a). The overhead vapor form the high pressured column is used to heat the reboiler from the low pressured column. There is only energy exchange between the two columns, so the heat duties of the feed splitting (FS) are approximately equal to the heat duty for single column with the same conditions divided by the number of effect.

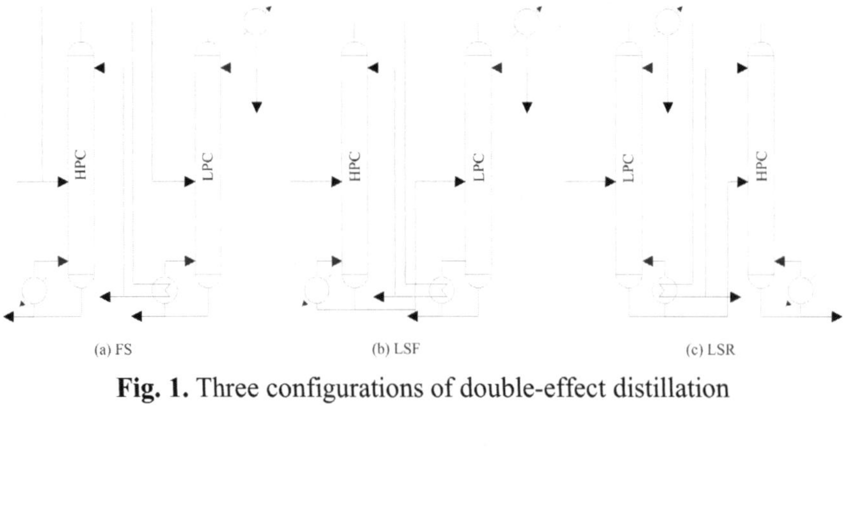

Fig. 1. Three configurations of double-effect distillation

Figure. 1 (b) and (c) respectively shows the configuration of the light split/forward heat integration (LSF) and the light split/reverse heat integration (LSR). Different from the feed splitting (FS), the total feed stream was sent to one column in the other two configurations. In the light split/forward (LSF), the total feed is pumped to the high-pressure column (HPC) and the bottom stream of HPC flows into the low-pressure column (LPC). In this way, all of the heavy components are from LPC and the light components from the top both of the two columns. Figure. 1 (b) indicates the direction of mass flow and heat integration is the same. The last configuration is the light split/reverse (LSR), as shown in figure. 1 (c), all of the feed is sent to the low-pressure column (LPC). It can remove the two kinds of light components from the top both of the two columns and one kind of heavy components from the bottom of HPC. Moreover, the direction of heat integration is in the reverse with the mass flow. These two kinds of configurations have not only energy exchange but also substantial exchange, therefore, when the operating parameters of one column are changed, the other column will also change.

Theoretically, the three kinds of configurations are empirically 30% less energy consumption than the single column distillation. As for choosing what kind of the configurations of the double-effect distillation, it depends on the properties of the multi-component mixture, the pressure of saturated steam, the temperature of cooling water, structure parameters of existing distillation column and so on.

2.2 Exergy analysis of the distillation process

According to the First Law of Thermodynamics, also called the Law of Energetic Balances, no matter what happens, the total amount of energy has not yet changed. However, when the various types of energy generate the mutual conversion, the quality of the energy is always decreased. The separation process in distillation is irreversible. In order to separate the mixture in distillation column, it needs the driving force which must consume a certain amount of external work, by this means, the decline of the quality of energy is inevitable. Therefore, we should choose a reasonable driving force in distillation process to avoid unnecessary loss of the quality of energy. It defined the concept of exergy (E_X) to compare the quality of the different types of energy. Exergy, which is a measure for evaluating the extent of available energy or the quality of energy, is based on the Second Law of Thermodynamics. Exergy analysis reveals the essence of the energy, and can be used as an important tool to evaluate the effect of energy saving. And it usually combines with process optimization studies, and as its guidance, to identify cost saving opportunities and amend the existing scheme.

Szargut [29] has summarized a comprehensive definition of the exergy, that is: 'Exergy is the amount of work obtainable when some matter is brought to a state thermodynamic equilibrium with the common components of the natural surroundings by means of reversible processes, involving interaction only with the above mentioned components of nature'. The state of one stream recorded as 1, according to the definition of exergy, the exergy of the stream can be expressed as:

$$E_x = (H_1 - H_0) - T_0(S_1 - S_0) \tag{2}$$

where, H_1 and S_1 is the enthalpy and the entropy of the stream 1, and T_0 is the temperature of the surrounding and H_0 and S_0 represent the enthalpy and the entropy of the surrounding.

The state of another stream recorded as 2, when the system stream change from state 1 (T_1, P_1) to state 2 (T_2, P_2), the variation of exergy (ΔE_X) is given by

$$\Delta E_x = E_{x,1} - E_{x,2} = \Delta H - T_0 \Delta S \tag{3}$$

In the conventional distillation process, it must input the heat or energy to drive the occurrence of separation process. The bottom of the column input the heat (Q_R) at this temperature (T_R), and the top of the column was sent to the heat (Q_C) at this temperature (T_C). The net work consumption (W) in distillation is

$$W = Q_R(1 - \frac{T_0}{T_R}) - Q_C(1 - \frac{T_0}{T_C}) \tag{4}$$

When the enthalpy difference between the product and the feed is very small, $Q_R=Q_C=Q$ can be approximately gotten, then Equation (4) is rewritten to be:

$$W = QT_0(\frac{1}{T_C} - \frac{1}{T_R}) \tag{5}$$

Then, the total exergy loss (Ex_{loss}) of the distillation column can be calculated by

$$Ex_{Loss} = W - \Delta E_x \tag{6}$$

3. *Prototype case: the prevalent distillation schemes*

3.1 The improved four-column methanol distillation scheme

In the methanol industry, it has gradually appeared two-column, three-column, four-column and five-column distillation scheme [20-23]. With the increase of the number of distillation column, the corresponding distillation scheme can not only increase the efficiency of the process, but also cut the energy consumption. At present, the four-column scheme has a dominant status to produce the purified methanol. However, the improved four-column distillation scheme can cut some more energy requirement with respect to the traditional four-column distillation scheme. The improved four-column distillation scheme is seen as the comparative reference in the paper.

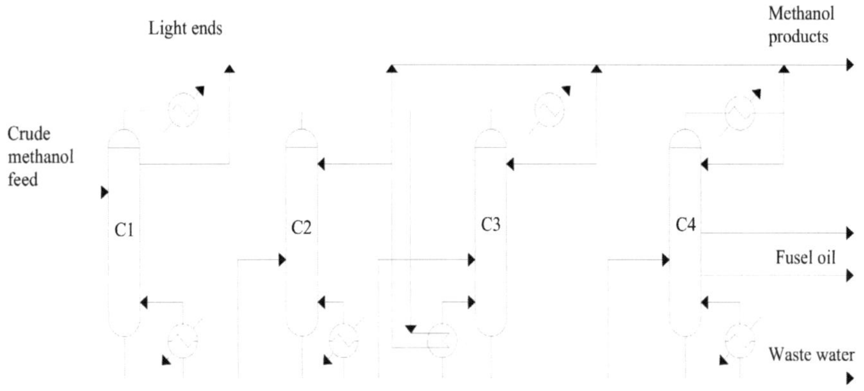

Fig. 2. The improved four-column double-effect methanol distillation scheme (C1, light ends column; C2, pressured column; C3, atmospheric column; C4, water column)

In the improved four-column distillation scheme, as shown in figure. 2 [24], the crude methanol flows into the light ends column (C1) in 87.5 ℃ for removing the light ends and the remnants of dissolved gas from the top of C1. Besides, in order to raise the methanol recovery rates, the light ends column (C1) designs a two-stage condenser system. The crude methanol was distillated successively from the pressured column (C2) and the atmospheric column (C3), and most of the methanol products can be received at the top both of C2 and C3. The bottom stream of C1, C2 and C3 is pumped respectively as the feed into C2, C3 and the water column (C4). Residual methanol, fusel oil and waste water are extracted from the corresponding position of the top, side stream and the bottom of the water column (C4).

In the improved scheme, the operation pressure of C2 is 0.9MPa, and the other three columns run under the atmospheric pressure. Due to the different of operating pressure, the overhead vapor of C2 can undertake thermal coupling connection with the reboiler of C3.

3.2 The current organosilicon monomer distillation scheme

According to the survey, there are several kinds of organosilicon distillation schemes. However, the 9-column organosilicon distillation scheme [9], in China, has an important role in the organosilicon industrial, as illustrated in Figure. 3, it is used as the base case.

Figure. 3 shows that the crude monomer is delivered to the heavies removal column (C1') and the bottom stream of C1' has a lot of the unsteady substances of high boiling points. These unsteady substances of high boiling points are pumped to the high boiling point column (C9') as the feed, and the products from the top of C9' continue to serve as the crude monomer by circumfluence. By this way, it can accomplish the separation between the unsteady substances of high boiling points and the useful substances at a relatively low temperature in the high boiling point column (C9') and avoid the unsteady substances of high boiling points from thermal cracking to generate many unknown substances which would greatly affect the subsequent separation of product. The low removal column (C2') provides trimethylchlorosilane (Me3), methyldichlorosilane (MeH), and other light components for lights removal column (C3'), MeH column (C4'), azeotrope removal column (C5') and Me3 column (C6'), which can finally achieve two products (MeH and Me3) and three wastes. It also provides methyltrichlorosilane (Me1) and dimethylchlorosilane (Me2) and other heavy components for Me1 column (C7') and Me2 column (C8'), which can finally get two products (Me1 and Me2).

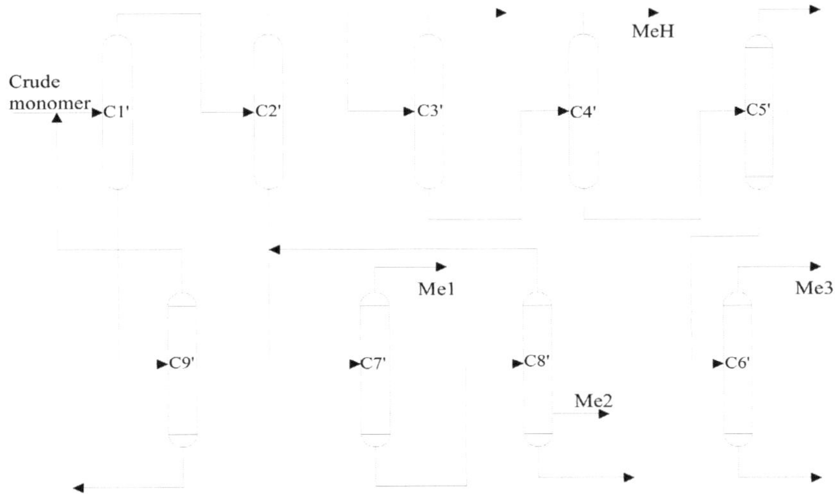

Fig. 3. The prevailing 9-column distillation scheme (C1', heavies removal column; C2', low removal column; C3', lights removal column; C4', MeH column; C5', azeotrope removal column; C6', Me3 column; C7', Me1 column; C8', Me2 column; C9', high boiling point column)

In this scheme, the operating pressure of all nine columns is relatively low, and the highest operating pressure is 260 kPa in the lights removal column (C3'). Although it has a different operating pressure between C3' and the other columns, but the temperature difference is not enough to make them use double-effect or heat integrated distillation.

4. Simulation and analysis both of the prevalent schemes and the optimized schemes

4.1 The novel five-column heat integrated methanol distillation scheme

Although the energy saving of the improved four-column methanol distillation scheme can reach 10% compared with the traditional four-column scheme, it still exist several shortages, which will be illustrated by the results of computer simulation to the improved four-column methanol distillation scheme. The flow rate of the crude methanol is 239230 kg/hr, and the temperature is 87.5 ℃, and the composition of the crude methanol is exhibited in table 1. The inlet and outlet temperature of the cooling water are 30 ℃ and 40 ℃.The methanol content is specified as 99.99 wt% in the methanol products and 10ppm in the wastewater of the water column. Table 2 shows the operating parameters of each column in the improved four-column scheme.

Table 1. Crude methanol feed data

Component	Mass fraction%	Component	Mass fraction%
H_2	0.0100	Pentanol	0.0338
N_2	0.0710	Formic acid	0.0030
Methane	0.0100	Acetic acid	0.0030
CO	0.0360	Methyl formate	0.0092
CO_2	0.6640	Acetone	0.0193
Methanol	95.0000	Methyl ethyl ketone	0.0030
Dimethyl ether	0.0309	Diisopropylether	0.0056
Trimethylamine	0.0010	Water	3.7100
Ethanol	0.0900	i-Pentane	0.0562
i-Propanol	0.0024	n-Pentane	0.0169
Propanol	0.0264	n-Hexane	0.0487
i-Butanol	0.0338	n-Heptane	0.0009
Butanol	0.0185	n-Octane	0.0050
3-Pentanol	0.1407	n-Nonane	0.0050

Table 2. The operating parameters of improved four-column scheme

The operating parameters	C1	C2	C3	C4
Top pressure (kPa)	150	900	110	110
Top temperature (℃)	81.0	132.6	67.7	67.7
Bottom temperature (℃)	83.8	139.1	96.5	104.5
Reflux ratio	0.58	2.44	1.68	11.57
Number of theoretical stage	33	52	58	61

The results of simulation indicate that the hot utility is still in a high level because the reboilers of C1, C2 and (C4) still need the saturated steam to transmit the heat. Table 2 shows the reflux ratio of C4 is 11.57 and this large reflux ratio can cause the increasing of the energy requirement of the water column (C4). Besides, according to the flowsheet of this scheme, the diameter of column is so large that the emergence of many urgent cases.

In view of above these defects in the improved four-column scheme, the novel five-column heat integrated methanol distillation scheme (shown as Fig. 4) is designed in this paper to reduce the impact of these factors as far as possible by using the paralleled or combined double-effect distillation. Juntao Zhang [23] and Jinsheng Sun [25] are also proposed a five-column methanol distillation scheme by using triple-effect distillation and double-effect distillation respectively, and the effect of energy saving of the two schemes is outstanding.

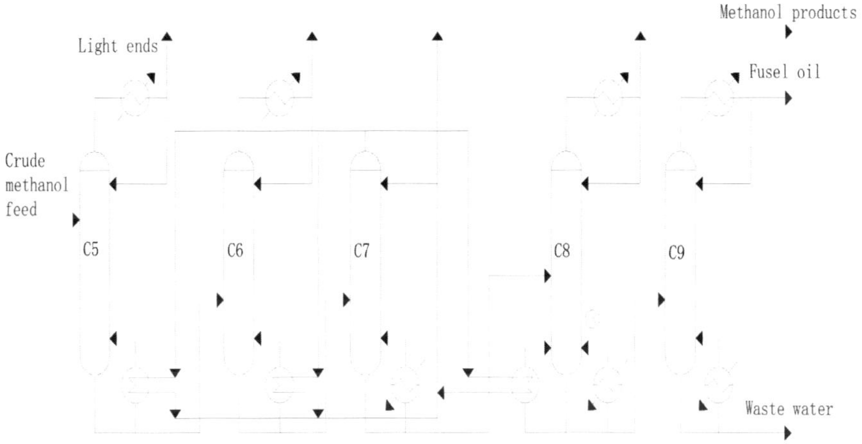

Fig. 4. The five-column heat integrated methanol distillation scheme (C5, light ends column; C6, prepositive atmospheric column; C7, pressured column; C8, postpositive atmospheric column; C9, water column)

According to the principle of the paralleled double-effect distillation, the four columns are unable to constitute the heat integration network in the improved four-column scheme. Through repeated verification, the prepositive atmospheric column (C6) is added after the light ends column (C5), meanwhile, the operating pressure of the pressured column (C7) is reduced to increase its throughput. In this way, the overhead vapor of C7 is split into three streams to respectively integrate with the light ends column (C5), the prepositive atmospheric column (C6) and the postpositive atmospheric column (C8). Due to use the paralleled double-effect distillation, the duties of the hot utility have decreased in a great margin. However, there are still part of heat from the background process is absorbed by the cooling water. So an inter-reboiler with a low temperature level is designed in the postpositive atmospheric column (C8) and it can share the heat with the background process streams, besides, we have proved the feasibility by using the column grand composite curve analysis of columns. Figure. 4 shows that the product of the top of the water column (C9) is fusel oil instead of purified methanol, which can not only reduce the reflux ratio of C9, but also make the diameter of column smaller to reduce the difficulty of operation.

Due to produce the fusel oil instead of purified methanol in the water column (C9) in the novel five-column distillation scheme, the methanol content of the feed for the water column have to limit at 0.016 wt%. In the five-column scheme, the operating pressure and the number of theoretical stage for each column are determined through the existing process, the onsite data and the requirement of the heat coupling. The proportion of the methanol products from the C6, C7 and C8 is determined by their own double-effect or heat integrated distillation system. The simulation results of the novel five-column scheme are listed in table 3.

Table 3. The operating parameters of five-column scheme

Column	C5	C6	C7	C8	C9
Top pressure (kPa)	160	110	400	110	110
Top temperature (℃)	83	66.6	104	66.6	86.5
Bottom temperature (℃)	85.6	84	118.8	104.5	103.2
Reflux ratio	1.46	1.49	2.23	1.87	4
Number of theoretical stage	33	54	54	65	29

4.2 Energy consumption and exergy loss analyses for the two methanol schemes

In the improved four-column scheme, the pressured column and the atmospheric column constitute the distillation system of heat coupling, there are three columns (C1, C2, C4) need the hot utility, and the entire duty of reboiler is 145.4×10^3 kw. As described above, the light ends column (C5) and the prepositive atmospheric column (C6) don't need the saturated steam in the five-column scheme and the total duty of reboiler is reduced to 122.7×10^3 kw. The detailed data of the reboiler duty as shown in table 4.

Table 4. The duty of reboiler in the two schemes

Column	C1	C2	C4	C7	C8	C9
Duty of reboiler (10^4 kw)	3.30	10.04	1.20	8.97	2.86	0.44

Obviously, the novel five-column scheme is much better than the improved four-column scheme in energy saving and the energy consumption of the novel scheme can be reduced by 15.61%. However, the recovery rate of methanol of the novel scheme is approximately equal with the improved four-column scheme.

Only the energy consumption analysis is not revealed the energy saving essence of the novel five-column scheme. Exergy analyses are used to further compare the two schemes, and the exergy flow rates are calculated through using the computer simulation data. The activity coefficient of the liquid is calculated using the Alcohol Package thermodynamic model. The calculation of the net work consumption of distillation column should pay more attention to the temperature of the cooling water. When the cooling water exchanges heat with other streams, the temperature of the cooling water is momentarily changed, so it is replaced the temperature of cooling water with the arithmetic average of inlet and outlet temperature of cooling water. Figure. 5 show the overall exergy loss of each column in the two methanol distillation schemes, in addition to the overall exergy loss of C6. Due to optimize the four-column scheme, the prepositive atmospheric column (C6) is added in the transformed scheme, and the overall exergy loss of C6 is 9157.6 kw. The exergy loss of each column, besides the pressured column (C7), in the novel five-column scheme is obviously less than their counterparts in the improved four-column scheme.

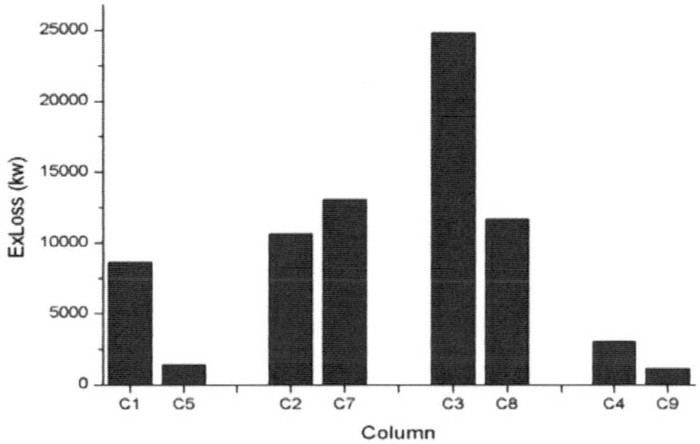

Fig. 5. The overall exergy loss analysis of each column in the two schemes

The total exergy loss of the improved four-column scheme and the transformed scheme is, respectively, 47100.1 kw and 36355.1 kw. The total exergy loss of the optimized scheme is dropped by 22.8%. Through the comparison between the exergy loss and energy consumption of the two schemes, it is indicated that the novel five-column scheme should be applied to produce the purified methanol in industry.

4.3 The novel heat coupling organosilicon distillation scheme

The primary source of the crude organosilicon monomers is the catalytic reaction products between chloromethane and silicon at high temperature and the table 5 displays the normal boiling points and mass percentage of the main components of the crude organosilicon monomers. Methyltrichlorosilane (Me1), dimethylchlorosilane (Me2), trimethylchlorosilane (Me3) and methyldichlorosilane (MeH) are the most valuable components in the crude organosilicon monomers. The purification requirements on organosilicon of these materials are shown in table 6.

Table 5. The crude organosilicon monomer feed data

Component	Normal boiling point at 1 atm/℃	Mass percentage %
LB		1
Methyldichlorosilane	41.8	4
Silicon tetrachloride	57.6	0.5
Trimethylchlorosilane	57.9	2.5
Methyltrichlorosilane	66.4	11
Dimethyldichlorosilane	70.2	80
HB		1
Temperature (℃)	40	
Pressure（kPa）	100	
Flow rate (kg/h)	26200	

Table 6. Purification demand on organosilicon

Monomer	Purification demand
MeH	99.5%
Me3	99.5%
Me1	99.5%
Me2	99.95%

The conventional nine-column scheme is simulated with the aid of existing process and onsite data to achieve the operating parameters of each column, as illustrated in table 7. The bottom temperature of the heavies removal column (C1') and the high boiling point column (C9') are obviously much bigger than that of the other seven columns in table 7. According to the principle of the double-effect distillation, the top temperature of the high pressured column is relatively high in comparison with the bottom temperature of the low pressured column, and the high pressured column can integrate with the low pressured column. Although the top temperature of C1' is less than the bottom temperature of the other seven columns, the overhead vapor of C1' still has a great opportunity to conduct heat with the other seven columns with the increasing of its top temperature.

Table 7. The simulation results of prevailing 9-column

Column	C1'	C2'	C3'	C4'	C5'	C6'	C7'	C8'	C9'
Top temperature (℃)	70.8	65.3	57.3	66.7	72.1	70.6	71.2	74.9	72.7
Bottom temperature (℃)	124.1	89.1	80.4	85.2	74.9	75.4	75.3	75.3	126.5
Reboiler duty (MW)	11.67	5.61	1.13	0.51	1.19	0.46	12.27	16.08	0.02
Condenser duty (MW)	-12.47	-5.53	-1.08	-0.53	-1.20	-0.47	-12.69	-16.12	-0.02
Top pressure (kPa)	110	180	260	220	170	150	110	110	110

According to the way of the paralleled double-effect distillation, the increasing operating pressure in the heavies removal column (C1') can make the top temperature of C1' ascending to integrate with the other seven columns. The idea is still in study to overcome some difficulties. However, we have got another feasible method, in which the operating pressure of C1' and the Me1 column (C7') is raised to 360 kPa and 280 kPa, respectively. In this way, the top temperature of C1' and C7' is increased to ensure they can integrate with the other columns, besides the high boiling point column, and the parallel double-effect or heat integrated organosilicon monomer distillation scheme is designed as shown in figure. 6 [26]. The feasibility of this heat coupling scheme has confirmed through three x-y diagrams including the heavies component and Me1, the heavies component and Me2, Me1 and Me2 [9].

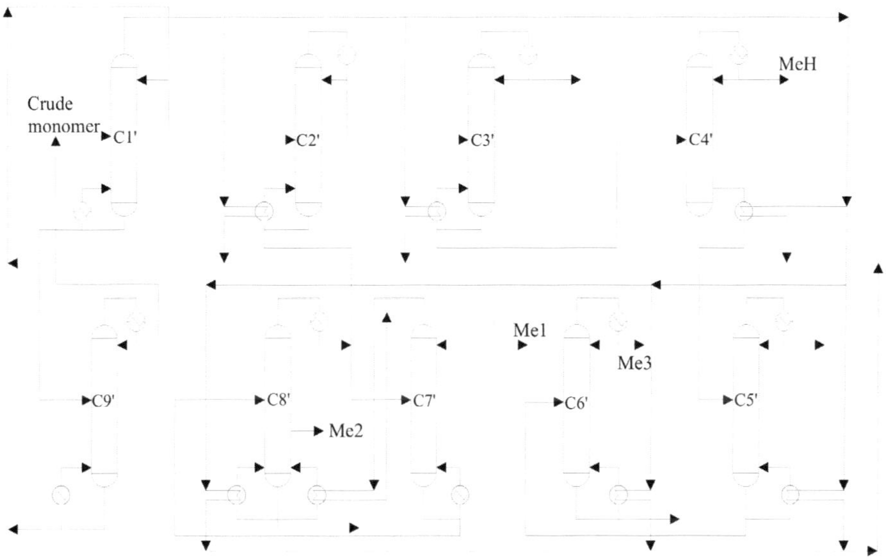

Fig. 6. The heat coupling 9-column scheme (C1', heavies removal column; C2', low removal column; C3', lights removal column; C4', MeH column; C5', azeotrope removal column; C6', Me3 column; C7', Me1 column; C8', Me2 column; C9', high boiling point column)

Figure. 6 shows the heavies removal column (C1') make the heat coupling with the low removal column (C2'), the lights removal column (C3'), the MeH column (C4'), the azeotrope removal column (C5'), the Me3 column (C6') and the Me2 column (C8'), successively. In order to gain a better energy saving effect, C8' is designed with two reboilers, the one integrated with C1' and the other integrated with the Me1 column (C7'). The operating parameters and the duty of reboiler and condenser of each column in the heat integrated 9-column are enumerated in table 8.

Table 8. The simulation results of heat integrated 9-column

Column	C1'	C2'	C3'	C4'	C5'	C6'	C7'	C8'	C9'
Top temperature (℃)	114.3	65.3	57.3	66.7	72.1	70.6	105.1	74.9	72.7
Bottom temperature (℃)	124.9	89.9	78.8	84.5	75.4	75.4	110.9	76.6	121.4
Reboiler duty (MW)	13.06	5.62	1.13	0.51	1.19	0.46	13.83	14.12	0.38
Condenser duty (MW)	-13.72	-5.54	-1.08	-0.53	-1.20	-0.47	-14.02	-14.43	-0.41
Top pressure (kPa)	360	180	260	220	170	150	280	110	110

4.4 Energy consumption and exergy loss analyses for the two organosilicon schemes

Figure. 4 indicate that each column of the prevalent nine-column scheme is not integrated with the other columns and the reboilers of all of these columns are required the saturated steam (600 kPa). According to the data from the table 7, the total duty of reboiler is 48.94 MW for the prevailing 9-column scheme. In the new nine-column scheme, C1' integrate with C2', C3', C4', C5', C6' and C8' successively and C7' integrate with C8', so C2', C3', C4', C5', C6' and C8' don't consume the saturated steam. The total duty of reboiler is 27.27 MW. Obviously, the heat integrated 9-column scheme has an absolute advantage over the prevailing 9-column scheme in energy saving, and the energy savings can reach 44.28%.

As the same with the methanol distillation scheme, the inlet and outlet temperature of cooling water are 30 ℃ and 40 ℃, and the total exergy loss is calculated through the results of computer simulation of the two schemes. The overall exergy loss comparison of each column in the two schemes is illustrated in figure. 7. In the normal scheme, the throughput of C9' is so small that the overall exergy loss of C9' is very near to zero, so it only have the exergy loss of C9' of the new scheme in figure. 7. The overall exergy loss of each column in the optimized scheme is declined in comparison with that of the corresponding column in the normal scheme except the high boiling point column (C9'). Because the crude organosilicon monomers contains some unstable components, in order to avoid these components happening thermal cracking in the heavies removal column (C1') in the novel scheme, more liquid from the bottom of C1' is pumped to C9' as the feed. That can cause the ascent of the throughput and the overall exergy loss of C9'.

The total exergy loss is 14035.5 kw and 8363.2 kw, correspondingly, in the conventional scheme and the novel scheme, and the exergy loss of the new scheme can be reduced by 40.41%. Through the above analysis and comparison, the heat integrated nine-column organosilicon distillation scheme should be a substitute for the prevalent nine-column scheme.

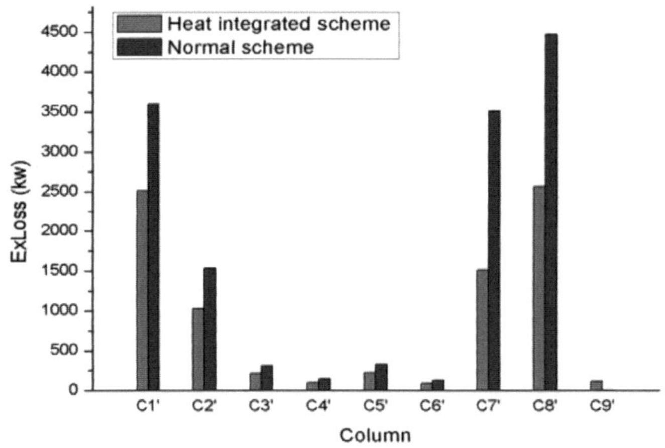

Fig. 7. The exergy loss comparison between the heat integrated scheme and the normal scheme

5. Conclusions

Paralleled double-effect or heat integrated distillation for energy saving is proposed, exampled by heat integrated five-column methanol distillation scheme and the heat coupling 9-column organosilicon monomer distillation scheme, with the aid of computer simulation. Besides, the optimized methanol distillation scheme and the transformed organosilicon scheme are comprehensively analyzed through the energy consumption and the exergy loss in comparison with the correspondingly prevalent scheme.

Adding a prepositive atmospheric column and decreasing the operating pressure of the pressured column in the novel five-column scheme can make the pressured column to simultaneously integrate with the light ends column, the prepositive atmospheric column and the postpositive atmospheric column. Besides, an inter-reboiler is added to integrate with the background process streams. The energy consumption and the exergy loss of the new scheme are 15.61% and 22.8% lower than the improved scheme.

Increasing the operating pressure of the heavies removal column and the MeI column can make them to integrate with the other columns, in addition to the high boiling point column, to constitute a paralleled double-effect distillation system in the optimized nine-column scheme. The energy consumption and the exergy loss of the novel scheme are 44.28% and 40.41% less than the normal scheme.

The two cases expound that the key of paralleled double-effect distillation process lies in the reasonable adjustment of the energy distribution, the operating parameters and the configuration of some columns. It not only reduces the quantity of energy consumption, but also avoids the decrease of the quality of energy. For factories with a large numbers of distillation schemes of high energy requirement, the paralleled schemes lead to improve economic benefit.

References

[1] Araujo AB, Brito RP, Vasconcelos LS, "Exergetic analysis of distillation processes-a case study," Energy 2007;32:1185-93.

[2] Michele Mascia, Francesca Ferrara, Annalisa Vacca, Giuseppe Tola, Massimiliano Errico, "Design of heat integrated distillation systems for a light ends separation plant," Applied Thermal Engineering 2007;27:1205-1211.

[3] Amiya K. Jana, "Heat integrated distillation operation," Applied Energy 2010;87:1477-1494.

[4] Hilde K. Engelien, Sigurd Skogestad, "Multi-effect distillation applied to an industrial case study," Chemical Engineering and Processing 2005;44:819-826.

[5] James G. Gebbie, Michael K. Jensen, Piotr A. Domanski, "Experimental transient performance of a heat pump equipped with a distillation column," International Journal of Refrigeration 2007;30:499-505.

[6] Massimiliano Errico,Giuseppe Tola, Michele Mascia, "Energy saving in a crude distillation unit by a preflash implementation," Applied Thermal Engineering2009;29:1642-1647.

[7] Massimiliano Erricoa, Giuseppe Tola, Ben-Guang Rong, Daniele Demurtas, Ilkka Turunen, "Energy saving and capital cost evaluation in distillation column sequences with a divided wall column," chemical engineering research and design 2009;87:1649-1657.

[8] Ivar J. Halvorsen, Sigurd Skogestad, "Energy efficient distillation," Journal of Natural Gas Science and Engineering 2011;3:571-580.

[9] Jinsheng Sun, Fan Wang, Tingting Ma, Hong Gao, Yanzhen Liu, Fang Cai, "Exergy analysis of a parallel double-effect organosilicon monomer distillation scheme," Energy 2012;47:498-504.

[10] Hilde K. Engelien, Sigurd Skogestad, "Multi-effect distillation applied to an industrial case study," Chemical Engineering and Processing 2005;44:819-826.

[11] Kazuo Matsuda, Kenichi Kawazuishi, Yasuki Kansha, Chihiro Fushimi, Masaki Nagao, Hiroshi Kunikiyo, Fusao Masuda, Atsushi Tsutsumi, "Advanced energy saving in distillation process with self-heat recuperation technology," Energy 2011;36:4640-4645.

[12] M.A. Sharaf, A.S. Nafey, Lourdes García-Rodríguez, "Exergy and thermo-economic analyses of a combined solar organic cycle with multi-effect distillation (MED) desalination process," Desalination2011;272:135-147.

[13] Weiwen Wang, Wenjian Wang, "Energy Saving in Methylchlorosilane Distillation," International Conference on Computer Distributed Control and Intelligent Enviromental Monitoring 2012:290-293.

[14] Myungwan Han, Sunwon Park, "Multivariable control of double-effect distillation

configurations," J. Proc. Cont. 1996:247-253.

[15] Hilde K. Engelien, Sigurd Skogestad, "Multi-effect distillation applied to an industrial case study," Chemical Engineering and Processing2005;44:819-826.

[16] Debadrita Maiti, Amiya K. Jana, Amar Nath Samanta, "A novel heat integrated batch distillation scheme," Applied Energy2011;88:5221-5225.

[17] Anton A. Kiss, Servando J. Flores Landaeta, Carlos A. Infante Ferreira, "Towards energy efficient distillation technologies-Making the right choice," Energy2012;47:531-542.

[18] Diego C. Alarcón-Padilla, Lourdes García-Rodríguez, Julián Blanco-Gálvez, "Experimental assessment of connection of an absorption heat pump to a multi-effect distillation unit," Desalination2010;250:500-505.

[19] Diego C. Alarcón-Padilla, Lourdes García-Rodríguez, Julián Blanco-Gálvez, "Design recommendations for a multi-effect distillation plant connected to a double-effect absorption heat pump: A solar desalination case study," Desalination2010;262:11-14.

[20] Fiedler E, Grossman G. Methanol, "Ullmann's Encyclopedia of Industrial Chemistry," Weinheim : Wiley-VCH Verlag/Gmbh & Co; 2002.

[21] Meyers RA, "Handbook of synfuels Technology," New York : McGraw-Hill; 1984.

[22] Chu YZ, Qin LP, "Optimum design of methanol distillation process and columns," Chem Ind Eng Prog 2008;(27):1659-62.

[23] Juntao Zhang, Shengrong Liang, Xiao Feng, "A novel multi-effect methanol distillation process," Chem Eng Pro 2010;(49):1031-37.

[24] Jinsheng Sun, Fan Wang, Tingting Ma, Hong Gao, Peng Wu, Lili Liu, "Energy and exergy analysis of a five-column methanol distillation scheme," Energy2012;45:696-703.

[25] J. S. Sun, P. Wu, Y. H. Wang, H. Gao, L.L. Liu.CN Patent102039058 A. 2011.

[26] J. S. Sun, P. Wu, Y. H. Wang, Y. X. Su, H. Gao, J. Li, D. Yan. CN Patent 101798322 A. 2010.

[27] Michele Mascia, Francesca Ferrara, Annalisa Vacca, Giuseppe Tola, Massimiliano Errico, "Design of heat integrated distillation systems for a light ends separation plant," Applied Thermal Engineering 2007;27:1205-1211.

[28] Olga A. Flores, J. Carlos Ca´ rdenas, Salvador Herna´ ndez, Vicente Rico-Ramı´rez, "Thermodynamic Analysis of Thermally Coupled Distillation Sequences," Ind. Eng. Chem. Res. 2003;42:5940-5945.

[29] Szargut J, "Exergy Analysis of Thermal, Chemical and Metallurgical Processes," New York: Hemisphere Publishing Corporation;1988:55-152.

(45n) In Situ Transesterification of Jatropha Curcas Seed Particles with Alkaline Phase Transfer Catalyst

Subbarao Duvvuri[1], Shuhaimi Mahadzir[2] and Sintayehu Hailegiorgis[2], (1)Department of Chemical Engineering, Universiti Teknologi PETRONAS, Tronoh, Malaysia, (2)Chemical Engineering, Universiti Teknologi PETRONAS, Tronoh, Malaysia

Economic viability of biodiesel technology is at stake due to the high cost of edible oil feedstock and slow reaction rates due to limited solubility of oil in methanol. In situ transesterification of non edible oil seed particles can minimize the cost of feedstock. Microwave pretreatment of seeds can make oil molecules more reactive. Use of phase transfer catalysts can enhance the reaction rates. In the present work, these three concepts are utilized together to investigate alkaline in situ transesterification of microwave pretreated jatropha curcas seed particles in presence of benzyltrimethylammonium hydroxide (BTMAOH) as a phase transfer catalyst (PTC). It was observed that addition of PTC to alkaline catalyst has improved conversion and reduced reaction time. Microwave pretreatment of jatropha seed particles reduced reaction time drastically. The quality of biodiesel produced was within the limits of the international standard requirements of ATME D6751 and EN-14214. Thus, combined use of microwave pretreatment of seed particles, phase transfer catalysis in the presence of alkaline catalyst can increase the rate of in situ transesterification and product yield.

(45o) Catalyzed-Assisted Manufacture of Olefins (CAMOL): Updated for Use in Naphtha Service

Maximilian Walter, Petrochemicals, BASF SE, Ludwigshafen, Germany and **Suzanne Rech**, BASF Qtech, BASF Corporation, Iselin, NJ

Since the commercialization of the olefins plant, operators have searched for ways to be more competitive. One area always targeted includes keeping their furnaces in operation longer between decoking steps. Over the years, many anti-coking methods/technologies have become commercially available using different approaches to prevent coils from coking. One of the latest available technologies is the novel CAMOL^{TM} (Catalyzed-assisted Manufacture of Olefins) Technology from BASF Qtech. As reported in the introduction of the technology to the EPC in 2008 & 2010, the CAMOL^{TM}technology differs from other available anti-coking technologies in that the technology not only eliminates the catalytic (or filamentous) coke make through a physical barrier, but also gasifies the thermal (or amorphous) coke for an even greater coking resistance. The coating process has generated a coating with almost identical thermo-mechanical properties as the base metal substrate (alloy). This allows bonding to the base metal similar to being welded with a minimum risk of spallation when going through thermal cycling. Commercially, the CAMOL Technology has been available since 2006 and is now in operation in furnaces of various designs.

With CAMOL^{TM}, operators have experienced extended run lengths, reduced energy requirement and GHG emissions, lower average TMT operating temperatures, reduced DMDS consumption, and less carburization. In addition, the technology has exhibited high operating temperature stability (>1130°C/ 2065°F).

At BASF SE in Ludwigshafen, Germany, the technology has been in operation since early 2010 in mostly naphtha service. After overcoming some initial challenges associated with a catalytic surface technology, such as reactivating the catalytic species after decoking, the coil is now successfully operating with run lengths significantly higher than uncoated coils. This paper will report on the results and learnings from the experiences gathered over the past three years along with the advancements made since the last report in 2010.

(45p) Extending Furnace Run Length and Coil Life by Optimizing Decoking

Vinod Mishra, Ingenero, Mumbai, India and **James Brigman**, Ingenero, Houston, TX

Abstract:

In the Ethylene manufacturing process, coke deposited in the radiant coils and Transfer Line Exchangers (TLEs) during normal operation is removed periodically by decoking the furnaces using a mixture of steam and air. There is a significant variation in operating parameters adopted by different plants for decoking. These are often either too conservative or aggressive, leading to production loss from increased off-stream time, TLE limitation and deterioration in coil life. This paper describes how the decoking procedure was modified based on furnace and TLE operating conditions and the decoking steps changed to obtain the best possible Ethylene yield, without adversely impacting coil life and TLE performance. This optimization has now been done in several major ethylene facilities and furnace run length has more than doubled and production has significantly increased.

SOLVING OPERATIONAL PROBLEMS IN A NAPHTHA SPLITTER

Flávio Weissheimer, GE Water & Process Technologies do Brasil Ltda, São Paulo/Brasil
Giorgio Franceschetti, Braskem, Santo André/Brasil
Rafael Henrique Camparin, GE Water & Process Technologies do Brasil Ltda, São Paulo/Brasil
Patrick Lucas, GE Water & Process Technologies, The Woodlands/US

Abstract

This paper sets out how the polymerization process of Reboilers of the (A-100) Crude Naphtha Fractionation Unit of Braskem UNIB 3 ABC can be controlled by applying polymerization inhibitors developed by GE Water & Process Technologies Ltda and Braskem´s engineering solutions.

These heat exchangers have a high degree of polymerization because of the content of olefinic compounds present in the feed composition which, when in contact with high temperatures in the heat exchange areas of reboilers, tend to polymerize. The joint solution of GE technology and Braskem engineering solutions applied to this system ensures regular operation of such equipment for periods of time greater than those hitherto recorded without manifesting loss of heat transfer efficiency.

In the 2012 shutdown while the opening of these reboilers was being inspected, it was observed that pipes had not been clogged nor was surface fouling present, which is a very positive state when compared to the historical records of previous campaigns.

The strategy adopted by GE/Braskem to ensure a longer campaign for these pieces of equipment was based on the quality control of feeding by adjusting dosages and monitoring performance using statistical analysis tools.

The use of statistical models based on Multivariable Regression Analysis (MRA) to evaluate performance, which is validated by inspecting reboilers internally, has been an efficient way to determine the degree to which the system has been fouled and also to schedule shut-downs for cleaning the mechanical parts.

Description of the Process

The (A-100) Naphta Fractionation Unit includes two distillation columns, the objective of which is to separate the light (C_6), medium (C_6 - C_8) and heavy (C_{9+}) cuts, included in the feed supply of crude naphtha.

The light cut and a portion of the heavy cut are sent to pyrolysis furnaces so as to produce ethylene. The medium cut is sent to the Hydrotreating and Catalytic Reforming Unit, with the objective of increasing the aromaticity of the system so as to produce BTX.

The A-100 processes 100 tons/h of crude naphtha, and approximately 35% of this current represents the average cut. Figure 1 below shows the schematic flow diagram of this fractionation process.

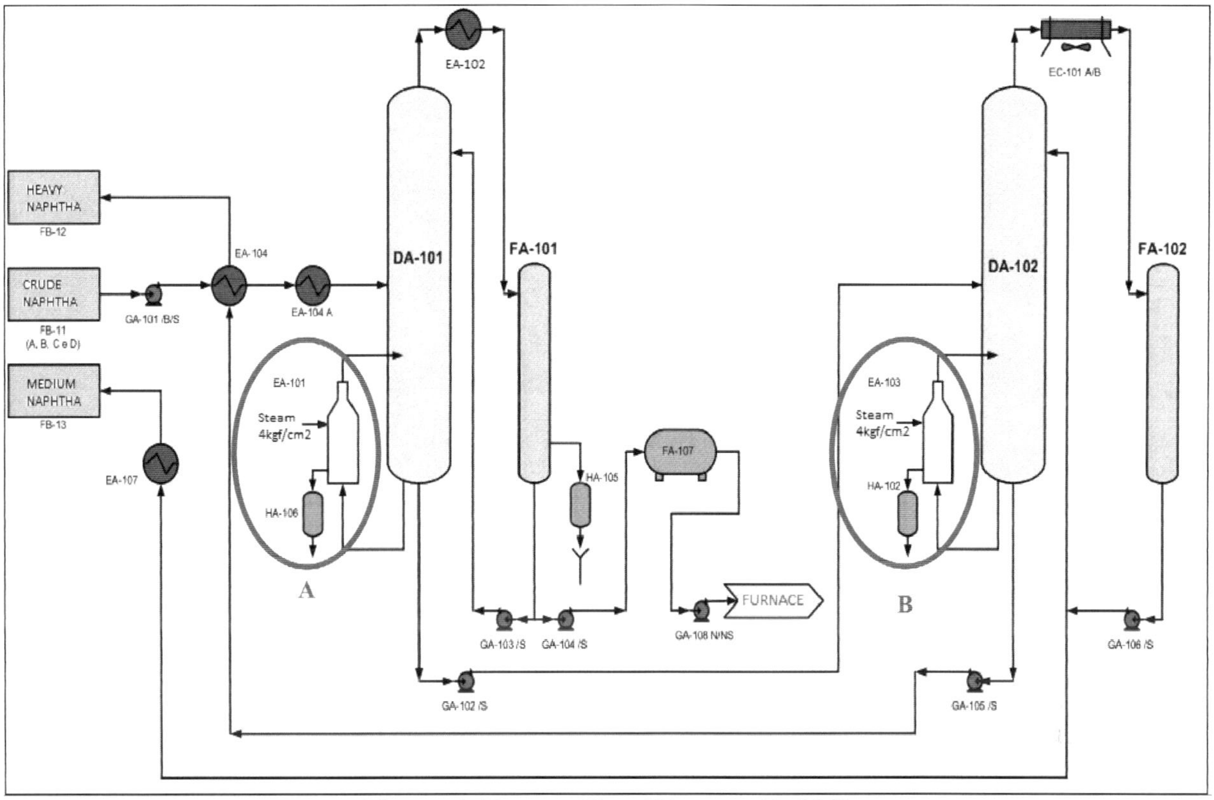

Figure 1. Process Flow Diagram (A-100)

The Reboilers, indicated by the letters A and B in Figure 1 above, operate with an injection of low steam (4kgf/cm2) to promote vaporization of the bottom stream of the A-100 distillation columns. These devices operate over a temperature range of about 150°C, which is sufficient to promote the polymerization of certain olefinic compounds present in the feed of crude naphtha.

Poor performance of these Reboilers can lead to production stoppages for maintenance and cleaning, which very often affects up to five days of production. The possibility of raising the times of campaigns and predicting the degree of fouling in downstream units helps to minimize production losses due to unscheduled stoppages.

This study will discuss the results of Reboiler A, which is the most critical equipment, with a view to synthesizing data, but gains in equipment B were also significant.

Evaluation System

Historical Record of the Campaign

Reboiler A was installed in 1972 but examination of the historical records did not reveal that the sheaf of tubes had ever been replaced. This exchanger has currently 11 tubes plugged into it.

The Campaign times of this equipment are shown in Table 1, with the last campaign being highlighted in blue so as to set it apart from the other ones. Table 1 gives a summary in the order of the timings of the campaigns, the average load processed in the period, the total load processed, the reason for operational stoppages etc.

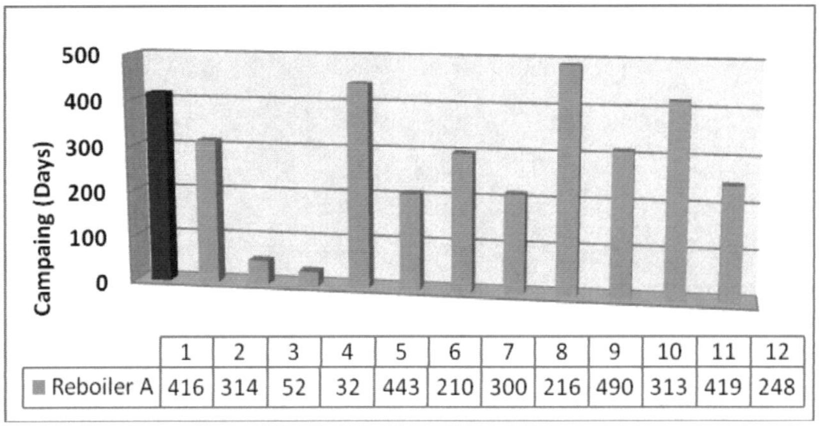

Figure 2. Campaign of the Reboiler A

Table 1. Historical Record of Reboiler A Campaigns

Date	Campaign *(days)*	Capacity *(ton/h)*	Capacity *(ton/campaign)*	Occurrence	Tubes Clogged *(%)*
13/06/12	416.0			Programmed Shutdown	0
17/03/11	314.0	93.1	701601.6	Heat exchange performance	30
03/05/10	52.0	90.8	113318.4	Stuck tube	-
11/03/10	32.0	71.1	54604.8	Heat exchange performance	50
09/02/10	443.0	72.0	765716.6	Heat exchange performance	20
16/11/08	210.0	98.1	494575.2	Heat exchange performance	-
16/04/08	300.0	104.0	748800.0	Programmed Shutdown	-
16/06/07	216.0	92.8	481075.2	Programmed Shutdown	-
10/11/06	490.0	-	-	Programmed Shutdown	-
30/06/05	313.0	-	-	Stuck tube	-
17/08/04	419.0	-	-	-	-
18/06/03	248.0	-	-	-	-
10/10/02	-	-	-	-	-

Conditions of Opening

In this review, a comparison is presented between two operational shut-downs of A-100, which occurred in March 2011 and May 2012, in order to observe the improvement in performance from the last campaign associated with the change in the GE Treatment strategy.

Campaign I: Production shutdown on 03/2011

Table 2 below summarizes the features observed during the inspection of the state of the inlet of this piece of this equipment during the shutdown of March 2011:

Table 2. Internal inspection of Reboiler A in 03/2011

Bottom Cover	Surface residue
Bottom Mirror	No deposits or internal blockage of the tubes were observed.
Top Mirror	Approximately 30% of the tubes were clogged, especially in the region of the vapor inlet, which has higher temperatures. Magnetic and non-magnetic deposits.

May/2011

Figure 3. The mirror of Reboiler A before it was cleaned.

May/2011

Figure 4. The mirror of Reboiler A at the vapor inlet before it was cleaned.

Note: It is difficult to carry out internal cleansing by Hydro jetting (pump pressure during cleaning is 20,000 psi). This reboiler has 11 tubes plugged into it, but none of them are plugged in during this shutdown.

Campaign II: Production shutdown on 03/2011

Table 3 below summarizes the features observed during the inspection of the state of the inlet of this piece of equipment during the shutdown in May 2012:

Table 3. Internal inspection of Reboiler A in 05/2012

Bottom Cover	Surface residue
Bottom Mirror	No deposits or internal blockage of the tubes were observed.
Top Mirror	Clogged tubes were not identified, not even in the region of the inlet vapor. There was only a thin layer of deposit that was easy to remove and which was evenly distributed over the upper surface of the mirror.
Spool	A thin deposit on the inner surface of the caisson.

May/2012

Figure 5. The mirror of Reboiler A before it was cleaned.

May/2012

Figure 6. The mirror of Reboiler A before it was cleaned.

Note: Hydro jetting was done quickly, without any obstructions due to deposits having been identified. This equipment has 11 tubes plugged into it, but none of them were during this shutdown.

Analysis of Deposits

The characterization of the deposits found in Reboiler A showed an 86.0% loss on ignition, thus indicating the strong presence of organic compounds in their composition, with the functional groups of carboxylic acid and ester standing out. The inorganic analysis of the sample indicated a high concentration of iron oxides and also sulfur oxides in the form of sulfides on the cover and spool of this reboiler. This is summarized in Table 4 below:

Table 4. Analysis of the Deposits of Reboiler A

Organic Compounds - LOI (%)	86.0
Water of hydration	Smaller
Sulfur compounds	Smaller
Carboxylic Acid and Ester	Greater
Dichloromethane soluble (%)	2.0
Elemental Sulfur	Greater
Dichloromethane insoluble (%)	98.0
Water of hydration	Smaller
Inorganics	Smaller
Carboxylic Acid and Ester	Greater
Composition of Dichloromethane insolubles (%)	
Carbon	56.0
Hydrogen	5.0
Nitrogen	3.0
Sulfur	8.0
Inorganic Compounds (%)	
Iron, Fe2O3 + Fe3O4	82.0
Sulfide, S2-	16.0
Phosphate, P2O5	1.0
Sodium, Na2O	1.0

Evaluation of Performance

The evaluation of performance of Reboiler A was prepared using the MRA tool, which is a GE good practice to predict fouling in critical systems. This statistical tool enables a dependent variable to be determined, in this case the percentage of opening the steam valve, by means of the numerical regression of the other operational variables of the column. This method generates a mathematical equation to predict the percentage of inlet steam when the equipment is clean, given that the equation obtained by regression includes data of this condition.

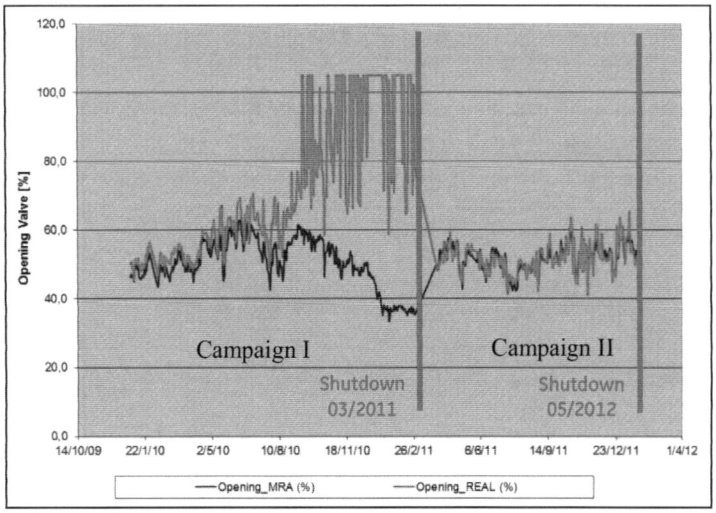

Figure 7. Performance Evaluation of Reboiler A

In Figure 7 the dark blue line represents the percentage of inlet steam when the equipment is clean, as per the equation defined by the MRA. The pink line represents the actual percentage of inlet steam. The sharp upward shift evidenced in the period between November 2010 and March 2011 shows the fouling of the system during Campaign I (prior to the 2011 operational shutdown).

After the 2011 shutdown, it can be seen that the two curves follow the same trend, without a shift, indicating that the unit has the equivalent performance of a clean piece of equipment. The change in the GE treatment strategy, taking the control of the dosages of the product Petroflo 20Y25 into account, was seen to have been effective in increasing the campaign time of this equipment.

Conclusion

The change of strategy in applying dispersant and controlling the dosages after the March 2011 shutdown, presented positive results, given the states of the inlet seen during the inspection of reboilers A and B in May 2012..

The low degree of fouling of these reboilers evidenced in the 2012 shutdown led to the production process being considered reliable, as the A-100 reboilers are no longer the pieces of equipment responsible for the loading limit of the unit. The downtime during the cleaning of the mechanical parts of the reboilers was about 10 times less than that observed in the previous shutdown, thus reducing Braskem´s maintenance costs.

In addition, the MRA statistical tool was able to produce a reliable representation of the degree of fouling of the system. This tool can be effective for scheduling shutdowns more than 03 months in advance.

References

1. Setal Lummus, "Manual de Operação da Unidade de Processamento de Nafta Bruta (Área 100)", Santo André, Brazil.
2. Ponce de Leon, Carolina; Weissheimer, Flavio and Camparin, Rafael H. (2011), "Proposta Técnica de Fornecimento de Especialidades Químicas, Equipamentos e Serviços para Tratamento Químico para a Unidade de Processamento de Nafta Bruta (Área 100)", Santo André, Brazil.
3. Kleinbaum, David G., Kupper, Lawrence L., Muller, Keith E. (2008) "Applied Regression Analysis and Other Multivariable Methods", Belmont, USA.

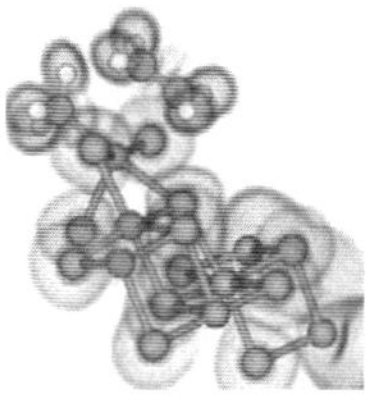

13AIChE
Spring Meeting, San Antonio, TX

SmartProcess® Application Strategy Achieves Faster and More Efficient Control of an Ethylene Complex

Patrick Truesdale
Principal Consultant
Industry Solution Group
Emerson Process Management

April 29, 2013
San Antonio, TX

EMERSON
Process Management

Speaker Introduction

- Principal Consultant
  Emerson Process Systems & Solutions
- BS Electrical Eng, PE
- 40+ Years experience with Refinery & Petrochem Process Control
- Joined Emerson in 2007
- Previous employers: duPont, Bonner & Moore, SETPOINT, AspenTech, Comcept Technology

Patrick Truesdale

E-mail: patrick.truesdale@emerson.com

13AIChE
Spring Meeting, San Antonio, TX

EMERSON
Process Management

Agenda

- Performance Package Review
- SmartProcess Ethylene Overview
  - Furnaces
  - Distillation
- Benefits

13 AIChE
Spring Meeting, San Antonio, TX

EMERSON
Process Management

SmartProcess Ethylene Packages

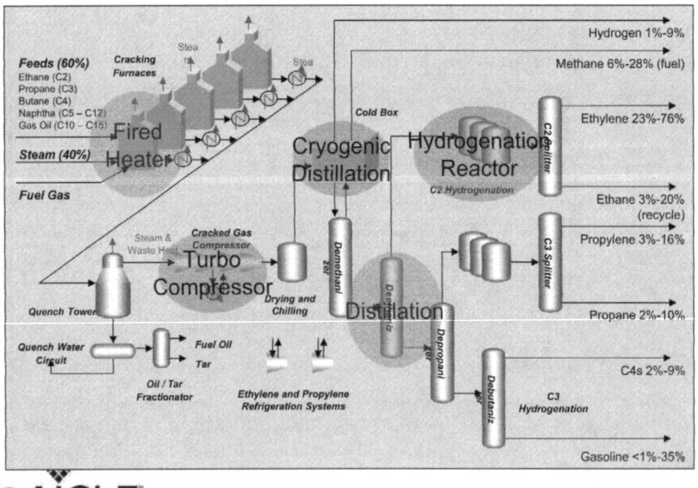

13 AIChE
Spring Meeting, San Antonio, TX

EMERSON
Process Management

SmartProcess Scope

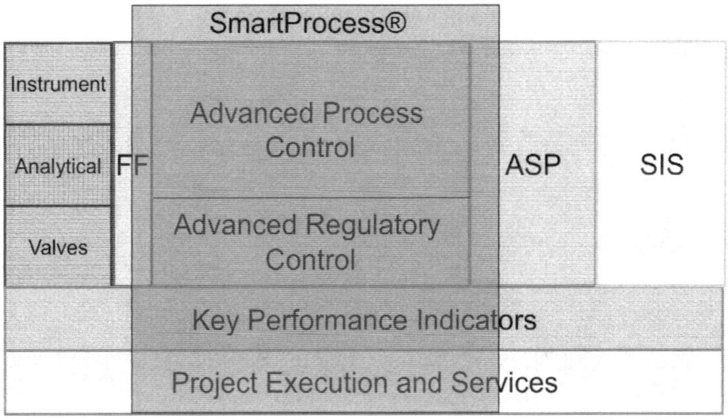

Basic Ethylene Furnace Variables

Key Operating Objectives:

- **Constant severity**
- **Maintain HC/Stm ratios**
- **Min excess air**
- **Maintain within constraints**
- **Max charge (if needed)**

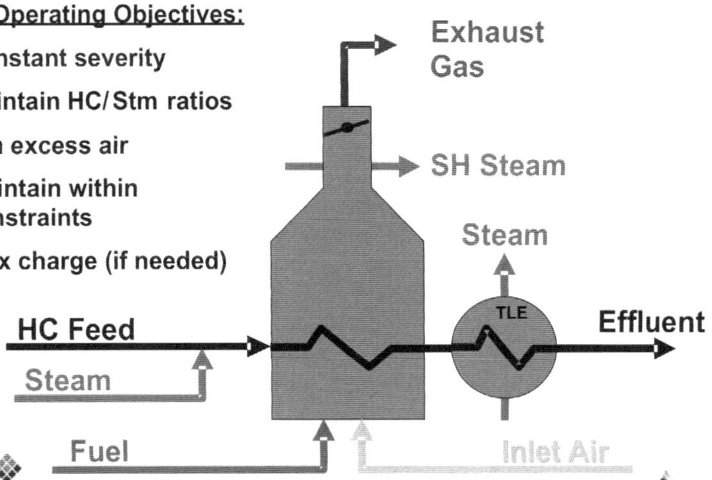

SmartProcess Furnace Module

- Design allows for different furnace configurations
  - Process Zones
  - Firing Zones
  - Multiple TLE analyzers
- Advanced Regulatory
  - Charge Control
  - Steam Control
  - Severity Calculation
  - Combustion Control
  - KPI's
- Advanced
  - Pass Balancing
  - Heat Balancing
  - Severity Control
  - O2 Control

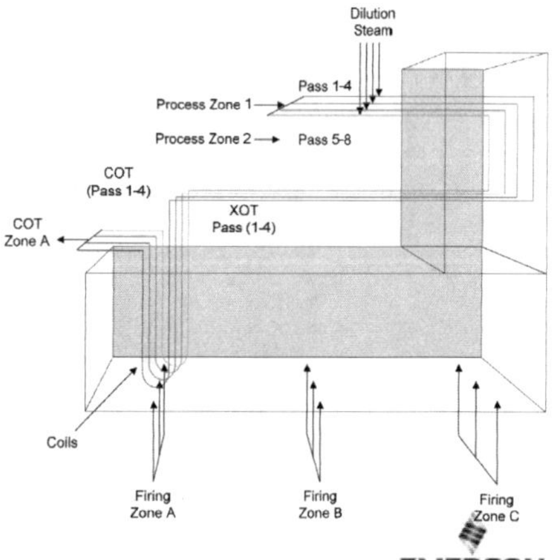

Standard Furnace Module Design

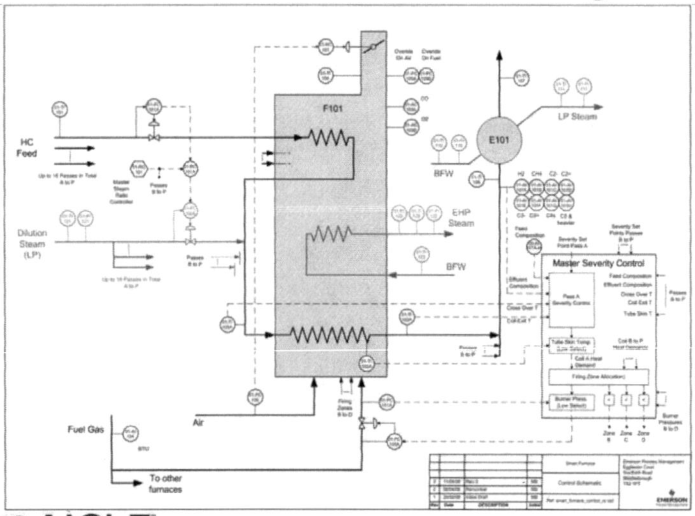

Charge Rate Control

Functions:
- Pass Balancing
- Constraints
- Feed Maximization

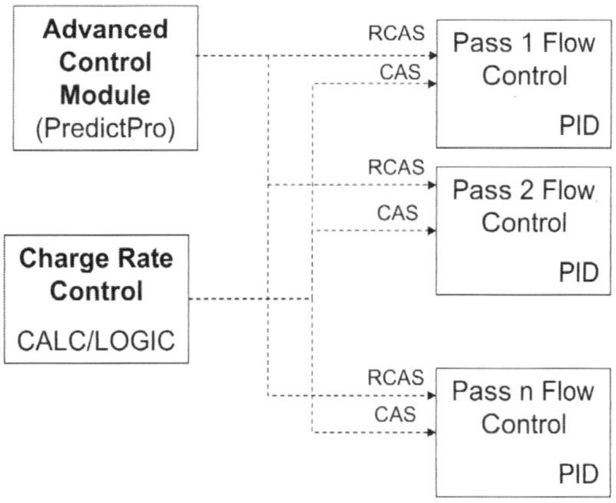

Functions:
- Ramping
- Limits & Alarms
- Multiple Feed Types
- Furnace Status
- Pass Offsets

13 AIChE Spring Meeting, San Antonio, TX

EMERSON. Process Management

Steam Ratio Control

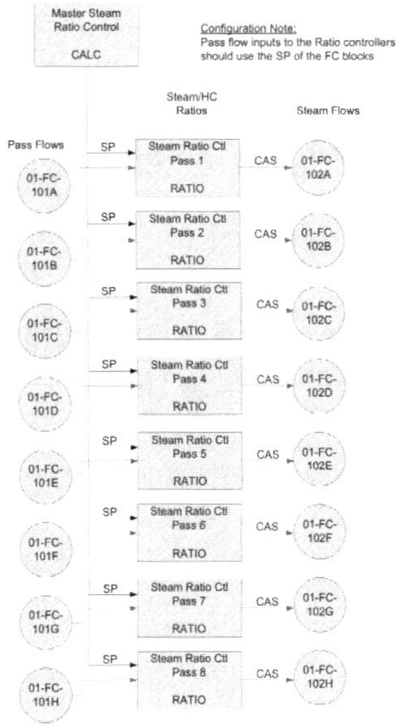

13 AIChE Spring Meeting, San Antonio, TX

EMERSON. Process Management

Severity Control

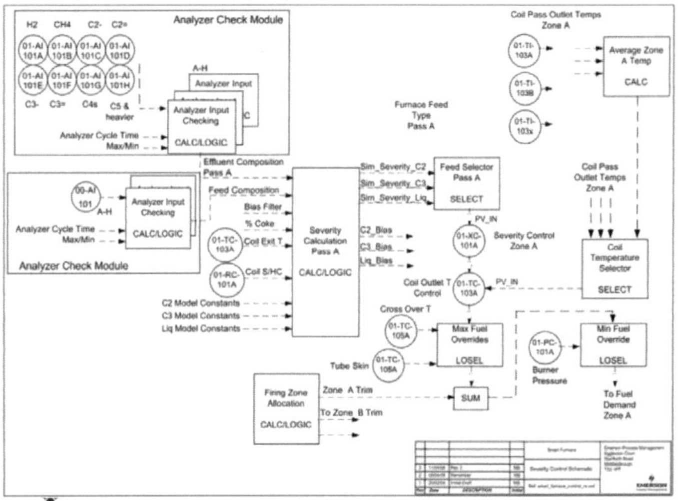

13 AIChE Spring Meeting, San Antonio, TX

EMERSON Process Management

Automated Furnace Sequences

- Standard module templates to be used for Furnace Startup, Shutdown and Decoking sequences.
- Utilizes preconfigured Sequence Function Charts (SFC) running in DeltaV controllers or App Station
- Standard sequence functions can be customized to match site requirements
- Manual operator steps with confirmation can be mixed with automated steps
- Sequence execution control graphics and faceplate templates

13 AIChE Spring Meeting, San Antonio, TX

EMERSON Process Management

Key Performance Calculations

- Furnace efficiency
  - Heat loss method
- Furnace Duty
- TLE Duty, Fouling factor
- Stack gas estimated composition
- Stack gas daily total tons exhaust for CO, CO2, S, N, etc.
- Stack gas NOx emissions (optional – based on CEMS)

13 AIChE
Spring Meeting, San Antonio, TX

EMERSON. Process Management

Distillation - Classic Multivariable Problem

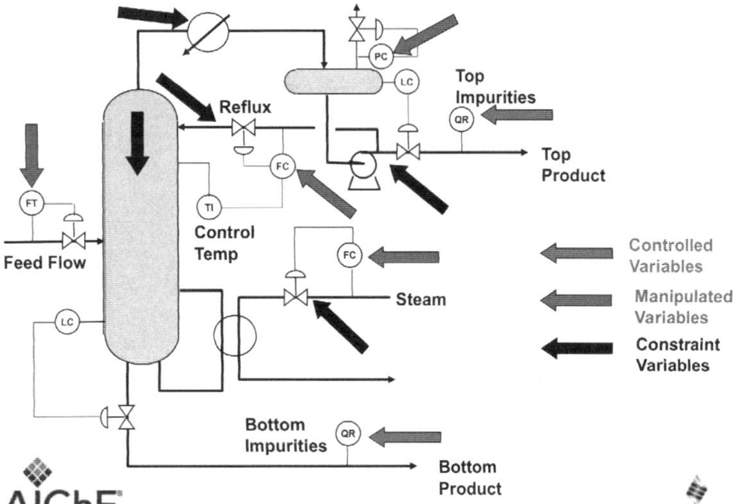

13 AIChE
Spring Meeting, San Antonio, TX

EMERSON. Process Management

Distillation Control Module

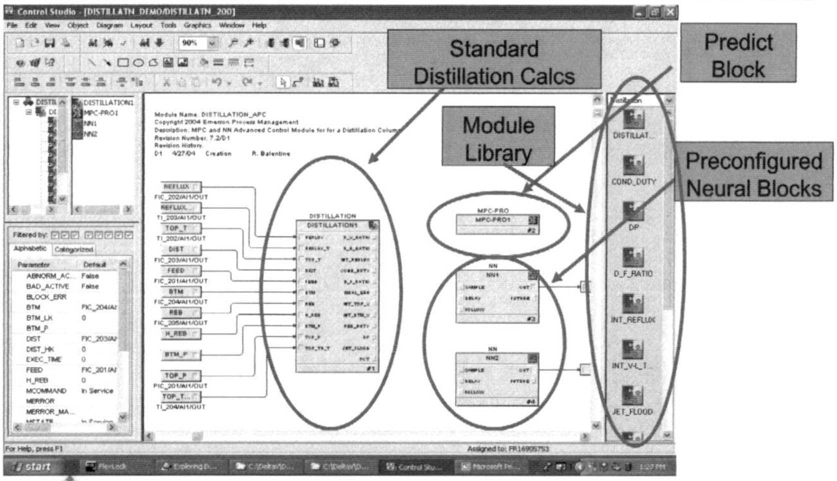

13 AIChE
Spring Meeting, San Antonio, TX

EMERSON
Process Management

SmartProcess Distillation

- DeltaV PredictPro with up to 4 MV's
- DeltaV Neural for up to 2 product qualities
- Standard set of calculations & constraints
  - Internal reflux
  - Pressure compensated temperatures
  - Jet flooding
  - Reboiler & condenser duty
- Engineering Services to design, implement and commission

13 AIChE
Spring Meeting, San Antonio, TX

EMERSON
Process Management

SmartProcess Applications

- Licensed industry applications for common processes that are:
  - Pre-engineered, tested, documented
  - Designed by APC Consultants with extensive practical experience
  - Supported & enhanced
- Bundled combination of:
  - DeltaV libraries: composite blocks, modules & templates
  - DeltaV simulation and demonstration example
  - APC tools (e.g. Predict, Neural)
  - Application code (e.g. VB, .Net)
  - Other databases and apps (e.g. SQL*Server, Excel, TankMaster)
  - Sample DeltaV Graphics
  - Documentation
- Templates to be used to configure for specific site requirements
- May be implemented by Emerson or Customer

Benefits

Ethylene Furnace High Temperature Imaging Cameras to Monitor Burner Flames ,Tube Temperature and Tube Failure Detection

Thomas M. Canty, P.E.
President
J.M. Canty Inc
Buffalo, NY USA

Abstract

The development of explosion proof, high temperature imaging cameras which operate in ethylene furnaces without cooling air now allows monitoring of ethylene furnaces to improve energy/operating efficiency along with providing important safety benefits. Imaging cameras work off the visual wavelength which provides 2 dimensional view and temperature measurement of the tube temperature which is more accurate than thermocouples and IR probes. In addition visual images of the furnace burners lets you detect flame detachment, smoking burners and flame shape in order to increase burner efficiency and save energy. The safety aspects of the Camera imaging systems is significant as it allows visual detection of pin hole leaks and tube failure in the furnaces.

Introduction

Ethylene furnaces, or any type of cracking unit, rely on thermal management in order to make product efficiently and to operate in a safe manner. Having 'eyes' in the furnace can help to identify problems in their early stages of development. High temperature camera technology is able to survive the environment and give a wide angle view of the tubes from which temperature and tube blockage may be determined. Achievable benefits from a camera system include:

1. Temperature analysis capacity for all areas of the image, not just single point. Multiple zone capability for user customization of analysis.
2. Flame detection and size measurement
3. Detection of tube movement and ability for software to automatically move measurement zones as tubes distorts.

Imaging Fundamentals

Basics of an imaging system include a high resolution sensor, proper illumination, appropriate hardware for the process and a robust software analysis package. For this particular application illumination is created by the process and need only be managed optimally for best results. A high density pixel array yields resolution of the image and measurement as well. The hardware and software are the more challenging aspects of the application and will be discussed in some detail.

Hardware

The design of the camera system is unique in that it's internally exposed lens assemblies are **designed to survive without cooling air**. When installed properly, the combination of system materials and heat transfer coefficients at the camera tip limit heat transfer to the system and allow it to run hot without chance of damage. This feature, when system is installed in a clean, gas burning environment, does not disrupt the temperature or air balances of the furnace because no cooling air is being introduced to the furnace. In addition, the design includes a disposable, quarts shield which protects the process exposed lenses from direct contact with the internal atmosphere. Should any particulate contact the system, the quartz shield is there to protect the optical lenses. Replacement of the protective quartz shield is inexpensive. Replacement of damaged lenses is very expensive and so the Canty design guards against that possibility.

Internally, the system is a two piece assembly consisting of the internal insertion lens and the external camera package. As mentioned earlier in this document, Canty employs a fused glass window as part of the electronic housing package. This has a dual purpose; Firstly, the camera housing assembly is a separate entity which carries WP and EXP ratings. Secondly, the camera housing bolts to the internal lens system and transmits the optical rays through the fused glass window to the CCD. This mounting also creates a hermetic barrier between the process and the electronic housing thereby preventing migration of hot, process gasses outside the pressure boundary of the system.

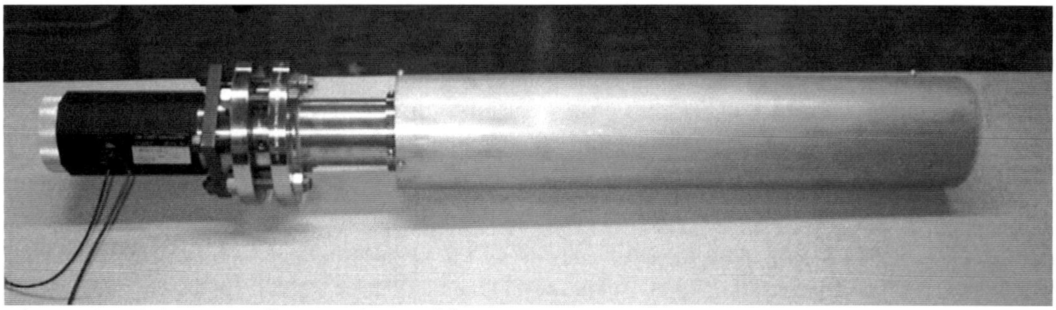

Figure 1. High Temp Camera Assembly

The camera electronics come with a full array of controls that allow it to be optimized to the software function to achieve accurate, consistent data output.

As previously mentioned, the installation is an important feature of the design. When installed correctly, the camera sits inside the refractory (see Figure 2) where its exposure to the process environment is minimized thus minimizing heat transfer into the internal lens.

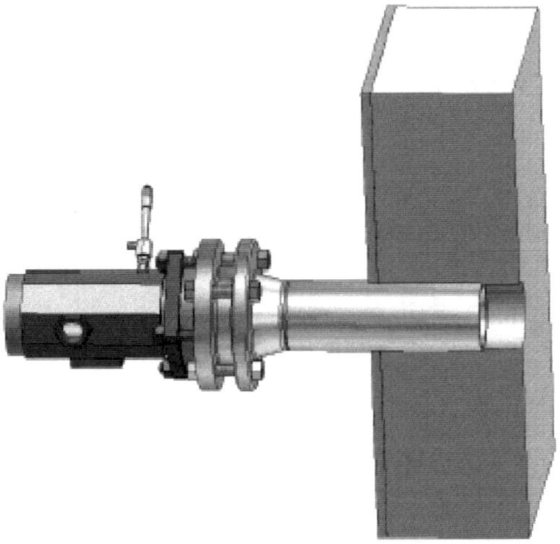

Figure 2. Camera Installation Schematic

Software

With respect to software design, the Canty software is fully flexible in analyzing any portion of the captured image. Multiple analysis zones can be drawn to monitor discreet areas of the view and each zone can run its own particular algorithm depending on need. The benefit here is that multiple tubes can be continually surveyed for changes and trends. The analysis data is also not single point, but is comprised of all the data points in the analysis zone. By choosing proper software filters various information can be obtained from the full array of data such as hot or cold spots. This data can also be archived in total or as snap shots at determined increments of time. The on screen data output can also be updated in the short or near term and can thus be used to judge what is happening in the furnace over time with just a glance at the screen. For instance, using the histogram feature with a 1x daily output a long term trend can be presented on screen in real time.

The basic theory of temperature measurement is based on the radiative emissions of a body at temperature – as the body temperature increases, the radiation curve shifts towards the shorter wavelengths and that is why as a body heats up the eye cannot detect it until it reaches 1000F which is the approximate visible point where emitted radiation is dark red in color. As the temperature increases, the color graduates to bright red, then orange then toward white. A pyrometer functions by measuring a narrow band of wavelength intensity and comparing it to the intensity of a black body to determine the temperature. From the graph below it can be seen that if a single wavelength is monitored the intensity of that measurement changes as the body temperature changes (curve shifts) which then corresponds to a different temperature on the calibration scale.

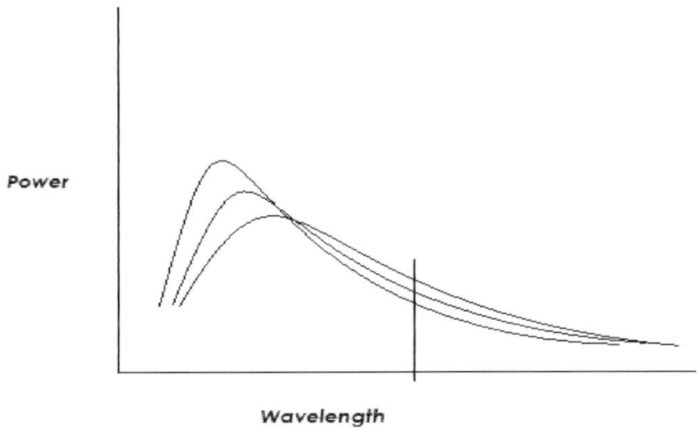

Figure 3. Power Curve at Varying Temperatures

The same is true of a camera system measuring the increasing intensity of an object as it radiates from approx 1000 F (dim, dark red color) on up the temperature scale to 3000+ F for instance which would be a bright white. From the graph above one can see that as color goes from red toward white the peak intensity is always increasing and it is the measure of this intensity which indicates the object temperature. Using an RGB camera the measurement can also be made by monitoring the individual color components of the signal impinging on the chip. For industrial uses, the measurement of intensity usually suffices.

Standard instrumentation used in industry include thermocouple type devices (contact) and thermal imaging and pyrometer systems (non contact). In severe temperatures and atmospheres, the contact devices deteriorate quickly and provide gradually less reliable information leaving the operator running blind. They also create a discontinuity on the surface they are welded to which, in the case of a furnace tube, can lead to premature tube failure.

Non contact systems have difficulties with calibration drift and emissivity adjustments to account for the fact that many items being measured are not black body emitters. In addition, these technologies are single point measurement systems and are best used as hand held devices which can be difficult to use and provide subjective measurement results based on operator preferences for targeting.

Vision technology offers several benefits these other technologies do not. The visual aspect of the system allows the operator to view the process and confirm the readings received. It also allows flexibility to set up measurement zones on screen at any time to assess any or all areas of the process in view should a change be required. Since vision systems are generally fixed installations, an added benefit is that the information is taken from the same process points and features within a given measurement zone at all times which means there is no error in sighting of the instrument.

The vision system is also unique in that it does not suffer calibration drift as almost all instruments do. Once the calibration is set the readings repeat themselves within a couple degrees F. The calibration is based on thermocouple or pyrometer readings initially, however after that calibration is finished it does not have to be redone in the future as long as the system remains in its location.

Process history and performance can also be best documented by a vision system. First off, the camera can analyze any area of process that is within the systems view. This affords the operator the capability of viewing many points of interest instead of just single point temperature readings. As an example, if the system is monitoring a tube bank it can assess the temperature of the tubes for the full length of view. If there is a local flow problem in a tube it will become overheated in one section. The camera can locate that section and indicate its temperature rise over the rest of the tube or adjacent tubes.
This information allows the tube location to be identified as well as the position in the tube that has encountered the upset condition. Over time this historical data, recorded in video or still image format by the control software, may be significant in improving the process. In addition to the ability to monitor the full area of the view, the full area of view can also be segmented into multiple zones of measurement, i.e. each tube having its own measurement zone for automated process control capabilities. This capability allows the tube to be operated at optimum efficiency since the addition of excess heat can be avoided thereby reducing energy draw to operate the furnace.

Thermal Analysis

Tube Temperature
The following figures illustrate the temperature analysis capability for tubes in a live test situation. Measurement zones are placed over the tubes (figure 4) as the furnace is heated and cooled. The temperature plots are shown in figures 5 and 6. The results are as expected and uniform from tube to tube, although a small number are included in this demonstration.

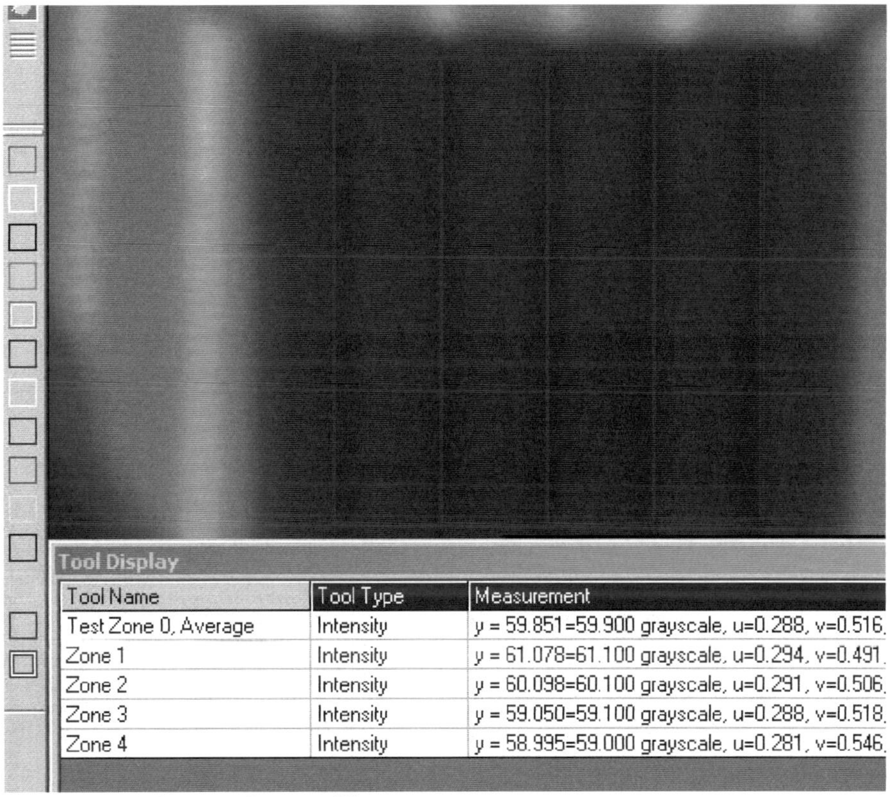

Figure 4. Software Screen Shot

As thermal loads are fluctuated up and down the following graphs track those changes. The cooling cycles are left to natural convection and thus are not linear in nature.

For the following curves, the vertical scale is gray scale intensity which is measured 0 – 255 (pitch black to bright white). The hotter the tube, the higher, or more intense, the gray scale reading. This value in process would be calibrated to the temperature. The horizontal scale is the image number, or time.

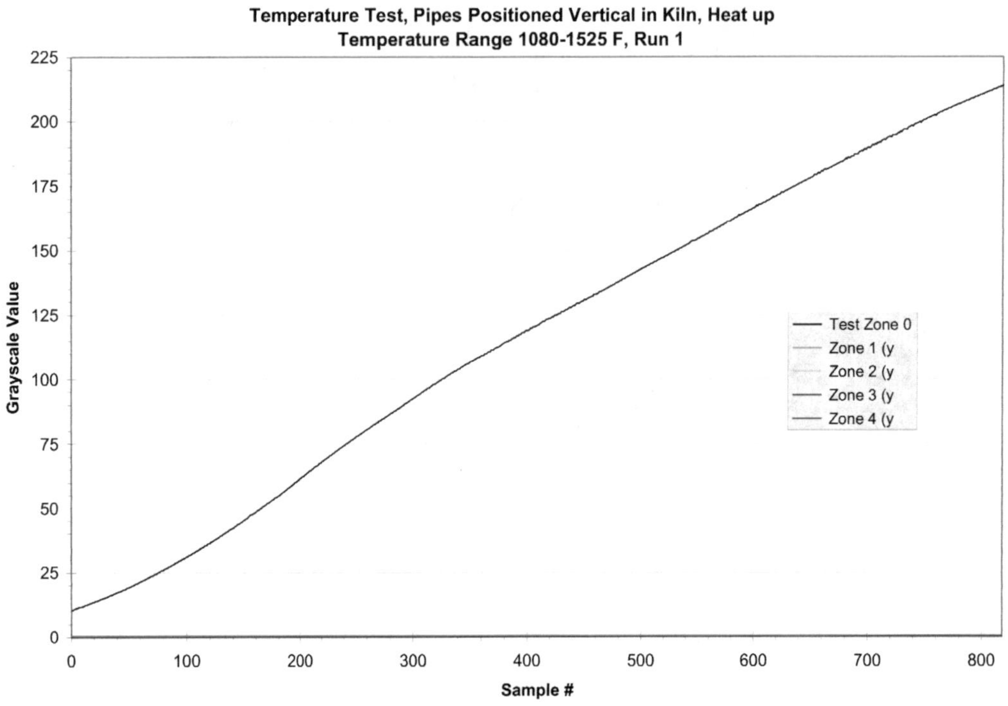

Figure 5. Heating Cycle

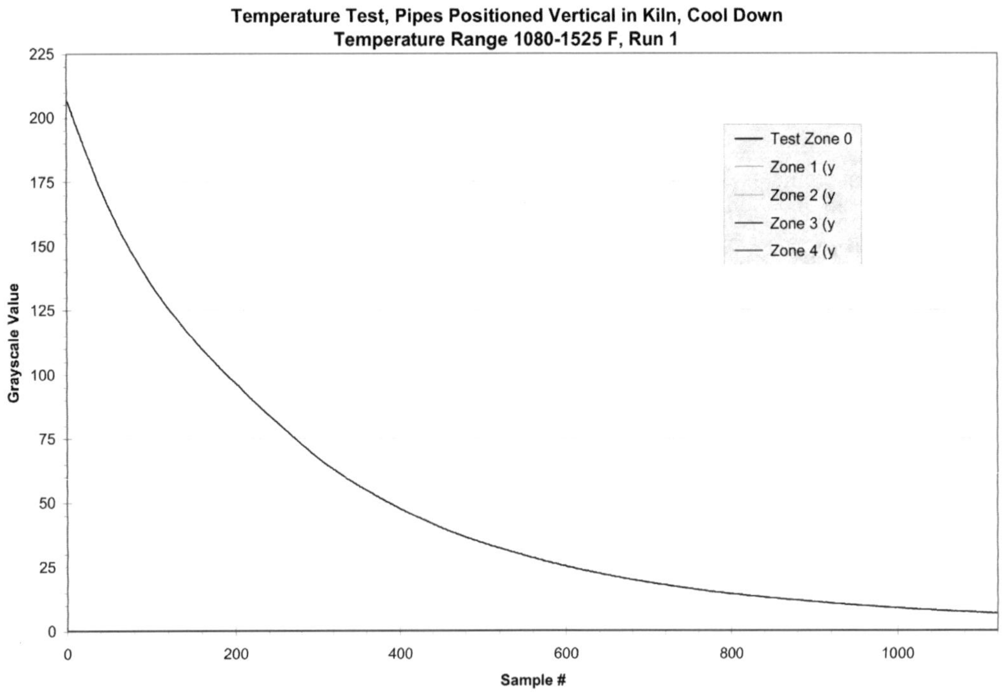

Figure 6. Cooling Cycle

Flame Measurement

Flame temperature can be monitored in the same way as described previously, however the flame size, or presence, can also be monitored. Figure 7 shows a live flame and Figure 8 shows a digital capture. Computing the flame area, or shape, can be done to track its changes over time.

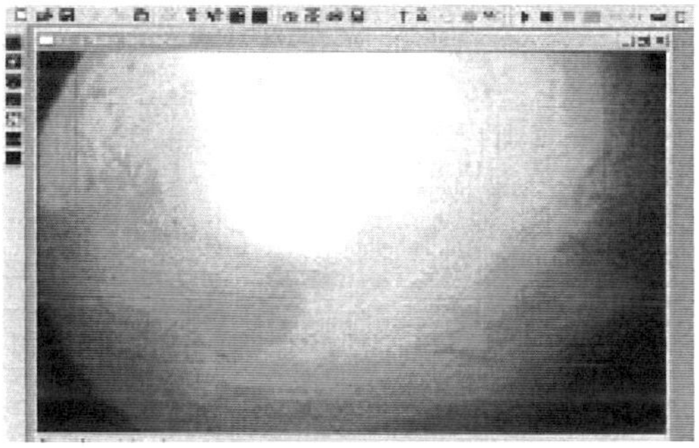

Figure 7. Kiln Flame (Live)

Figure 8. Kiln Flame (Digital)

Conclusion

Vision Technology presents an opportunity to increase effectiveness of furnace monitoring through hardware and software capabilities. The thermal design at the core of the hardware enables the system to provide the operator a view into the furnace and feed back on its thermal performance over time. This data can assist in detecting building upset conditions, but also provide a record for study to assist in improving the operational method of the furnace.

(45v) Straighten up! Non-Linear Compensation Using Feed Forward Control

Barbara Hamilton, Process Systems and Solutions, Emerson Process Management, Hampden , ME

Non Linear processes occur throughout the petrochemical plant. While the PID loop is surprisingly robust for most situations, there are cases where compensation of these non-linearity's is critical for stable control. Similarly, disturbances can cause major disruptions in control stability.

A method of feed forward control, using data based models, can provide robust control throughout the process range. Historical data are used to build models to drive the output toward the solution with PID control providing the final drive to zero error.

The model is also used to develop dynamic tuning. The result is a simple to configure, robust solution to manage non-linear loop and/or disturbance variations.

(45w) Adsorption of CO2 by Alkaline Activated Ca-Al2O3-SiO2 Material

Naim Faqir, King Fahd University of Petroleum & Minerals, Dhahran, Saudi Arabia and Reyad Shawabkeh, Chemical Engineering, King Fahd University of Petroleum & Minerals, Dhahran, Saudi Arabia

Natural gas contains a slight level of carbon dioxide which leads to several industrial and domestic heating problems. In addition, it is one of main greenhouse gases that contribute for global warming, where its pollution is a wide-spread problem arising from a wide range of industries, leading to complex contaminant in atmosphere. Given the increasing pressures worldwide to minimize the emission of greenhouse gases into atmosphere, CO_2 must be remediated efficiently and cost effectively.

In this short abstract a chemically treated limestone and alumino-silicate material were utilized for removal of CO_2 from natural gas using a combined adsorption-chemical reaction processes. This functionalized Ca-Al_2O_3-SiO_2 material was characterized according to its crystalline phase, surface area, pore size distribution and surface morphology. Then it was tested against different concentrations of CO_2 in CO_2-N_2 gas streams.

Pressure Gauges: Hidden Danger or Visible Safety

by Jeff Placek, WIKA Instrument, LP, Lawrenceville, GA

Introduction

In recent years, refining plants, petrochemical plants and chemical plants have been required to increase capacity, but have been provided with very few upgrades to existing equipment to accommodate the increased capacity requirements. This has caused plants to have increased stress and strain on their systems and operating units. For example, higher pulsating pressures and heavy mechanically-induced vibrations are translating to all areas of the processing equipment, especially pressure instrumentation.

Pressure gauges are one of the most important, yet most overlooked or ignored keys to ensuring the continued safety, productivity and profitability of process plants. Process pressure gauges are used to read pressure on transfer lines, discharge pumps, and other places where it is important to be able to read pressures.

Problems Associated with Misapplied, Faulty or Ignored Gauges

Incorrect, misapplied, damaged, faulty or ignored gauges can lead to a number of issues, including safety problems, downtime, and costly regulatory agency fines and other penalties.

Process Safety:

One group of problems relates to process safety issues, such as loss of containment that leads to fugitive emissions, fires and explosions. Bourdon tubes in malfunctioning gauges can rupture, causing the release of media, such as gasoline, oil or gas. Because this media is often volatile and corrosive, it can cause fires, explosions, or other serious safety issues, such as a toxic gas releases or acids spraying onto workers.

Media releases are common results of misapplied, faulty or ignored gauges. Fires and explosions are less common but they do occur. When reviewing plant instrumentation, findings have revealed instances where the wrong gauges have been installed in locations on pumps. In one instance, a gauge vibrated so severely when the pump came out of alignment that it caused the Bourdon tube in the gauge to crack. The media coming out of the tube leaked onto the pump, which was excessively hot because of age and misalignment, causing a fire.

Expensive Downtime:

A second group of problems relates to expensive downtime, which can result from the breakdown of process safety just mentioned, or simply from having to shut down a process in order to replace incorrect, misapplied, broken or otherwise damaged gauges.

1 - Safety-Related Downtime:

As a result of the problems identified above (emissions, fires, explosions, etc.), there can be any number of causes of subsequent downtime, including fire suppression activities, released media containment, line shutdown, personnel evacuation, and then eventually having to rebuild or repair the lines, replace equipment, etc. In addition, if there is a media release with no injuries or fatalities, the plant is required to notify the EPA. However, if there is a media release that does cause injuries or deaths, the plant is required to notify OSHA. These can lead to very long process shutdowns, anywhere from a day to a month or more. (More on OSHA and the EPA below.)

2 - Non-Safety-Related Downtime:

When a line fails, causing a gauge to fail, even if there is no release or safety incident, downtime can still occur as a result of the process needing to be shut down in order to replace the gauge. Gauge failure rarely causes other parts of the process to fail, although gauge failure is usually an alert that there is a problem. In most cases, a plant worker will simply replace the pressure gauge and move on. When another failure occurs, the worker will do the same, until the root cause of the failure, such as a misaligned pump, causes the plant to shut down the line, resulting in more costly, unplanned downtime and a major maintenance time.

Regulatory Agency Fines and Penalties:

A third group of problems relates to the potential for significant fines and other penalties from federal regulatory agencies (primarily OSHA and the EPA), as well as from state and local jurisdictions, that result from compromised process safety.

1 - OSHA:

OSHA addresses many process-related safety issues in 1910 Subpart H, "Hazardous Materials." More specific is 1910.119, "Process safety management of highly hazardous chemicals." In its Process Safety Management section, OSHA notes that, "Unexpected releases of toxic, reactive, or flammable liquids and gases in processes involving highly hazardous chemicals have been reported for many years in various industries that use chemicals with such properties. Regardless of the industry that uses these hazardous chemicals, there is a potential for an accidental release at times they are not properly controlled, creating the possibility of disaster."

OSHA goes on to note that: "To help ensure safe and healthful workforces, OSHA has issued the Process Safety Management of Highly Hazardous Chemicals standard (29 CFR 1910.119), which contains requirements for the management of hazards associated with processes using highly hazardous chemicals."

OSHA's standard emphasizes the management of hazards associated with highly hazardous chemicals and establishes a comprehensive management program that integrates technologies, procedures, and management practices.

While OSHA does not monitor gauges in specifically, they will enforce guidelines for worker safety. OSHA investigates injuries and deaths, and may levy fines or other penalties.

2 - EPA:

According to Greg Neneman, director of business development solutions for Valin Corporation (San Jose, Calif.), fugitive emission regulations from the Federal level have tightened up over the past several years. In many states, such as California, Oregon and Washington, the regulations are even tighter.

"One challenge companies face is ensuring compliance with emission regulations when they upgrade or expand their processes," points out Neneman. "They are not allowed to produce additional emissions beyond the original limit set by regulations."

From the standpoint of federal EPA regulations, each company that produces or emits volatile organic compounds (VOCs) is required to come up with a leak detection and repair program. "In some areas, county and local requirements are even tighter than federal requirements," continues Neneman.

In certain areas, companies that produce or emit VOCs are required by regulation to do audits on valves, flanges, gauges, and all other process connections on a quarterly basis. "And in some jurisdictions, the regulatory agencies can give plants 24 hours notice before they descend on that plant to check all of their records and conduct other investigations," adds Neneman.

In May 2012, the U.S. Environmental Protection Agency (EPA) levied a $1 million civil penalty on a plastics company for alleged violations of the Clean Air Act at two of its chemical manufacturing plants. In addition to paying the penalty, the company agreed to implement a comprehensive program to reduce fugitive emissions of hazardous air pollutants (HAPs) from leaking equipment such as valves and pumps. The settlement also required the company to implement enhanced work practices, including more frequent leak monitoring, better repair practices, and new efforts designed to prevent leaks. After being fined by the EPA, the company realized that instrumentation was one of the areas where they having fugitive emissions. Until that time, workers had been replacing gauges left and right, because they weren't aware of the problems that could occur. The gauges weren't a big expense to them, so they considered the replacement activity to be more of a nuisance than anything. It's when they had the rupture that they realized there was a real problem.

An audit of the plant, which involved reviewing at all of the instrumentation, found that in many applications, they were using the wrong instrumentation. To correct the issues, the right instrumentation was specified for them, and the correct gauges were ordered. The plant now has its instrumentation under control.

While conforming to EPA regulations has always been a challenge for process facilities, it is going to become even more challenging in the future as a result of EPA's recent introduction of new requirements related to fugitive emissions and leak detection and repair (LDAR) strategies. On April 17, 2012, the EPA Administrator signed a suite of final air quality regulations that affect both upstream and midstream oil and gas sectors - the production/processing and transmission/storage sectors of the oil and natural gas industry.

For the first time, EPA will regulate air emissions from natural gas wells that are hydraulically fractured, as well as other emission sources associated with exploration, production, processing, and transportation of oil and natural gas.

Promulgation of the final rules will result in new control equipment requirements, as well as increased requirements for monitoring, recordkeeping, and reporting for owners, operators, contractors and vendors.

The final rules require the use of emission controls, as well as work practices, through two different air regulatory programs. One is the New Source Performance Standards (NSPS). The other is the National Emissions Standards for Hazardous Air Pollutants (NESHAP).

EPA finalized a new NSPS at 40 CFR Part 60, Subpart OOOO, which regulates emissions of volatile organic compounds (VOC) and sulfur dioxide (SO2) from a variety of oil and gas exploration, production, processing, and transportation facilities, including natural gas wells, certain compressors and pneumatic controllers, storage vessels, and sweetening units at onshore natural gas processing plants. Subpart OOOO also includes startup, shutdown and malfunction provisions.

The EPA has also made it clear that oil and gas production and processing sources will be the focus of its National Enforcement Initiative (NEI) for fiscal years 2012 and 2013. With the new NSPS and revised NESHAPs in place, EPA now has the tools it needs to enforce in a comprehensive manner.

As a result of these new regulations, plants should look at everything more closely, including gauges.

How Problems Can Occur With Gauges

Technical problems can occur with process pressure gauges in a number of ways, which can be conveniently categorized into seven broad categories with specialties or subsets. The most common reason for gauge problems is that workers often install gauges without knowing how they should be installed or accessorized. As a result, the gauges can fail because they were never designed to function in those specific situations. They can also fail because they were installed incorrectly, or because they were not properly maintained.

One way to eliminate or minimize these problems is to consider specific pressure gauge accessories. These can prove to be the extra step needed to ensure a successful, proactive preventive maintenance program, and for extending the life of the gauges. Accessories can provide additional protection for the gauges during their lifetimes.

1 - Mechanical Vibration.

This is oscillating, reciprocating or other periodic motion of a rigid or elastic body or medium forced from a position or state of equilibrium. Typically, vibration is the main cause of pressure gauge failure in process plants. Pump applications produce most of this vibration.

Gauges that aren't designed to withstand vibration may be installed in applications where vibration occurs, causing the gauges to come out of accuracy. Even more serious, a gauge can vibrate so much that the Bourdon tube inside it can crack, releasing media."

One way to prevent vibration-related problems is to install liquid-filled gauges, which help to dampen the vibration and also protect the internal components. The liquid case filling also lubricates the internals and prevents corrosion.

2 - Pulsation.

When media rapidly cycles through the pressure system of the gauge, pulsation occurs. If the pointer pulsation increments are greater than 5 percent of the full scale value, you must intervene to prevent damage to the gauge.

A restrictor or throttle screw can solve most pulsation problems. It will slow down the media by reducing the size of the orifice and averaging out pressure fluctuations. For extreme pulsation, such as spiking and surging, a snubber works best. It allows more control over the amount of snubbing and the flow of media into the pressure system.

3 - Temperature:

If a gauge is exposed to a media with very high temperature, the metal alloy in the gauge may not be able to handle the temperature, causing the Bourdon tube to fail and release media.

Long pipes that separate the pressure gauge from the process, or small cooling elements with radiator fins, protect the pressure instrument from high process temperatures by allowing ambient air to flow over the surface area. In environments with steam, using a siphon with the pressure gauge cools the steam down and drives it back to its liquid state before it gets to the gauge. With very high-temperature media, a diaphragm seal with a cooling element will isolate and prevent the hot media from reaching the gauge.

4 - Spikes or Overpressure.

In some cases, process media may be under extreme pressure. Pumps also can cause spikes and over-pressure, both of which shorten gauge life. Gauges can fail as a result of these pressure spikes. When workers turn pumps on or off, or open or close valves, a surge of media or oil goes through the pipe and hits the pressure gauge, which causes a huge spike. For example, if a line is pumping 100 psi of oil, someone may install a 100 psi gauge, not realizing that this leaves no allowance for any overpressure, surging, or spiking. When pressure does increase, it causes overpressure on the gauge, sometimes to the point of where the Bourdon tube ruptures and leaks media.

One way to prevent this problem is to use a different type of gauge, such as 200 psi gauge for a 100 psi line. Another option is to accessorize the gauge with an over-pressure protector, which would shut off the pressure if there were a surge or spike. Overpressure protectors are designed to avert spikes and the resulting overpressure damage. These protectors provide protection for pressure gauges against the effects of pressures exceeding their maximum pressure ratings.

5 - Corrosion.

Corrosion problems can occur with gauges if the process media is not compatible with the pressure gauges' wetted parts material. Media can sometimes be so corrosive that it corrodes through the Bourdon tube and leaks onto the ground.

When the media is not compatible with the wetted parts material of the pressure gauge, a diaphragm seal is required to isolate the gauge from the media. The benefit of a diaphragm seal is that it can be made of special alloys that are designed to withstand corrosion. It is essential to select a diaphragm material that is compatible to the process media.

6 - Clogging.

In some processes, the media may eventually crystallize or may otherwise harden. If a standard gauge is used in such an application, it will eventually get clogged and will not be able to read the pressure accurately. For example, if a gauge gets clogged and ends up reading zero pressure, when in fact it is under tremendous pressure, there will be an unexpected and dangerous release of media if and when the piping system is opened. Diaphragm seals protect the pressure gauge from solids, particles and viscous media by creating a barrier between the process and the pressure inlet of the gauge. There is also an option to equip the diaphragm seal with flushing ports to constantly flush material out from the diaphragm surface.

7 - Mishandling and Misapplication.

Gauges can also fail prematurely if they are misapplied. For example, a gauge may have been ideal for the application when it was installed. However, if there are equipment modifications or more components subsequently added, this can change process parameters, resulting in the gauge not longer being right for that application. Misapplication can lead to serious problems associated with failure and subsequent media release.

It is also possible to accidentally damage gauges during installation. Gauges should be installed using the proper sized wrench on the flats. Gauges should never be installed by turning the case. After installation, gauges can also fail if they are mishandled or used improperly.

Why Problems Can Occur With Gauges

In a typical process plant, about 25 percent of pressure gauges have failed or are about to fail.[1]
A lack of management focus and worker education and experience are key reasons plants experience issues with gauges.

Management Focus.

Gauges are often viewed as inexpensive commodities, with management failing to realize that they are very critical parts of the processes. Often, plants will replace failed gauges with the same replacement gauges, which will also fail, only to have to be replaced again. Eventually, their luck may run out and a serious failure may occur because the incorrect gauge has been installed.

Why do gauges tend to receive such little attention from management? Management tends to focus on the expensive equipment - $100,000 or million dollars items. They often don't understand the importance of focusing attention on $100 gauges. This is not the fault of management. The problem is that no one has explained to them just how important gauges are.

As noted earlier, though, ignoring instrumentation can cause serious safety and financial problems. Plant process gauges can be thought of in the same way as vehicle gauges. If the temperature gauge on an automobile is defective, it is possible to destroy the engine as a result of overheating that is not registered by the defective temperature gauge. Similarly, if the oil pressure gauge on an automobile is defective, it is again possible to destroy the engine as a result of seized components resulting from loss of oil that is not indicated by the defective oil pressure gauge.

Worker Education and Experience.

A lack of worker education and experience also contributes to gauge issues. In many instances, when maintenance technicians look at gauges, they don't pay much attention to them unless they leaking or look as if they are going to explode. They don't realize all of the information that that gauge can provide. In addition, technicians may not realize that a gauge has failed or is about to fail. Again, as long as it is not leaking media, they will often just keep walking. This is rarely the fault of the technicians. Many have not been properly trained to understand process pressure gauge performance.

Along the same lines, the industry has been experiencing a "brain drain" for several years, with large numbers of experienced technical workers and engineers retiring and being replaced by younger people who lack the education and experience to understand what gauges are telling them, or, worse, not being replaced at all.

According to research conducted by Cambridge Energy Research Associates in 2007, more than half of all oil-field professionals will reach retirement age during the next decade. The Interstate Oil & Gas Compact Commission found that the average age of production and service company employees is 46 to 49 years, and the average retirement age in the oil and gas industry is 55 years. [2]

Historically, plants had specialists whose jobs focused exclusively on instrumentation. They spec'd the gauges, maintained them, and made sure they were operating correctly. However, as plants got leaner in recent years, they started letting these people go. Plants first shifted gauge responsibility to the pipefitter whose main responsibility was to make sure the pipes were put together correctly and not leaking.

[1]WIKA data aggregated from more than 250 instrument audits.
[2]Peter Peek, Jocelyn Fenard, Patrick Gantes, Christine Theiler. December 2008, "Skills Shortage in the Global Oil and Gas Industry: How to close the gap" Part I. Pages 14-15.

Then, as plants began letting pipefitters go, they shifted the responsibility to the unit operators or technicians, who now install the gauges. These people usually don't have any experience or knowledge of how to install instrumentation. They usually just replace old gauges with similar gauges. The problem is that the gauges are usually the wrong ones to begin with. The result is that, these days, few plants have people who have much if any education and experience with instrumentation.

Taking Stock of Process Pressure Gauges

When it comes to improving the safety, productivity and profitability of process plants through an effective gauge management program, companies can take a couple of steps. First, an understanding of the vital importance of gauges must be established. Second, workers must understand what kind of gauges will be effective.

Focus on Gauges:

World-class process plants maintain their assets proactively. In such plants, pressure gauges receive as much attention as pumps and control valves. Management understands the importance of proactively maintaining their assets, and they realize that one of the best ways to maintain their most expensive assets is to properly source and maintain the gauges that protect these assets. That is, they understand the value of spending $100 for a gauge that can subsequently prevent damage or destruction to a $100,000 or $500,000 piece of equipment and the resulting negative safety, downtime and regulatory implications.

There is an interest in predictive maintenance these days in process plants, which is a good thing. A lot of this, though, involves expensive software. In addition, when something is out of kilter, the system may be set up such that technicians may have to wait for it to get far enough out of the normal range before it will generate an alert on the computer. However, properly-maintained pressure gauges, studied by people who know what to look for, can be very inexpensive but reliable and useful mechanical predictive maintenance tools.

Gauge Procurement Considerations:

Over time, as noted earlier, the status of pressure gauges has unfortunately diminished to that of commodities in many process plants. In order to ensure process plant safety, productivity and profitability, it is essential that plant professionals understand the importance of selecting the right gauges for applications.

It is also important that these be high-quality gauges, not "commodity" gauges. That is, not all process gauges are created equal. The so-called "lowest price" gauges appear attractive at first, but are rarely the "lowest cost" in the long run, because they usually have to be replaced more often, leading to excessive downtime and inconvenience. Worse, they may fail unexpectedly, leading to catastrophic results.

Consider these 10 steps to select the right gauges that will help to ensure safe, productive and profitable plant processes.

1 - Know what is in the pipe. This way, you can select a gauge alloy that will be compatible with the process media.

2 - Next, know the temperature of the media. If it is very hot, for example, it will be unsafe to use a standard pressure gauge. A long pipe, siphon or diaphragm seal assembly may be needed to cool the media sufficiently before it reaches the gauge internals.

3 – Third, know the ambient conditions surrounding the gauge. For example, if it is corrosive around the gauge, you need to make sure that the external components of the gauge such as the case and window can handle the environmental conditions.

4 - Next, know the pressure in the piping that the gauge is meant to measure. A best practice is using a gauge that is designed for two times the normal operating pressure. For example, if normal pressure is 100 psi, use a 200 psi gauge.

5 – Know the conditions that affect gauge wear. These could include mechanical vibration, pulsation, pressure spikes, etc. For example, if vibration exists, a gauge with a liquid-filled case is recommended. If there is pulsation, use a throttle plug or restrictor accessory.

6 - Select gauges that are designed for a particular service. Process type gauges are designed with strong Bourdon tubes and other heavy duty components. If a gauge is misapplied, the internals of the gauge may not be rugged enough to resist failure and subsequent media release in these applications.

7 - The gauge case must be rugged and have the correct safety features. In some case designs, if there is a rupture of the Bourdon tube, media will be released out the front of the gauge, which can spray workers. Select a gauge where media would release out from the back of the gauge.

8 - Accuracy is important. In most process plants, +/-0.5% accuracy is the norm, and +/-1% is acceptable.

9 - Size is also important. The gauge dial needs to be large enough that operators can read it from a distance – four inches or larger.

10 - Finally, the gauge should be mounted properly. A gauge is usually mounted by its stem, but it can also be mounted in a panel or against a flat surface.

Looking Outside for Assistance

As noted earlier, it can be difficult for plant management to devote enough time and attention to an effective process pressure gauge program. Similarly, it can be difficult for technical personnel to get up-to-speed in terms of education and experience related to process pressure gauges.

There has been a growing trend over the last 30 years in virtually all industries for companies to outsource services that are not considered "core competencies." That is, companies have come to realize that it makes sense to focus their attention on their missions and what they do best, and leave secondary and peripheral activities to third parties that specialize in these other activities and services. Becoming experts in pressure gauges is certainly not a core competency for refining plants, petrochemical, and chemical plants, so this is an area that can easily be outsourced to an experienced third-party audit team.

If a plant elects not to gain the experience that it needs on its own, free assistance is available. To ensure that gauges are working properly and are able to protect the valuable assets that they were designed to protect, a five-phase instrument review is recommended. A best practices review identifies gauges with unwanted conditions and recommends corrective actions to ensure optimum performance. The review identifies issues that affect reliability, longevity and safety. Such a review can uncover evidence of pulsation, vibration, pressure spikes, excess process or ambient temperature, installation damage, clogging, chemical incompatibility, and other problems that can lead to serious problems.

Phase 1 - Interview. This involves discussions between the third-party audit team and plant personnel. During the interview, plant personnel learn the purpose of the instrument audit, how it will be conducted, and how they will benefit from the audit. The interview also helps determine the best way to match the audit to the needs of that specific plant.

Phase 2 - Pre-Meeting. This involves determining the units or sections of the plant to audit, as well as outline roles and responsibilities. It also identifies prohibited and exempt areas of the plant. The third-party team reviews problematic instrument installations and discusses safety procedures with plant personnel.

Phase 3 - Instrument Application Audit. This involves having key plant personnel lead the third-party team on an actual plant walk-through. The audit team reviews the target gauge installations, identifies gauges that demonstrate unwanted conditions, as well as those that demonstrate installation best practices.

Phase 4 - Post-Meeting. This involves explaining the instrument audit results to key plant personnel, specifically noting gauge installations that pose immediate safety concerns. Together, the audit team and plant personnel determine which installations to investigate immediately and which to block out. The audit team and plant personnel then set schedules for the follow-up meeting.

Phase 5 - Follow-up Meeting and Presentation. The third-party team presents its general findings and the executive summary resulting from the instrument audit. The team also presents a spreadsheet that lists details about the gauges from the instrument review, color-coded by severity. The team also provides recommendations for improvements based on experience and best practices. Together, the audit team and plant personnel develop a strategy for moving forward.

The plant should also consider leveraging the audit team in other ways. For example, over time a plant's storeroom inventory can get out of hand. A qualified third-party audit team can help plant personnel rationalize, consolidate and manage their inventory of pressure gauges. For example, if a plant stocks spares for 10 or more ranges, the audit team may be able to help reduce that number to five or six ranges.

Essentially, when used properly, gauges play a significant role in plant safety, productivity and profitability. They act like silent alarms, indicating when something is seriously wrong or helping to diagnose deeper issues within systems and processes that can eventually lead to serious safety issues. To ensure safety and reliability at your plant, make certain the proper gauges are in place by reviewing your processes, gauges and your inventory. Everything is at stake.

REFERENCES

1. WIKA data aggregated from more than 250 instrument audits.

2. Peter Peek, Jocelyn Fenard, Patrick Gantes, Christine Theiler. December 2008, "Skills Shortage in the Global Oil and Gas Industry: How to close the gap" Part I. Pages 14-15.

(62a) Distillation of Extremely Hazardous Materials

Andrew (Jack) Eckles, National Thermodynamics Laboratory, Galveston, TX

Frequently, engineers are faced with design of systems that contain or process extremely hazardous materials. We will discuss and review both common and extraordinary precautions and design aspects to mitigate risks and address potential process hazards. Examples of designs and systems will be provided as well as methodologies to address processing of these materials. Hazards addressed will include extremes of operating conditions as well as materials with extreme toxicities, pyrophoric properties, energetic compounds, and even nuclear materials.

(62b) Plastic, Ceramic and Metal! Translating Efficiency with Confidence?

Samrat Mukherjee, DuPont Engineering Research & Technology, DuPont, Wilmington, DE, Jose Bravo, Shell Global Solutions (US) Inc, Houston, TX and Ruth Sands, Heat, Mass & Momentum Transfer Group, DuPont Engineering Technology, Wilmington, DE

Translating data from one packing material or type to another is a common challenge for mass transfer engineers. Techniques for accomplishing this will be reviewed, and the validity of these existing techniques will be examined using representative systems available in the literature.

(62c) Application of the New FRI Valve Tray Efficiency Model in the Simulation and Design of Distillation Columns

Ross Taylor, Clarkson University, Potsdam, NY and Attilio J. Praderio, LNG Technology Development, ConocoPhillips, Houston, TN

Models for the estimation of the efficiency of distillation trays have been part of the chemical engineers toolbox for many decades. Efficiency models fall into a number of distinct classes that include:

1. Empirical correlations of tray efficiency data
2. Semi-empirical that combine correlations for Numbers of Transfer Units with a mechanistic model of mass transfer.
3. Fundamental models based on a simplified picture of the hydrodynamics on distillation trays.

In this paper we are concerned with the latter class of model.

There are two quite different regimes of operation of distillation trays, the froth regime and the spray regime. Detailed models of the hydrodynamic and mass transfer character of these regimes have been developed by several investigators (Lockett and Plaka, 1983; Lockett, 1986; Kaltenbacher, 1982; Hofer, 1983; Prado and Fair, 1990, Syeda et al. 2007). In this class of model the froth regime on a distillation tray typically consists of three zones:

1. Zone I - Jetting-bubble formation region.
2. Zone II - Free bubbling zone.
3. Zone III - The splash zone.

In Zone I, the jetting-bubbling formation zone, the vapor issues through the perforations in the tray in the form of jets, breaking into bubbles. In the free bubbling zone a distribution of bubble sizes is assumed. Available experimental data show that the assumption of a bimodal bubble size distribution is a good approximation. Thus, we have fast-rising "large" bubbles and slowrising "small" bubbles in the dispersion and this picture has been incorporated in the class of model inspired by this picture. The splash zone above the free bubbling zone consists of entrained droplets. This zone – if it is considered at all – may be modeled as being made up of droplets of uniform size rising in plug flow through the splash zone.

Recently, a new member of this class of fundamental model has been developed by Fractionatation Reesearch Incorporated.

The FRI model has been implemented in the rate-based column simulation program ChemSep.

In this paper we use this new model to model distillation columns for which performance data is available. We show how a column performance model can be tuned to match operational data by fitting model parameters that have physical significance. Column profiles calculated with the FRI model are compared to profiles calculated with standard efficiency models such as that published by the AIChE.

13 AIChE®
Spring Meeting, San Antonio, TX

Use of Distillation Simulation and Process Analysis Techniques In Toll Manufacturing:

A Case Study

Jeffrey L. Silverstein, PhD
Cytec Industries, Inc.
5 Garret Mountain Plaza
Woodland Park, NJ 07424
jeff.silverstein@cytec.com

Prepared for Presentation at
American Institute of Chemical Engineers
2013 Spring Meeting
San Antonio, Texas
April 28 – May 2, 2013

UNPUBLISHED

Use of Distillation Simulation and Process Analysis Techniques In Toll Manufacturing:

A Case Study

Jeffrey L. Silverstein, PhD
Cytec Industries, Inc.
5 Garret Mountain Plaza
Woodland Park, NJ 07424
jeff.silverstein@cytec.com

Keywords: simulation, process analysis, distillation, process engineering, toll distillation

Abstract

Because of scheduling issues or lack of appropriate equipment, chemical companies often use the services of small toll manufacturers to produce products for customer trials or to meet unexpected product demand. In 2012, Cytec Industries required the distillation services of a toller to upgrade the quality of one of its raw materials containing two closely boiling isomers. Since the process is carried out in a batch reactor, the mass of the batch could then be increased and the batch reaction cycle time reduced to meet an unexpected demand for the final product.

This paper is a case study in the use of distillation simulation and process analysis techniques to guide the selection process for an appropriate toller, develop the operating parameters for the distillation campaign, and provide day-to-day direction for the campaign to overcome unexpected timeline challenges caused by a delay in start-up, lower than anticipated column efficiency, a power outage, a pump seal failure, and underperformance of a rented cooling water chiller.

Introduction

Because of scheduling issues or lack of appropriate equipment, chemical companies often use the services of small toll manufacturers to produce products for customer trials or to meet unexpected product demand. In 2012, Cytec Industries required the distillation services of a toller to upgrade the quality of one of its raw materials containing two closely boiling isomers. Since the process is carried out in a batch reactor, the mass of the batch could then be increased, the initial concentration of the reactant would be higher, and the batch reaction cycle time reduced to meet an unexpected demand for the final product. Cytec requires that the raw material not be identified. Henceforth, its principle components are designated as ISOM-1and ISOM-2.

The reactant in the process is ISOM-1 which is purchased as a crude raw material nominally containing 78% ISOM-1 and 22% ISOM-2. The cost of the crude raw material was about

\$0.79/lb. The highest quality ISOM-1 commercially available was 95%/5% ISOM-1/ISOM-2 at a cost of about \$2.57/lb. Purchasing high quality ISOM-1 was not economically justifiable.

Consequently, a study was undertaken to technically and economically evaluate ways that the crude 78% ISOM-1 might be upgraded by Cytec to at least 97% ISOM-1. Three methods were evaluated:

1. In-house isomerization and distillation process
2. In-house semi-continuous distillation
3. Continuous distillation using an outside toll distiller

Isomerization & Distillation Process Concept

A process concept was developed by computer simulation for a semi-batch acid catalyzed isomerization reaction and distillation process.

The reaction is the equilibrium between the isomers:

$$\text{ISOM-1} \rightleftarrows \text{ISOM-2}$$

Literature data are available that permit the reaction rate constants to be calculated for a particular catalyst. The distillation removes ISOM-1 from the reactor which shifts the reaction towards more ISOM-1 and depletes the ISOM-2 in the system. The reactor is continuously fed with the crude isomer mixture at the same rate that the product is taken overhead in the distillation column until the required amount of high quality ISOM-1 is accumulated. The reactor contents remain at or near the equilibrium mixture, 78%/22% ISOM-1/ISOM-2 throughout the distillation. The reactor residue at this concentration is retained for charging the reactor in the next campaign.

The simulations provided the basis for lab reaction experiments to evaluate several likely catalysts. The lab experiments were short duration, lasting about 4 hours. The data obtained were regressed to obtain kinetic rate constants for the isomerization reaction and for a dimerization reaction that was observed to occur in Cytec's experiments, but was not described in the literature. The experimental study focused on finding a catalyst and reaction conditions that reduce drastically or eliminate the side reaction. Unfortunately none were found. Computer simulations of the plant-scale equipment for a 14 day campaign using the rate constants for the best catalyst studied showed that the side reaction produced so much dimer that the process became too inefficient to consider implementation. The efficiency of the distillation deteriorated rapidly as the dimer built up in the reactor.

In-House Semi-Continuous Distillation

The plant equipment evaluated for the isomerization and distillation process was re-evaluated for a semi-continuous distillation without reaction. The goal was to determine the minimum fixed reflux ratio that would produce the maximum average rate of 97 - 98% ISOM-1 with the lowest batch cycle time.

Making a theoretical calculation for a column with the same diameter and column packing as in the plant and an infinite number of theoretical stages it was found that the minimum reflux ratio is 6.17 and that a 6 day distillation cycle would produce a theoretical maximum of 27,450 lbs. of 98% ISOM-1. Assuming 24 hours for equipment start-up and shutdown for each batch, the average hourly production rate is 163.4 pounds. The product requirement in 2013 was projected to be 239,207 pounds at 98% purity (234,423 pounds real ISOM-1). Thus the number of operating days would be about 61, with the best theoretical production capability. It was clear that the equipment would not be available for the required time. Evaluation of other distillation equipment in Cytec revealed similar occupancy challenges. So attention shifted to distillation by an outside toll manufacturer.

VLE Relationship for ISOM-1 – ISOM-2

The normal boiling point for ISOM-1 is 101.53 C and for ISOM-2 it is 104.89 C. No VLE data for this system were found in the literature, so UNIFAC [1] was used to predict it. The predicted X-Y diagram at 760 mm Hg is given in Figure 1. The relative volatility is quite low. The T-X-Y and liquid phase activity coefficient curves at 760 mm Hg are given in Figures 2 and 3, respectively. The system is very nearly ideal for pressures $\leq$ 760 mm Hg.

The system relative volatility is very sensitive to operating pressure. Figure 4 shows the average relative volatility as a function of system pressure. The average relative volatility is determined by the average of the relative volatilities calculated at 19 points on the x-axis, from 0.05 to 0.95 mole fraction ISOM-1, of the X-Y diagram at a given system pressure. The curve shows that the distillation is less difficult with decreasing pressure. Vapor pressure for each component and relative volatility as a function of temperature are given in Figure 5.

Selection of Toll Distiller

The objective of the project was to recover 96 – 98% ISOM-1 as the overhead product with 20% or less ISOM-1 in the bottoms stream. In addition to the column efficiency requirement implicit in this material balance requirement, the column had to have sufficient throughput capacity so that the required 234,423 pounds of real ISOM-1 could be produced in a reasonable time to keep the toll cost low enough to make the economics of the project attractive.

The minimum number of theoretical stages to accomplish the separation was calculated using the Fenske equation [2]. Results are provided in Figure 6. This simple curve was used to screen the efficiency of distillation columns of candidate tollers and to determine the required operating pressure. All of the columns that were possible candidates were packed and had 25 – 40 theoretical stages. Thus, the operating pressure would have to be 30 – 200 mm Hg. Only 2 tollers were found that met both the column efficiency and column throughput requirements. Only one of these columns was available for toll manufacturing in the time period that was compatible with the production requirements of Cytec's plant.

Description of Column

The continuous column is 42" ID and contains 6 packed beds. Each bed has a feed nozzle with an associated liquid distributor. Details are provided in Table 1. The top two beds contain Sulzer 352Y structured packing, 12 feet and 10.5 feet high, respectively. The remaining four beds contain Norton Intalox 2T structured packing and are each 8.25 feet tall. The specific surface areas are obtained from vendor bulletins [3, 4]. The packing factors are obtained from the databank compiled by Kister, et.al. for Norton Intalox 2T [5] and for Sulzer MellapakPlus 252Y, 452Y and 752Y [6] . The packing factor for 352Y, which is in the column, was obtained by fitting the plot of packing factor and specific surface area with a second order polynomial for the three sizes provide by Kister [6]. The resulting polynomial is:

$$F_p = 0.0000366675\, a_{pm}^2 + 0.0191864\, a_{pm} + 4.836547 \qquad (1)$$

where: F_p is the packing factor and a_{pm} is the specific surface area of the packing in m^2 / m^3. The PF for Sulzer 352Y is estimated at 16.13.

Kister's conservative rule of thumb [5] for HETP for structured packing was used to estimate the number of theoretical stages:

$$HETP = \frac{1200}{a_{pf}} + 4 \qquad (2)$$

where: $HETP$ is in inches and a_{pf} is the specific surface area of the packing in ft^2 / ft^3. The calculated values are provided in Table 1.

Since Chemcad [7], the simulation software used for this work, only permits a whole number of theoretical stages, a Murphee stage efficiency was applied to each bed to meet this condition. The HETP and the stage efficiency are 2 adjustable parameters that can be used to calibrate the mathematical model of the column so that the simulation matches the column performance. The column is estimated to have 34.55 theoretical stages using Kister's equation and 39 when efficiency values are applied to obtain whole numbers. So from Figure 6, to obtain a 98% product, the column must operate below 105 mm Hg, for a 97% product, below 140 mm Hg, and for a 96% product, below 170 mm Hg.

Establishing Initial Operating Conditions for the Campaign

Simulations of the column performance were made to establish operating conditions for the campaign, which was scheduled to begin at 8:00 on Monday June 4, 2012. The toller established a nonnegotiable end date and time of 23:59 on Saturday, June 16. So the goal was to establish operating conditions that would produce the required amount of real ISOM-1 in a maximum of 304 operating hours. The amount of real ISOM-1 was established by the product manager at 234,423 pounds, with a target quality of 97%. The average 97% ISOM-1 product rate over the course of the campaign needed to be 795 lb/hr or greater.

For this system, the relative volatility between ISOM-1 and ISOM-2 increases with decreasing pressure, making the separation easier and more energy efficient. However, at low operating pressure the overhead condensing system must be efficient and significantly subcool the overhead condensate to minimize product loss to the vacuum system due to air leakage. In addition, at lower operating pressures, although offset by higher relative volatility, the column feed rate may be unacceptably constrained due to hydraulic limitations of the packing resulting from high volumetric vapor rates. A simulation study was made to evaluate the relationship of column operating pressure with all of these factors.

Effect of Operating Pressure

Table 2 summarizes the results of this study. Operating pressures over the range 30 – 150 mm Hg were evaluated. The initial case evaluated was at 30 mm Hg. The constraints on the simulations were: feed rate = 1250 lb/hr, reflux ratio = 9.75, 16% ISOM-1 in the bottoms, condenser pressure drop = 0.10 psia, process-side condenser outlet temperature = 7 ^{O}C, and air leakage = 5 lb/hr.

The table shows that as the pressure increases, the quality of the overhead product decreases, but that the actual yield of ISOM-1 in the overhead product increases. The latter is due principally to the decrease of ISOM-1 loss in the vacuum system due to air leakage as the system pressure increases. So, although the ability of the column to achieve a high quality separation is improved with reduced pressure, product yield can be significantly impacted by even low rates of air leakage if the overhead condensing system is incapable of sufficiently subcooling the overhead column vapor stream and the downstream vacuum and air pollution control equipment are not capable of recovering the organic components. Both of these competing effects of pressure need to be taken into account when choosing operating conditions for the toller's column.

Overhead Condensing System

The choice of 7 ^{O}C as the condensate temperature in the cases presented in Table 2 is conservative. The 2000 ton chiller that was rented by the toller was rated for 2.0 MBtu/hr, with a 0 ^{O}C coolant supply temperature at 500 gpm flow rate to the condenser. If this rating were accurate, operation at 30 mm Hg would have been feasible, because the loss due to air leakage would have been acceptably small.

The Primary Condenser surface area is 1400 ft^2. Simulating the condenser performance with Aspen Exchanger Design & Rating software [8] and Chemcad [7] confirmed that this is more than sufficient HTA under these process conditions to cause pinching and achieve a process-side outlet temperature about 0.2 C^{O} higher than the coolant inlet temperature, well under the target process-side outlet temperature of 7 ^{O}C.

The results of this operating pressure study were reviewed with the toller. He expressed concern about the capability of their air pollution control devices to handle any upsets that might result in higher than anticipated air leakage rates and condenser outlet temperature. Consequently, Cytec recommended operation at 75 mm Hg as a reasonable compromise between achievable product quality and losses to the vacuum system. From Cytec's point of view, the slight reduction in product quality would be offset by the reduction in product loss to the vacuum system, the resulting increase of real ISOM-1 recovery in the overhead product and the confidence the toller manifested in the decision to operate at the higher pressure. However, his request to operate still higher at 150 mm Hg was parried by the results given in Table 2 and the process simulations that showed an unacceptable reduction in product quality below the target of 97% ISOM-1.

Initial Operating Conditions Selected For Start of Campaign

It was decided to initially operate the column at 75 mm Hg, with a feed rate of 1250 lb/hr, a reboiler duty of 1.70 MBtu/hr, and a reflux ratio of 10.0. This corresponds to the case given in the last line of Table 2. Increasing the ISOM-1 content of the bottoms stream to 20% permits the overhead quality to exceed the minimum target of 97%.

Campaign Begins

The toller was scheduled to have the chiller installed and fully operational by 8:00 on Monday, June 4, 2012 and the column started up on total recycle with the agreed upon initial operating conditions.

As might be expected, the chiller was still being installed on Monday morning. The chiller was finally fully installed and operational sometime after I left the plant at 17:30 on Monday June 4. Overnight, the column was started up under total recycle with the feed rate of 1200 lb/hr and the desired column top pressure of 75 mm Hg. The overhead product flow rate was set at 900 lb/hr and the hot oil flow to the reboiler was adjusted until the product quality was measured at 97.08%. The feed rate was 50 lb/hr lower and the product rate was 36 lb/hr lower than the desired rates. This feed rate was agreed to at the request of the toller for initial start-up.

It became apparent immediately, that the target coolant temperature into the condenser of 32 ^{O}F could not be achieved. The temperature was measured at 44.6 ^{O}F. The condenser performed as expected, however, achieving a process-side outlet temperature of 45 ^{O}F. The loss of ISOM-1 to the vacuum system was larger than expected as a result. Analyses of the feed, the overhead product and the bottoms stream were made every 30 minutes to check for steady state. The decision was made at 11:00 on Tuesday, June 5, to take the column off of total recycle and begin accumulation of overhead and bottoms product. Thus 27 hours of production time that were expected to be available were lost. The column material balance is given in Table 3. As shown in Table 4, at the current operation, the target product mass would be accumulated in 268.3 hours, with only 8.7 hours safety margin until the absolute end of the campaign.

Two tasks required immediate attention. The first task was to calibrate the column by simulation of its current operation to determine the actual number of theoretical stages. The next task was to use this new model of the column to establish a higher feed rate and overhead product rate to reduce the time to meet the production objectives and increase the safety margin to a comfortable level for the time to complete the campaign.

Column Performance Analysis

A simulation analysis of the column, which included the effect of the underperforming chiller, was begun while the column was operating in total recycle. This first simulation used the assumed efficiency of the column based on Kister's correlation and an air leakage rate of 5 lb/hr. Results are shown in Table 3. The expected product quality is 97.84% compared to an analyzed 97.08%. The expected bottoms ISOM-1 content is 18.05% compared to an analyzed 19.92%. The column efficiency was clearly lower than expected.

The HETP was reduced by 2 inches across the entire column in the next set of simulations. This resulted in 33 theoretical stages with efficiencies applied as described previously. It was also apparent that the ISOM-1 loss to the vacuum system due to air leakage was higher than the expected 5 lb/hr air equivalent. The air leakage was adjusted upward to 6.5 lb/hr. These adjustments resulted in a simulation that closely matched the actual column performance as shown in Table 3.

Since the actual air leakage rate could not be measured, this became an adjustable parameter in the simulation to permit the column performance to be matched. The vendor conducted leak tests prior to the start of the campaign and declared that the "system was tight." A post-campaign analysis by the vendor showed that there was measurable air leakage into the system at the feed nozzle.

Feed Location Change

On Tuesday morning at 9:00, after about several hours of steady satisfactory column operation in total recycle, the VP of Operations and the Technical Manager for the toller requested a meeting to discuss moving the feed location down one packed bed from the current bed #3 to bed #4. Based on their experience with other systems and different operating conditions, they believed that product quality would improve with no other change in operating conditions. They believed that the reflux ratio could ultimately be reduced and the feed rate could be increased without affecting the current product quality.

I pointed out that I had made a sensitivity analysis for feed location and that the current feed location to the top of bed #3 is optimum. Despite the results of the simulation analysis, which it was now evident they did not trust, and to feel like they were in control of their own plant operation, they firmly reiterated their desire to move the feed to bed #4.

I insisted on simulating this new case to determine the result, which is given in Table 3. Moving the feed would result in a modest, but measurable reduction in product quality. However, it was still acceptable for Cytec's purposes. I showed them the results. They were skeptical and still believed that the product quality would improve. So, I authorized the change to take place at noon, one hour after the column was taken off total recycle. I insisted that the overhead and bottoms streams be sampled and analyzed hourly. I decided to authorize the change to demonstrate the validity of the simulation model and to eliminate any skepticism about it during the remainder of the campaign if an unexpected event arose that required a change in operating conditions to keep the campaign on track for product quality and quantity. The feed was moved down one packed section to the top of bed #4 on June 5 at 12:00.

The graph in Figure 7 shows how the overhead product quality varied with time after the feed location was changed from bed #3 to bed #4. After 18 hours, when it became clear that the product quality was declining, the VP of Operations and the Technical Manager admitted that the simulation was accurate and they agreed to move the feed back to bed #3. This move was made at 6:00 on Wednesday, June 6, 18 hours after lowering the feed location.

An examination of the overhead product quality shows an initial increase during the first 2 hours, followed by a steady decrease for the remaining 16 hours. This initial "wrong way effect" occurs because the number of enriching stages increases. As a result, the overhead product quality initially increases, but because the number of stages below the feed has decreased, after a while the effect of reduced stripping in the column catches up and the net effect is a drop in product quality as the new steady state is established.

Increase Feed Rate to 1260 lb/hr

The status of the campaign and results of the simulation were discussed in a conference call with the Cytec project manager, the business manager and the Cytec plant technical director on Wednesday morning, June 6. I requested that the minimum acceptable ISOM-1content in the product be established. Then, based on a simulation study, I recommended modifications in the column operating conditions to achieve at least this acceptable product quality at a production rate that would meet our accumulated product goal well before the midnight June 16 deadline for terminating the campaign. New operating conditions were agreed to and involved the following:

a. Increasing the feed rate to 1260 lb/hr to provide more safety margin for achieving the target product production.
b. Reducing the target product quality from 97% ISOM-1 minimum to 96% ISOM minimum.

The simulation results of these new operating conditions are given in Table 3, corresponding to the line labeled 6/6 13:00. As shown in Table 4, as a result of the increase in feed rate, the time to meet the product production objective was reduced by 10.5 hours, providing a safety margin increase from 8.7 to 19.2 hours. The additional time available could be used to process any excess feedstock that was available, since more than the minimum requirement was ordered.

After the feed rate was increased the column reached steady state with the material balance given in Table 3. The actual analyzed overhead product quality was 96.43% ISOM-1 in the overhead and 18.13% in the bottoms. The simulation predicted 96.94% and 20.63%, respectively. When the model was adjusted to increase the air leakage from 6.5 lb/hr to 14.5 lb/hr, these results changed to 96.49% and 18.09%, respectively. It is not unreasonable to expect that the air leakage could have increased during the course of the campaign.

As shown in Table 4, using plant measurements, the time to meet the product production objective was reduced by 9.24 hours, providing a safety margin increase from 8.7 to 17.94 hours.

Power Outage and Pump Seal Failure

On Thursday, June 7, as a result of a severe thunderstorm, a power outage occurred at 23:30. The column was down for 4.6 hours before it came back on stream. The operating conditions were unchanged. At 10:25 on Friday, June 8, the column was shut down again due to a pump seal failure. The column was down for 4.9 hours before it came back on stream. So production was lost for a total of 9.5 hours. The operating conditions remained unchanged when the column resumed production at 15:20 on Friday, June 8.

Increase Feed Rate to 1320 lb/hr

Calculations were made to determine if the current feed rate was adequate to meet the product demand in the remaining available operating hours. Results are given in Table 4. At this point 61,852 lb of 96.68% product had been collected. The mass of real ISOM-1 was 59,796 lb. With 234,423 lb required, 174,627 lb remained to be collected. There was no change in the measured column material balance after it came back on stream. It was being fed at 1260 lb/hr and producing 942 lb/hr of 96.43% ISOM-1 overhead product. At this rate it would require 192.24 operating hours to achieve the production goal. The remaining available hours in the campaign were 200.68. Given the distinct possibility of other unforeseen events occurring that would shut the column down, it was decided to increase the feed rate to reduce the required operating time to achieve the production goal. In addition, it was desirable to have enough available extra time to process the excess feedstock purchased beyond that required to meet the production objective.

The constraint on the column operation from the beginning of the campaign was the performance of the chiller. Given the chiller limitations, it was determined that operating the reboiler at a maximum of about 1.95 – 2.0 MBtu/hr allowed the overhead condensing system to operate in a stable manner with minimal loss of product to the vacuum system.

With these operating constraints, any proportional increase in feed and product rate with fixed reboiler duty would necessitate a reduction in the reflux ratio and a drop-off in overhead product quality. If the resulting product quality is too low, the only remedy available is to reduce the overhead product flow to cause an increase in reflux ratio and accept additional ISOM-1 loss in the column bottoms.

A series of column simulations were made to arrive at operating conditions with increased feed rate that would meet the dual objectives of meeting the production goal and maintaining acceptable product quality at all times.

It was decided to set the feed rate to 1320 lb/hr and the overhead product rate to 990 lb/hr. As shown in Table 3, the simulation predicts that this would result in an overhead product quality of about 96.41% ISOM-1 and a bottoms product containing 18.59% ISOM-1. At 13:00 on June 9, the operating parameters were changed accordingly. Under these conditions, as shown in Table 4, it would require 162.33 hours to achieve the target production with 179 available hours remaining in the campaign, providing a safety margin of 16.67 hours.

Overnight Shift Operators Make "A Small Change"

At the resumption of column operations after the pump seal was replaced, a running plot was begun of overhead product quality with time to monitor performance until the end of the campaign. The graph is provided as Figure 8, and time = 0, corresponds to 7:40 on Saturday, June 9.

At this time, the overhead product quality was 96.75% ISOM-1, slightly higher than plant analyses and simulation predictions that were developed prior to the shutdowns. As shown in the graph, the measured product quality continued to rise through the morning, reaching a value of 97.23% at 12:05. From this trend, it was deduced that a reduction in overhead product rate was made the previous night. A check of the flow meter by the plant operations manager confirmed that the product rate was indeed reduced 12 lb/hr by the operators to 930 lb/hr. As often occurs with night shift operators everywhere, they believed it would be desirable to increase the product quality and simply made this "small" change without informing the operations manager. Rather than increase the product rate immediately, it was decided to maintain these operating conditions for less than an hour, until the increase in feed rate to 1320 lb/hr and product rate to 990 lb/hr was made at 13:00. The expected material balance, developed by simulation is given in Table 3.

Watching Product Quality Drop Off After Feed Rate Increase

It was expected that a small drop off in product quality would occur as a result of the feed rate increase. As the plot in Figure 8 shows, the quality appears to level off at about 96.1% ISOM-1, a little low, but still acceptable. The decision was made at this point to wait to see if steady state had been reached and then to reduce the overhead draw rate by 6 lb/hr to bump the product quality up a bit. The further deterioration in overhead quality triggered the decrease in product rate to be made, but by 12 lb/hr to a flow rate of 978 lb/hr. This change caused the product quality to improve and stabilize at an average quality of about 96.37% ISOM-1.

The sensitivity of the product quality to small changes in product rate is depicted in the plot. Starting at 22:57 on Wednesday, June 13, (t =107 hr in Figure 8), there is an increase in product quality that peaks at about 96.58%. The overhead product rate dropped from 978 lb/hr at 22:57 to 936 lb/hr at 1:00 on June 14. It increased to 948 lb/hr at 3:13, and was back to 978 lb/hr by 5:00. When the rate was returned to the desired 978 lb/hr the quality dropped back into the previous range. According to the toller's technical manager, this product rate excursion was caused by an increase in the product tank pressure from 1 psig to 15 psig during a transfer of product to a tank truck. Either the flow controller in the product line from the column or the product pump was not capable of compensating for this increase in downstream pressure.

Product Volume Objective Met

At approximately 8:15 on Saturday, June 16 (at 165.5 hours in Figure 8), the required target accumulation of 234,423 pounds real ISOM-1 was reached. The actual safety margin was 15.75 hours, consistent with the estimate in Table 4 of 15.14 hours.

Process "Off-Spec Material" Accumulated During Start-Up

The column continued to operate for about 2 hours past this point at the same operating conditions until approximately 5000 pounds of unprocessed feedstock remained and was blended with 10,000 pounds of "off-spec material" that had been accumulated during start-up, principally from column rinses.

The "off-spec material" contained 81.36% ISOM-1 compared to 77.63% ISOM-1 measured in the crude feed to the column. Combining these in the feed tank to the column produced a new feed containing 80.12% ISOM-1. Simulations were made to establish new operating conditions. Holding the feed rate constant at 1320 lb/hr and the reboiler duty at 1.955 MBtu/hr, at 10:30 the overhead product rate was increased to 1014 lb/hr to maintain the product quality at a predicted 96.49% ISOM-1. The simulation results of these conditions are given in Table 3. The reflux ratio dropped from 11.355 to 11.143. This material was processed to near completion and the campaign ended at 21:40 on Saturday, June 16, 2 hours and 20 minutes earlier than the mandatory completion time. The graph in Figure 8 shows that the average product quality during this combined feed phase of the campaign was about 96.64%.

Material Balance Summary

The quality of the accumulated ISOM-1 overhead product was 96.48%. To meet the target of 234,423 pounds of real ISOM-1, 242,976 pounds of product was accumulated. The campaign actually produced 255,728 pounds as measured by the flow totalizer in the line to the storage tank. The bottoms accumulation was 81,577 pounds containing 21.08% ISOM-1, 76.06% ISOM-2 and the remainder were higher boilers. A customer was found for the bottoms product who was willing to take it free of charge and with no cost to Cytec. The feed total was 339,629 pounds containing 77.57% ISOM-1, 21.78& ISOM-2, 0.65% heavies and about 100 ppm water. These results do not include losses in the vacuum system.

Conclusions

The use of computer simulations and process analysis tools to guide the distillation campaign were important reasons that the production and product quality targets were met. Continuous monitoring of the mass and quality of accumulated product provided signals that the column operating conditions needed to be changed to ensure that production rates would meet the campaign objectives in the fixed amount of time available for the campaign.

These process engineering techniques were used to successfully provide initial and mid-course corrections that overcame the following unforeseen issues that contributed to reducing the effective number of operating hours of the campaign:

- Delay in commencement of campaign
- Lower than expected column efficiency
- Underperformance of the rented chiller
- A power outage

- A pump seal failure
- Operator meddling with product rates
- Flow controller malfunction

The accumulation of the target quantity of purified ISOM-1 was completed sufficiently in advance of the fixed termination date of the campaign to permit all of the additional purchased feedstock to be successfully processed through the column to produce additional product grade ISOM-1.

References

[1] Fredenslund, A., Jones, R. L., Prausnitz, J. M., "Group Contribution Estimation of Activity coefficients in Nonideal Liquid Mixtures", AIChE Journal, Volume 21, No. 6, 1086, 1975

[2] Aerstein, F. and Street, G., *Applied Chemical Process Design*, Plenum Press, New York, 1978

[3] Koch-Glitsch, "Intalox High Capacity Structured Packing", Bulletin KGSP-1, Wichita, KS, 2003

[4] Sulzer Chemtech, "Structured Packings for Distillation, Absorption and Reactive Distillation", Bulletin 22.13.06.40-V.08 – 50, Winterhur, Switzerland

[5] Kister, H. Z., *Distillation Design*, McGraw-Hill, New York, 1992

[6] Kister, H. Z., Scherffius, J., Afshar, K., and Abkar, E., "GPDC Interpolation: Use, Misuse and Updates", presented at the AIChE Spring National Meeting, Houston, Texas, April, 2007

[7] Chemcad, version 6.5.1.5866, Chemstations, Inc., Houston, Texas

[8] Aspen Exchanger Design & Rating software, version 7.3.1, Aspen Technology, Inc., Burlington, MA

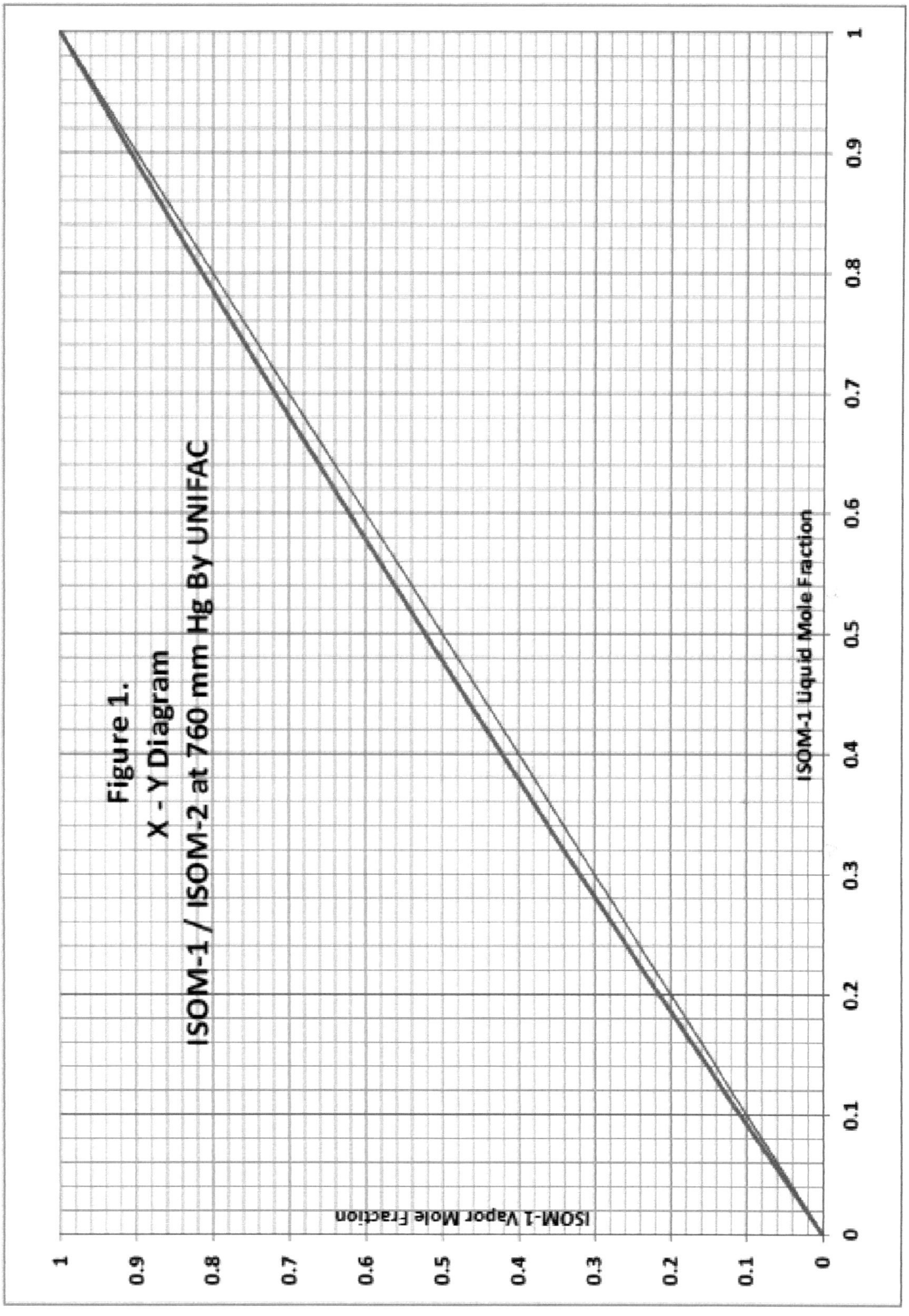

Figure 1.
X - Y Diagram
ISOM-1 / ISOM-2 at 760 mm Hg By UNIFAC
ISOM-1 Liquid Mole Fraction
ISOM-1 Vapor Mole Fraction
0
0.1
0.2
0.3
0.4
0.5
0.6
0.7
0.8
0.9
1

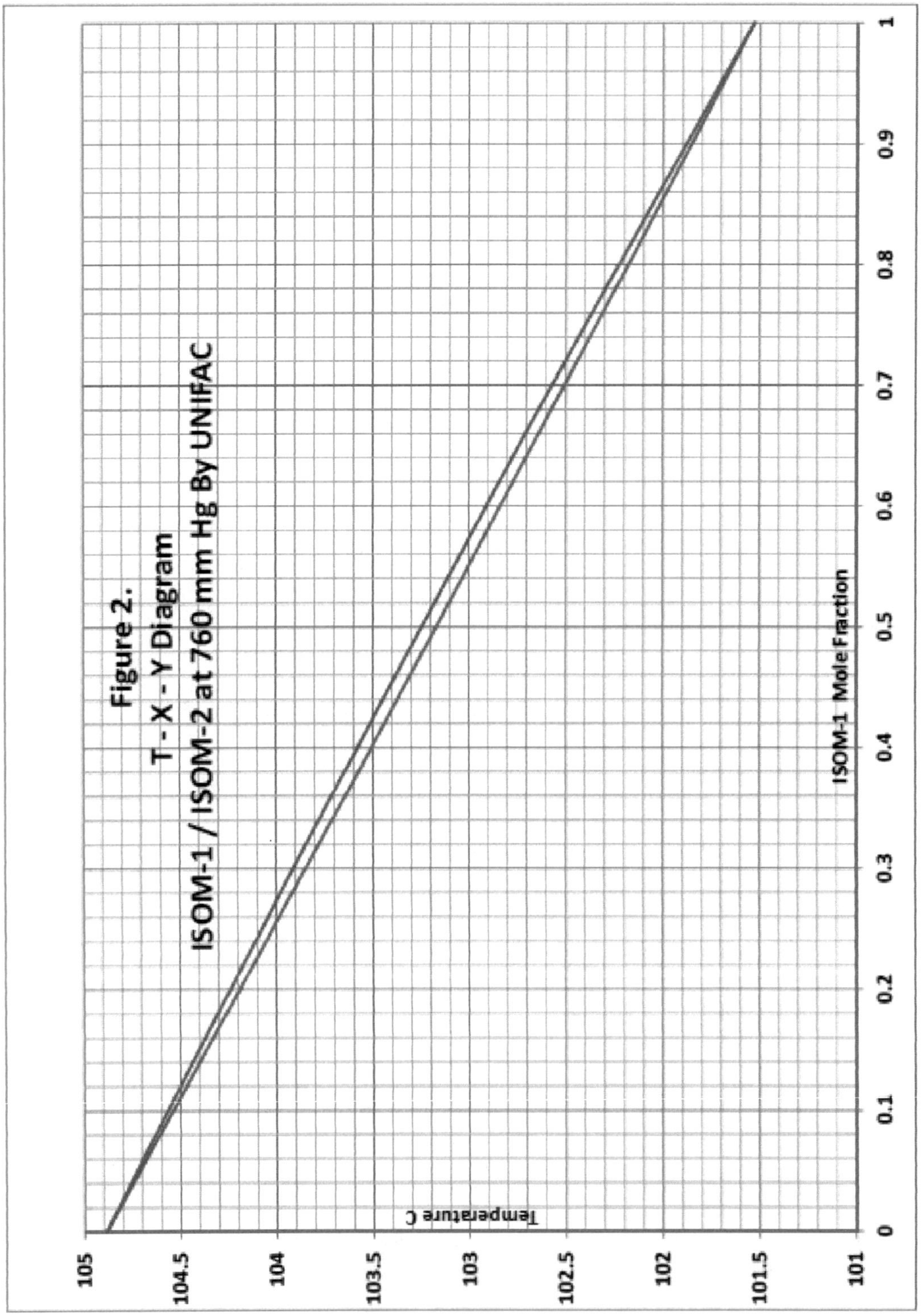
Figure 2.
T - X - Y Diagram
ISOM-1 / ISOM-2 at 760 mm Hg By UNIFAC
Temperature C
ISOM-1 Mole Fraction
105
104.5
104
103.5
103
102.5
102
101.5
101
0
0.1
0.2
0.3
0.4
0.5
0.6
0.7
0.8
0.9
1

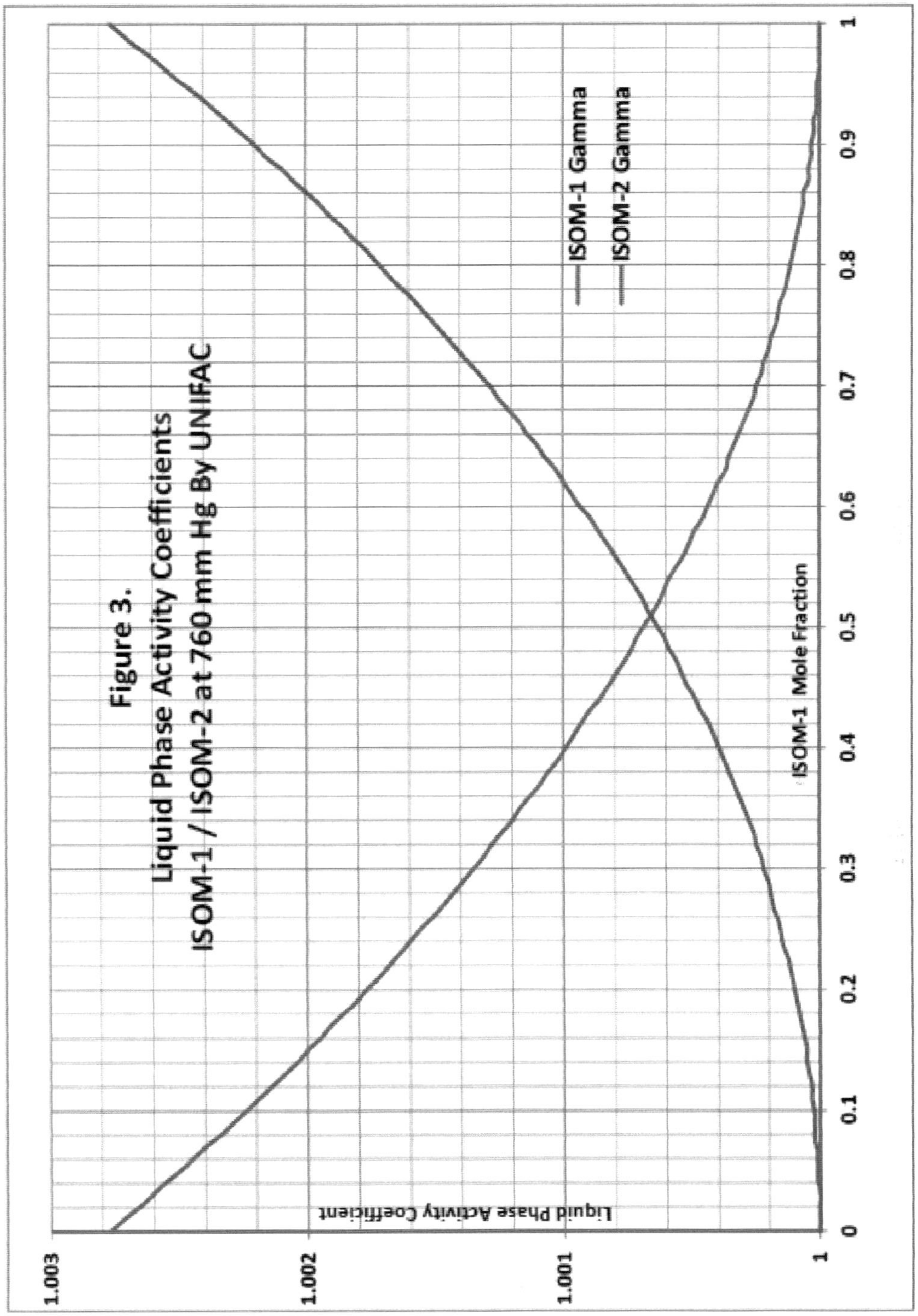
Figure 3.
Liquid Phase Activity Coefficients
ISOM-1 / ISOM-2 at 760 mm Hg By UNIFAC
ISOM-1 Gamma
ISOM-2 Gamma
Liquid Phase Activity Coefficient
ISOM-1 Mole Fraction
1.003
1.002
1.001
1
0
0.1
0.2
0.3
0.4
0.5
0.6
0.7
0.8
0.9
1

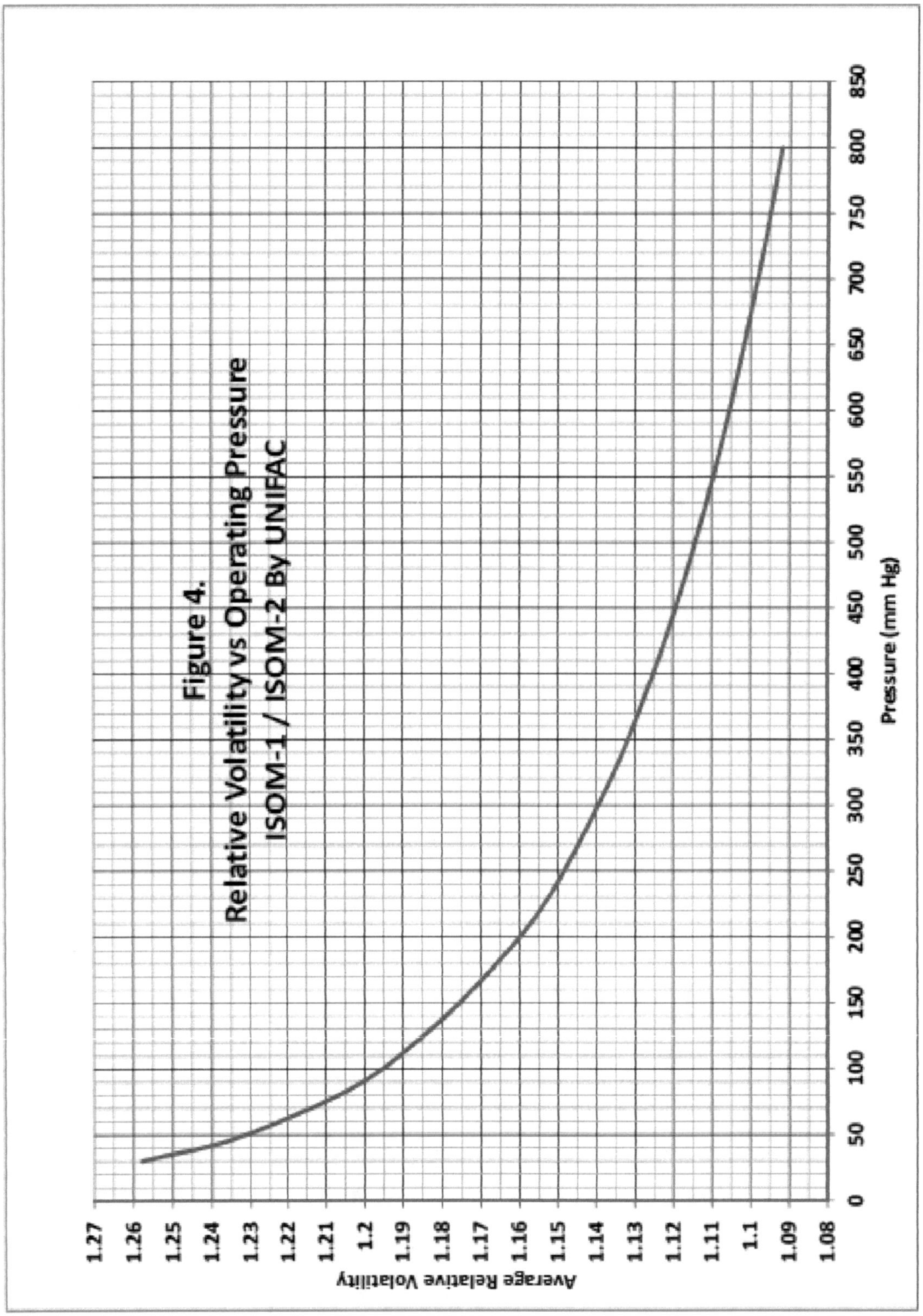
Figure 4.
Relative Volatility vs Operating Pressure
ISOM-1 / ISOM-2 By UNIFAC
Pressure (mm Hg)
Average Relative Volatility
0 50 100 150 200 250 300 350 400 450 500 550 600 650 700 750 800 850
1.27 1.26 1.25 1.24 1.23 1.22 1.21 1.2 1.19 1.18 1.17 1.16 1.15 1.14 1.13 1.12 1.11 1.1 1.09 1.08

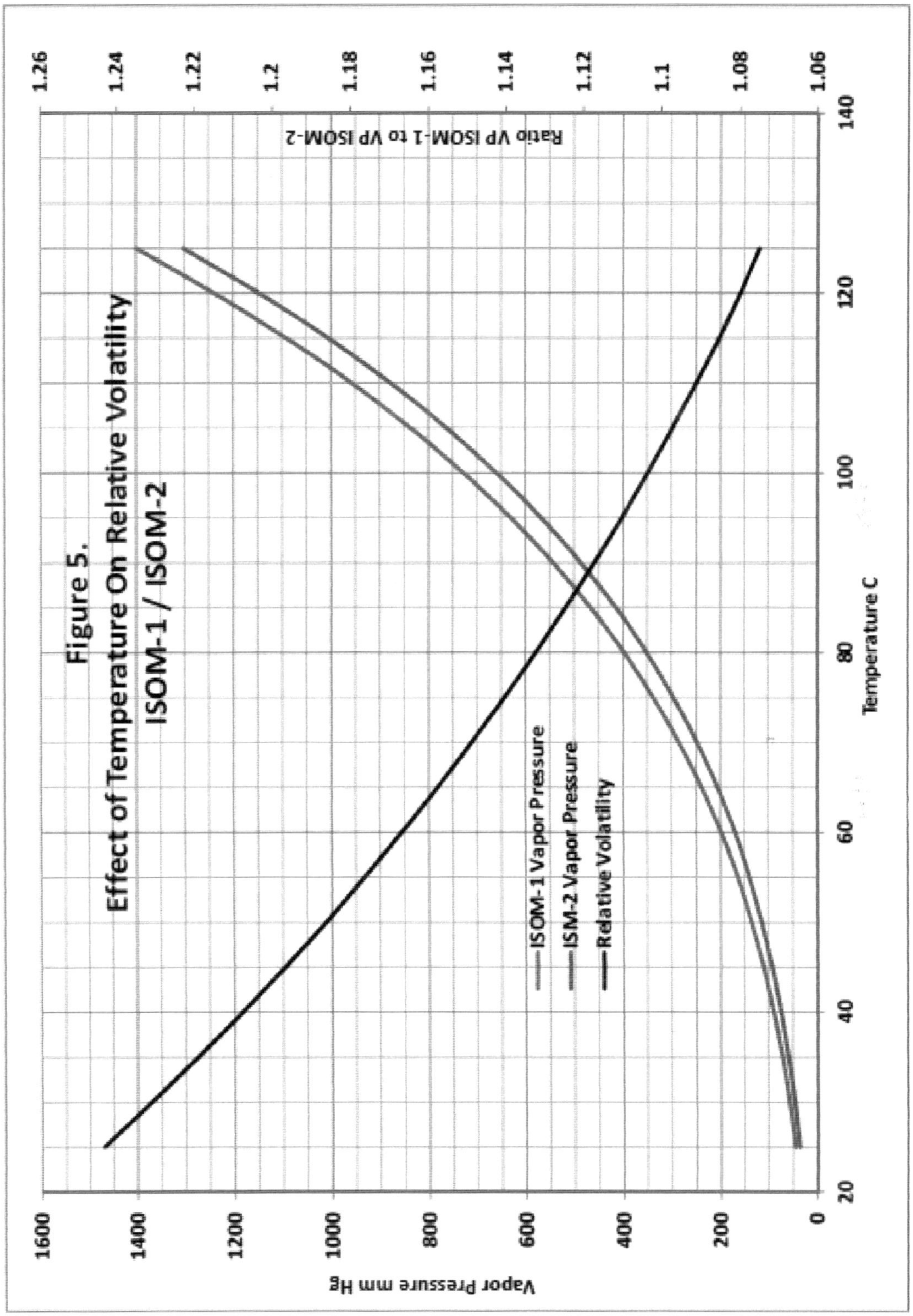
Figure 5.
Effect of Temperature On Relative Volatility
ISOM-1 / ISOM-2
ISOM-1 Vapor Pressure
ISM-2 Vapor Pressure
Relative Volatility
Vapor Pressure mm Hg
Ratio VP ISOM-1 to VP ISOM-2
Temperature C
0
200
400
600
800
1000
1200
1400
1600
20
40
60
80
100
120
140
1.06
1.08
1.1
1.12
1.14
1.16
1.18
1.2
1.22
1.24
1.26

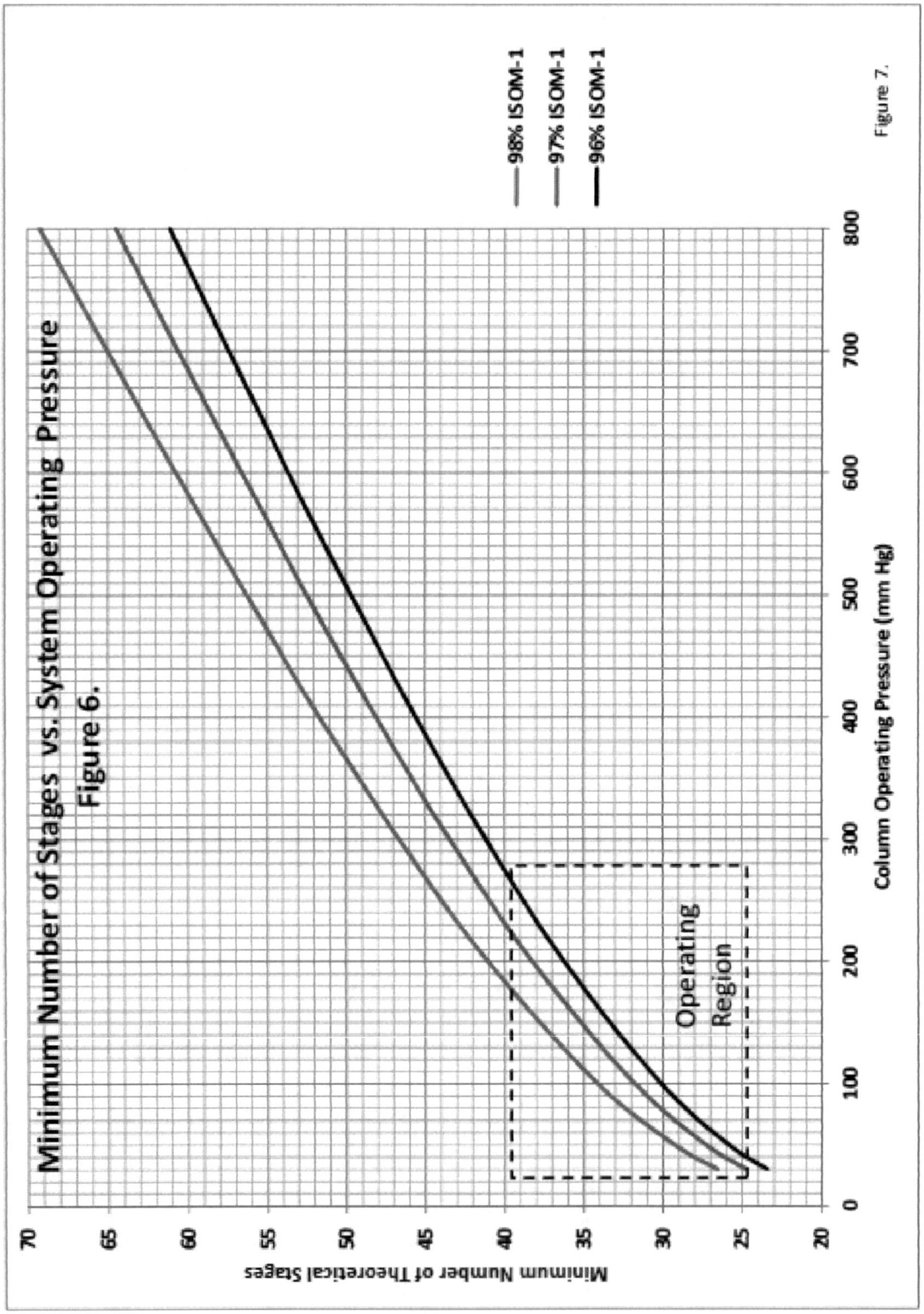
Minimum Number of Stages vs. System Operating Pressure
Figure 6.
Minimum Number of Theoretical Stages
Column Operating Pressure (mm Hg)
98% ISOM-1
97% ISOM-1
96% ISOM-1
Operating Region
70
65
60
55
50
45
40
35
30
25
20
0
100
200
300
400
500
600
700
800
Figure 7.

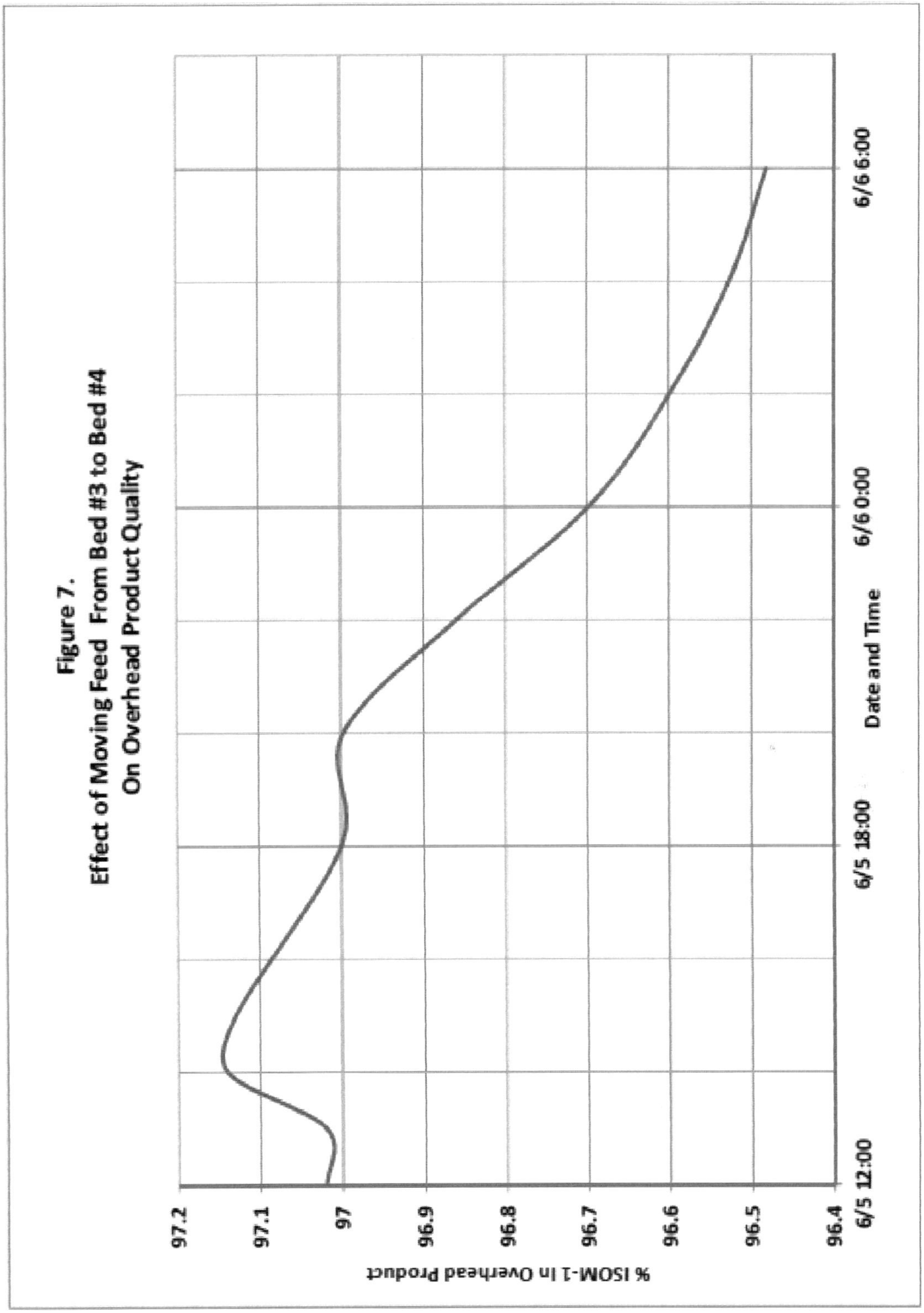

Figure 7.
Effect of Moving Feed From Bed #3 to Bed #4
On Overhead Product Quality

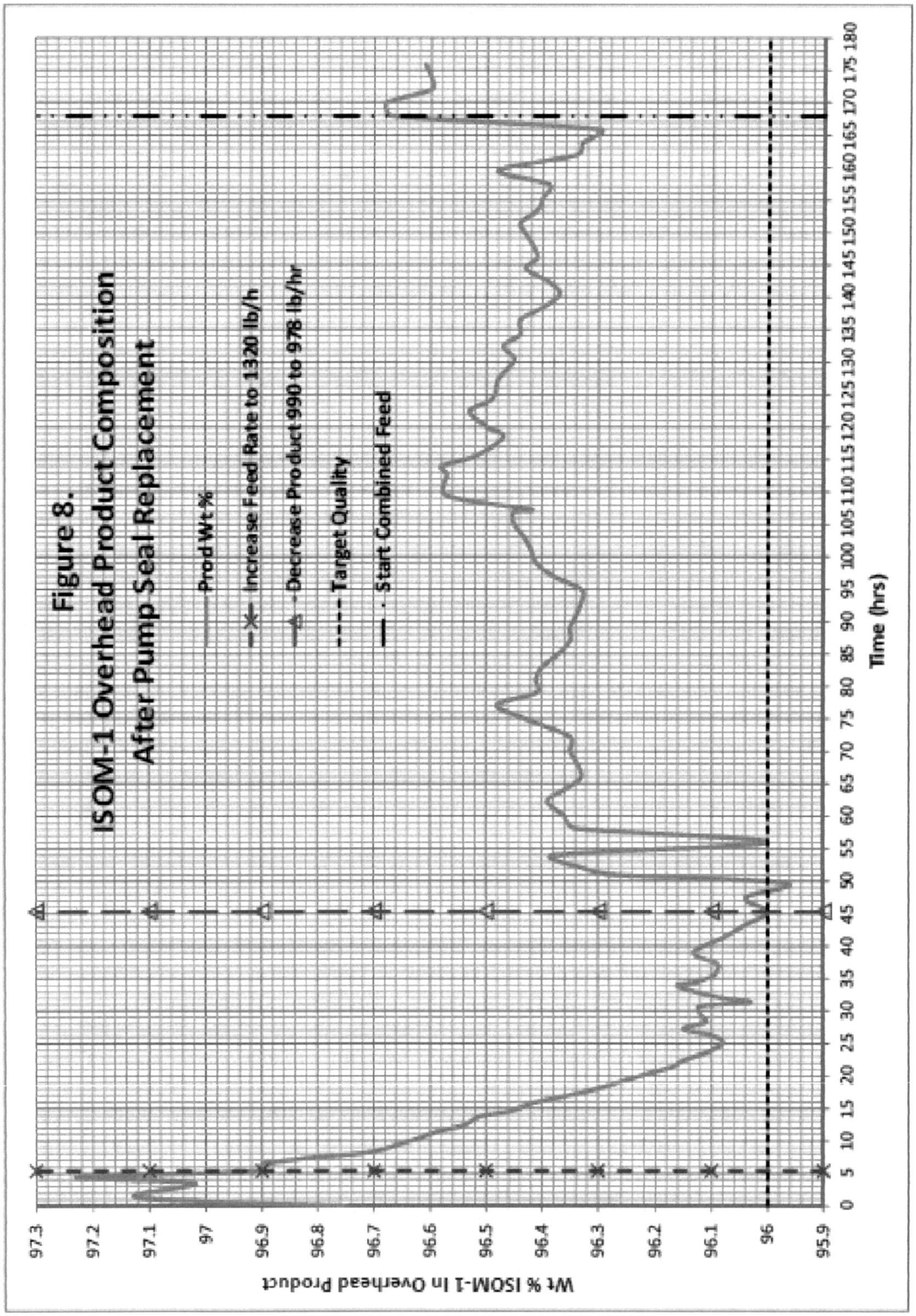
Figure 8.
ISOM-1 Overhead Product Composition
After Pump Seal Replacement
Prod Wt %
Increase Feed Rate to 1320 lb/h
Decrease Product 990 to 978 lb/hr
Target Quality
Start Combined Feed
Time (hrs)
Wt % ISOM-1 In Overhead Product
97.3
97.2
97.1
97
96.9
96.8
96.7
96.6
96.5
96.4
96.3
96.2
96.1
96
95.9
0 5 10 15 20 25 30 35 40 45 50 55 60 65 70 75 80 85 90 95 100 105 110 115 120 125 130 135 140 145 150 155 160 165 170 175 180

Table 1.

Column Configuration

Column Diameter = 42" ID

	Packed Section	Packing Type	Height feet	Packing Factor	Specific Surface Area m^3/m^2	Specific Surface Area ft^3/ft^2	Estimated HETP inches	Number of Stages	Simulation Stages	Stage Efficiency
Top	1	Sulzer 352Y	12.00	16	352	107.29	15.18	9.49	10	0.9486
	2	Sulzer 352Y	10.50	16	352	107.29	15.18	8.30	9	0.9223
	3	Norton 2T	8.25	17	215	65.53	22.31	4.44	5	0.8875
	4	Norton 2T	8.25	17	215	65.53	22.31	4.44	5	0.8875
	5	Norton 2T	8.25	17	215	65.53	22.31	4.44	5	0.8875
Bottom	6	Norton 2T	8.25	17	215	65.53	22.31	4.44	5	0.8875
								TOTAL	39	

Table 2.
Effect of Operating Pressure
On Column Performance

Pressure	Feed Rate	Reflux	Ovhd Product	Ovhd Product	Btms Product	Btms Product	Column ΔP	Reboiler Duty	ISOM-1Vac Syst Loss*
mm Hg	lb/hr	Ratio	lb/hr	wt% ISOM-1	lb/hr	wt% ISOM-1	mm Hg	MBtu/hr	lb/hr
30	1250	9.75	905.6	98.02	300.2	16	37.62	1.438	43.61
75	1250	9.75	958.4	96.77	285.1	16	30.04	1.655	6.20
100	1250	9.75	970.6	95.94	275.1	16	27.09	1.725	4.17
150	1250	9.75	993.8	94.21	253.5	16	22.96	1.845	2.49
75	1250	9.75	936.6	97.34	307.0	20	28.84	1.617	6.24
100	1250	9.75	946.8	96.67	298.8	20	25.84	1.683	4.20
150	1250	9.75	966.0	95.29	281.3	20	21.65	1.793	2.52
75	1250	10.00	935.7	97.41	307.8	20	30.07	1.654	6.25

* Based on 7 C process-side condenser outlet temperature, 0.10 psia condenser pressure drop, and 5 lb/hr air leakage rate

Table 3.

Column Operating Parameters During Campaign

Operating Pressure = 75 mmHg

Date and Time	Description	Theoret Stages	Feed Stage	Feed lb/hr	Reflux Flow gpm	Ovhd Product lb/hr	Ovhd Product wt% ISOM-1	Btms Product lb/hr	Btms Product wt% ISOM-1	Cndnsate Temp °F	Reboiler Duty MBtu/hr	Air Leak lb/hr
6/4 8:00	Simulation of Proposed Conditions	39	20	1250	25.74	935.7	97.41	307.8	20	44.6	1.654	5.00
6/5 11:00	Plant Operation			1200	31.00	900.0	97.08		19.92	45	1.955	
	Simulation With Expected Colm Effic	39	20	1200	31.09	900.0	97.84	291.7	18.05	45	1.955	5.00
	Column Calibration Simulation	33	18	1200	31.07	900.0	97.00	289.3	20	45	1.955	6.50
6/5 12:00	Plant Op - Move Fd To Bed #4											
	Simulation of Feed to Bed #4	33	22	1200	31.07	900.0	96.72	289.3	20.9	45	1.955	6.50
6/6 6:10	Actual plant Operation Feed To Bed #4			1200	31.00	900.0	96.48			45	1.955	
	Simulation To Set Operating Conditions	33	18	1260	31.35	942	96.94	305.5	20.63	48.4	1.955	6.5
	Actual Plant Operation			1260	32.1	942	96.43		18.13	48.4	1.955	
	Adjusted Simulation To Match Plant	33	18	1260	31.32	942	96.49	290.2	18.09	48.4	1.955	14.5
6/9 13:00	Increase Feed rate to 1320 lb/hr											
	Simulation To Set Operating Conditions	33	18	1320	30.95	990	96.41	304.87	18.59	46.2	1.955	14.5
6/16 10:30	Combined Feed Remaining 15,000 lbs											
	Simulation To Set Operating Conditions	33	18	1320	31.13	1014	96.49	279.6	19.17	47.2	1.955	14.5

Table 4.

Campaign Timeline

Effect of Events On Column Production Requirements

Date	Hours	Product Accmulated	Product To Be Accmulated	Feed Rate	Required Product Rate	Overhead Flow Rate	ISOM-1 In Overhead	Actual Product Rate	Hours to	Safety Margin
Day, Time	Remaining	lb	lb	lb/hr	lb/hr	lb/r	wt%	lb/hr	Completion	hr
6/4 8:00	304.00	0	234,423	1,250	771.13	936	97.41	911.5	257.19	46.81
	Corresponds to column operation with expected efficiency and 1250 lb/hr feed rate									
6/5 11:00	277.00	0	234,423	1,200	846.29	900	97.08	873.7	268.30	8.70
	Actual production start time. Feed =1200 lb/hr with product rate = 900 lb/hr.									
6/6 13:00	251.00	22,717	211,706	1,260	843.45	942	96.95	913.3	231.81	19.19
	Increase feed rate to 1260 lb/hr with 942 lb/hr product. Results based on simulation prediction.									
6/6 13:00	251.00	22,717	211,706	1,260	843.45	942	96.43	908.4	233.06	17.94
	Increase feed rate to 1260 lb/hr with 942 lb/hr product. Results based on plant measurements									
6/7 23:30	216.50	54,056	180,367	0	833.11	0	0.00	0.0	198.56	17.94
	Power outage begins; plant down for 4.6 hours									
6/8 4:06	211.90	54,056	180,367	1,260	851.19	942	96.43	908.4	198.56	13.34
	Plant resumes product takeoff at 942 lb/hr with feed rate = 1260 lb/hr									
6/8 10:25	205.58	59,796	174,627	0	849.43	0	0.00	0.0	192.24	13.34
	Plant shutdown due to pump seal failure. Plant down for 4.9 hrs									
6/8 15:20	200.68	59,796	174,627	1,260	870.17	942	96.43	908.4	192.24	8.44
	Plant resumes product takeoff at 942 lb/hr with feed rate = 1260 lb/hr									
6/9 13:00	179.00	79,490	154,933	1,320	865.55	990	96.41	954.5	162.33	16.67
	Increase feed rate to 1320 lb/hr with 990 lb/hr product. Results based on simulation prediction									
6/11 6:06	137.90	118,718	115,705	1,320	839.05	978	96.37	942.5	122.76	15.14
	Reduce product rate to 978 lb/hr									

(62e) Comparison of Batch and Continuous Industrial Distillation

Zbigniew T. Fidkowski, Air Products, Allentown, PA

Batch distillation is still widely used in chemical industry. Many smaller chemical plants operate in batch mode, with batch reactors and batch rectifiers. The main reason is production flexibility where different chemicals are being made in the same equipment in short (one- or two-week) campaigns. This is in fact the way how majority of specialty chemicals are made.

Despite great flexibility, batch distillation has many disadvantages when compared with continuous process. Some of them are related to operational difficulties:

- It may be difficult to establish when to switch from one cut to another.
- It may be difficult to establish optimal reflux ratio, since the pot composition continuously changes
- Making significant changes in process parameters may cause operating instabilities.

Other disadvantages are related to inherent inefficiencies of batch distillation. For example, it is impractical and in many cases impossible to switch directly from a "waste cut" to pure cut. There is usually a slop cut ("front cut") in between, containing off-spec product. This front cut needs to be reprocessed during a separate batch or mixed with the next batch crude (feed). Reprocessing means waste, due to reduced capacity and increased energy consumption. Sometimes, when the product content in the crude is too small, we may not be able to recover it in one batch distillation. That means that the product needs to be concentrated and then reprocessed in a separate batch distillation.

The question arises then: is batch distillation really more economical than continuous distillation, even for small production rate?

Several typical case studies will be presented to compare these two processes.

(62f) Trace Contaminant Control in High Purity Distillation Processes

Andrew (Jack) Eckles, National Thermodynamics Laboratory, Galveston, TX

Control of trace contaminants is a common problem in most processes to one degree or another. Contamination issues affect not only product quality but often can affect functionality as well as health and safety. System design, assembly and construction, component and equipment selection, sealing methods, cleaning, operating parameters, control strategies and materials of construction can all affect contribution or reduction of trace contaminants to the distillation process. Strategies and procedures will be presented for trace contaminant control in distillation systems.

RHODIUS
Safety and Environmental Solutions

Foam decay with knitted wire mesh

Overview about characterization, decay and prevention of foam

Armin Pitzer - Rhodius GmbH

AIChE Conference April 28th – May 3rd 2013 San Antonio

RHODIUS
Safety and Environmental Solutions

My special thanks to:

- Mr. Stefan Hegner (Rhodius), partner and specialist in our lab
- Mr. Wolfgang Heikamp (Rhodius), for support and trust

Content

RHODIUS
Safety and Environmental Solutions

- Introduction – Company Rhodius
- Intention to prevent foaming by knitted wire mesh
- Characterization and formation of foam
- Test equipment – experimental studies
- Results
- Conclusions - further steps for Rhodius research

Introduction

RHODIUS
Safety and Environmental Solutions

Locations

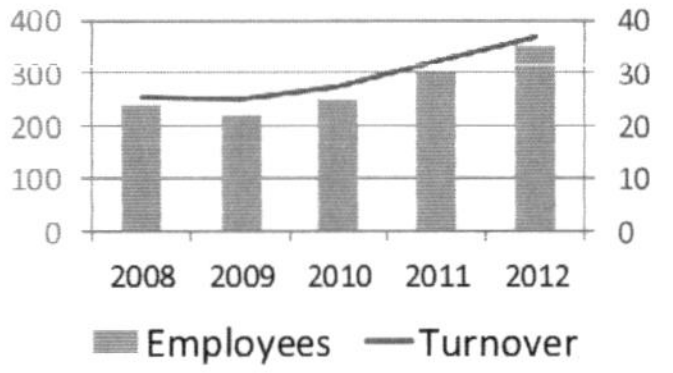

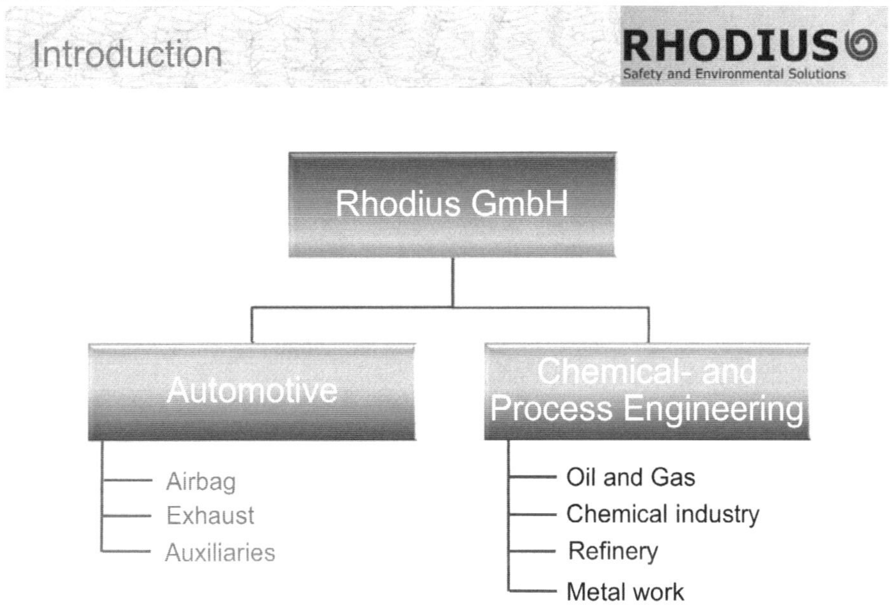
Introduction
RHODIUS
Safety and Environmental Solutions
Rhodius GmbH
Automotive
Airbag
Exhaust
Auxiliaries
Chemical- and Process Engineering
Oil and Gas
Chemical industry
Refinery
Metal work

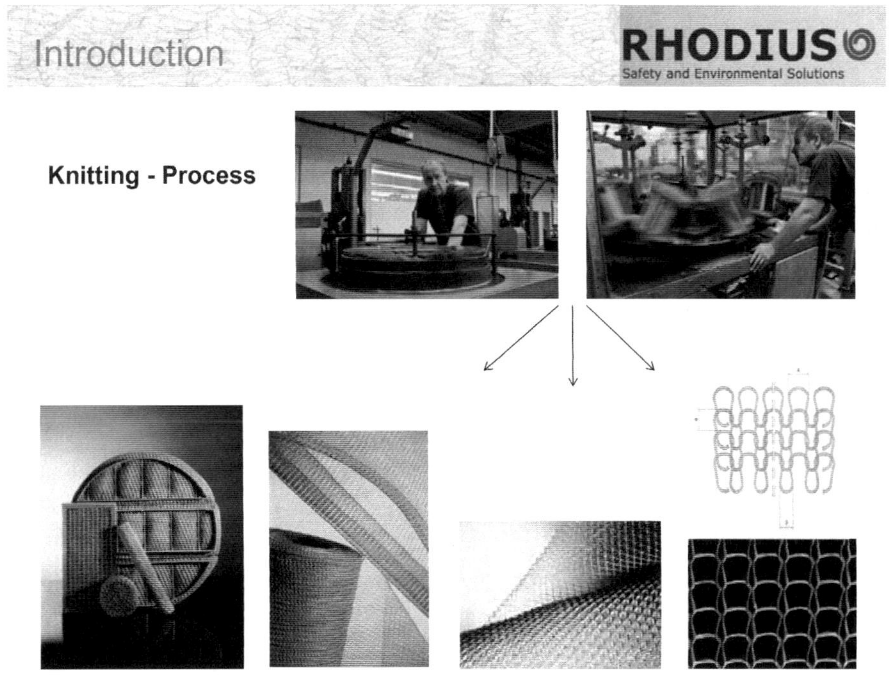
Introduction
RHODIUS
Safety and Environmental Solutions
Knitting - Process

Introduction
Automotive - Division

RHODIUS
Safety and Environmental Solutions

Airbag - components

Gasgenerator-complete system: Knitted wire mesh as filter + cooler

Generator pre-mounting: Filter made out of knitted wire mesh

Half-finished products for cold gas containers

Generator pre-mounting: Filter made out of knitted wire mesh

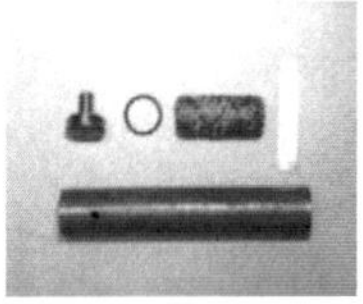

Components of an airbag module with knitted wire mesh filter

Mounting parts without knitted wire mesh

Generator pre-mounting: filters, coolers, volume balance made out of knitted wire mesh

Introduction
Automotive - Division

RHODIUS
Safety and Environmental Solutions

Exhaust gas system

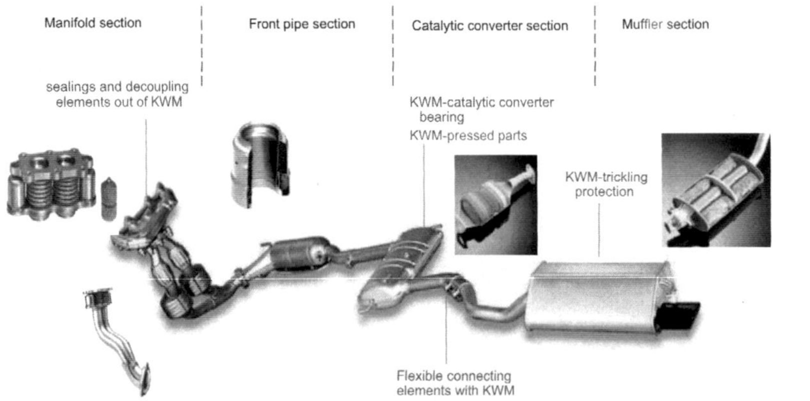

KWM = Knitted wire mesh

Introduction

Chemical- and Process Engineering

RHODIUS
Safety and Environmental Solutions

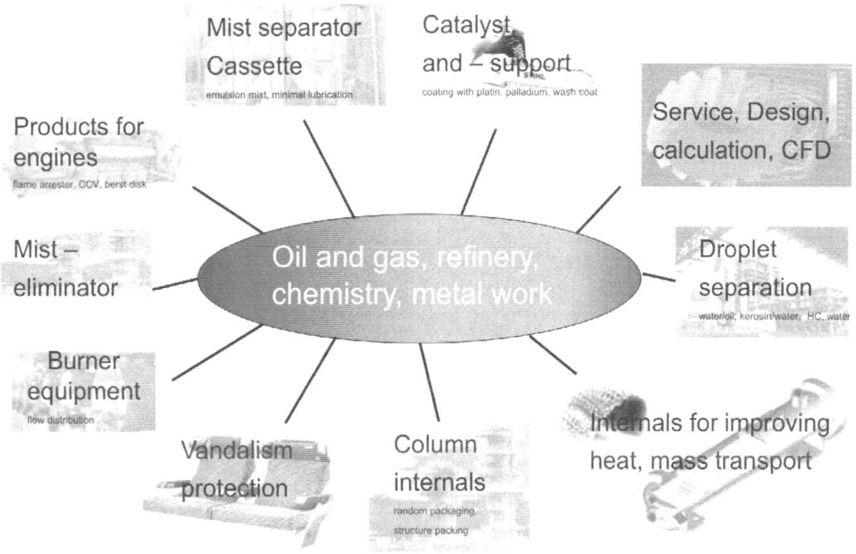

Intention to prevent foaming by knitted wire mesh

RHODIUS
Safety and Environmental Solutions

The occurrence of foam in chemical industry, e.g. in packing columns, can cause many problems:

- additional pressure drop
- low mass transfer
- low capacity

Prevent foaming in general by use of

- *chemical*
- *thermal*
- *mechanical*

measurements.

→ *Intention of Rhodius is to destroy foam by mechanical forces using knitted wire mesh*

Characterization and formation of foam

RHODIUS
Safety and Environmental Solutions

Definition of foaming:

→ Dispersion of gas in liquids generated by gasing, beating or the drop down of liquids.

3 common methods to define foamability:

Method	Origin	Generation
Ross-Miles	tensid	liquid drop-down
Bartsch	tensid	shaking
Bikerman	tensid	gasing

Characterization and formation of foam

RHODIUS
Safety and Environmental Solutions

General splitting of dispersions in 3 different parts:

- ***top:*** metastable / polyhedron foam
  $\phi \geq 0{,}74$
- ***middle:*** spherical foam
  $0{,}52 \leq \phi \leq 0{,}74$
- ***bottom:*** gas/liquid dispersion
  $\phi \leq 0{,}52$

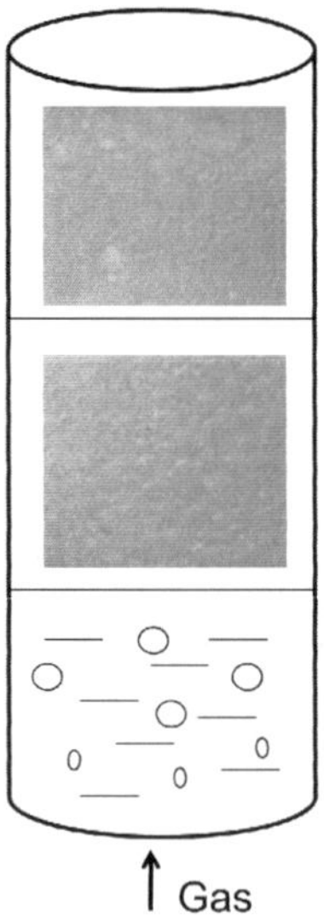

Characterization and formation of foam

RHODIUS Safety and Environmental Solutions

Foams – thermodynamically instable systems

→ Driving force for the instability is the growth of interface surface energy.

3 main processes lead to the minimization of free energy:

- diffusion of gas between neighboring bubbles
- drainage of interlamellar fluid
- the coalescence of bubbles

Characterization and formation of foam

RHODIUS Safety and Environmental Solutions

Applications in thermal process technology which can cause foaming:

- acid and amine gas treatment (e.g. CO_2-sequestration)
- rectification of organic acids and alcohol
- heterogene azeotropic - rectification
- crude oil – rectification
- extractive / reactive – rectification
- natural gas drying

Test equipment – experimental studies

RHODIUS
Safety and Environmental Solutions

Glass – column
(built-up in style of Bikerman)

- Diameter 100 [mm]
- Length 1000 [mm]
- Heating system with thermostat
- Pressure drop ($P_1 - P_3$)
- Glass - Frit

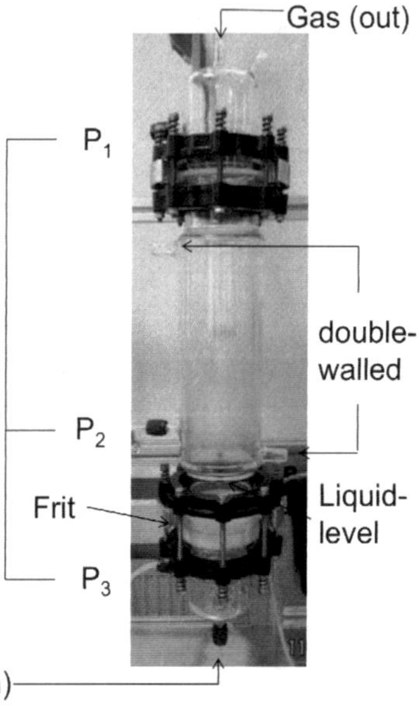

Test equipment – experimental studies

RHODIUS
Safety and Environmental Solutions

Test system

1. Fill in the fluid butanol + water

2. Open the gas flow in steps and register the height of foaming

→ Register maximum stable foaming height
→ Pressure drop

3. Put in different knitted wire mesh types and observe flooding point by increasing gas flow rate

→ Determine flooding point
→ Pressure drop over packing and whole column

Test equipment – experimental studies

RHODIUS

Safety and Environmental Solutions

Substance data - *butanol / water + clean air*

	Butanol	Water	
Density	809	998	kg/m³
	50.5	62.25	lb/ft³
Viscosity	2,9	1	mPas
	2.9	1	cP
Surface tension	25	72,5	mN/m
	25	72.5	dyne/cm
Vapor pressure	5,6	31,7	mbar
	0.081	0.46	psi

Test equipment – experimental studies

RHODIUS

Safety and Environmental Solutions

Test - program

→ Knitted wire mesh parts – DN 100

Number	Material	Style
I	metal	10
II	metal	20
III	co-knit	10
IV	co-knit	20
V	plastic	10

Results

RHODIUS
Safety and Environmental Solutions

Video - Foam decay

near flooding *_high foaming load*

Number I - Style 10

stable
high foaming height

instable
polyhedron foam
at bottom of KWM

Results

RHODIUS
Safety and Environmental Solutions

Video - Foam decay

no flooding *_low foaming load*

Number I - Style 10

stable
low foaming height

instable
polyhedron foam
at bottom of KWM

Results

RHODIUS
Safety and Environmental Solutions

Foam decay with KWM → low foaming load
→ high foaming load

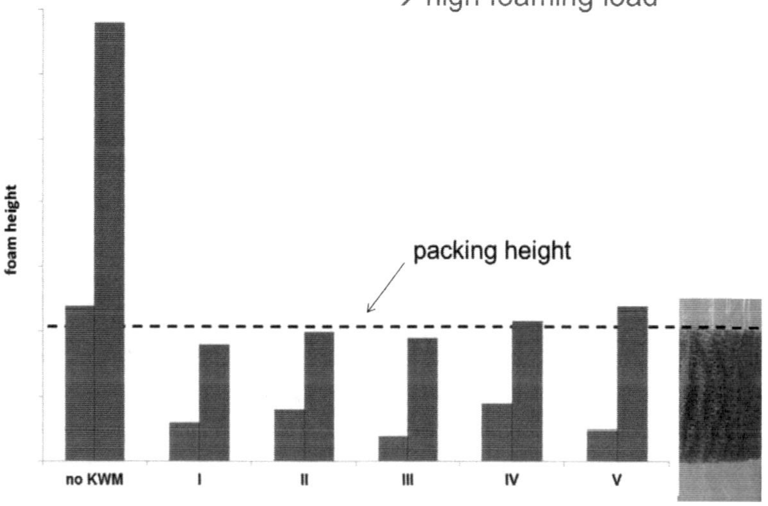

Results

RHODIUS
Safety and Environmental Solutions

Foam decay with KWM → low foaming load
→ high foaming load

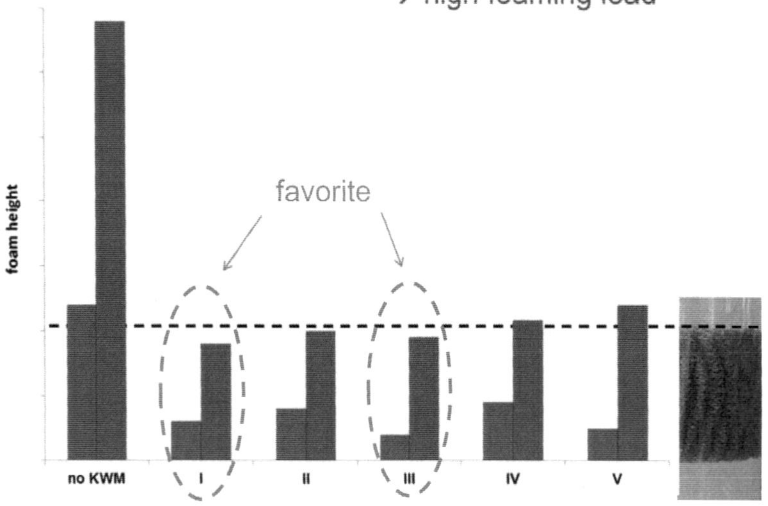

Conclusions

RHODIUS
Safety and Environmental Solutions

- Test rig is able to reproduce the results
- KWM is suitable for : foam destruction
  droplet separation
  coalescence (liquid / liquid)
- Metal and co-knit styles are the favorite types for foam destruction
- KWM – advantages: low pressure drop (high porosity)
  cleanable
  unlimited lifetime

Further steps for Rhodius research

RHODIUS
Safety and Environmental Solutions

- Different liquid / gas systems
  e.g. tensid: *MDEA, Triton X-100*
- Detail research on wettability - co-knit styles
- Influence of temperature
- Improve design of KWM to avoid wall effects

RHODIUS
Safety and Environmental Solutions

Thank you

for your

kind attention!

Questions ?

(85a) Tower Scanning – Now a Quantitative Tool, Part A

Lowell Pless, Tracerco, Pasadena, TX and Izak Nieuwoudt, Koch-Glitsch, Wichita, KS

Tower gamma scanning and neutron backscatter techniques are well established in the process industries as qualitative tools to help troubleshoot separation towers, both trayed and packed towers. In this presentation it will be shown that through careful and innovative application of gamma scanning and neutron backscatter techniques, quantitative information about separation towers can be extracted.

Part A: For trayed towers this will be illustrated by comparing data taken under very controlled conditions on a hydrocarbon pilot plant tower equipped with windows, with the data obtained from scans of the tray downcomer and active area. Three sets of trays, designed to flood under jet flood, downcomer backup or choke flood mechanisms, were tested and studied. The experimental observations, tray model predictions, and scan data for each of these sets of trays will be compared. Based on this information it can be concluded that tower scanning can be used to generate quantitative information about tray performance.

(85b) Performance Comparison of Structured Packing Liquid Holdup and Pressure Drop Correlations

Anil Krishna Jammula[1], James R. Whiteley[1], Tony Cai[2] and Mike Resetarits[2], (1)School of Chemical Engineering, Oklahoma State University, Stillwater, OK, (2)Fractionation Research, Inc., Stillwater, OK

This paper presents an analysis and comparison of popular liquid holdup and pressure drop correlations developed for structured packing. Correlation performance was analyzed using a screened database containing the literature liquid holdup and pressure drop data. This database includes FRI pressure drop data which have been released to the public. The screening process excludes data points from very high pressure test runs and flooding data. The full database consists of 938 pressure drop data points and 368 liquid holdup data points obtained from 7 different sources. A total of 24 different structured packings are present in this database. This database consists of data from both hydrocarbon and aqueous system.

Liquid holdup performance results are documented for following correlations: Billet (1999), Del Carlo (2006), Delft (2011), Gualito (1997), Mackowiak (1991), Pondebat (1992), Spiegel (1992), SRP (1993) and Stichlmair (1989).

Pressure drop performance results are documented for following correlations: Billet (1999), Brunazzi (1997), Delft (2011), Gualito (1997), Iliuta (2001), Mackowiak (1991), Pondebat (1992), Spiegel (1992), SRP (1993) and Stichlmair (1989). Performance of the correlations was evaluated based on different factors e.g., types of packings and types of systems, etc. The paper concludes with an overall summary of the findings.

(85c) Mass Transfer Characterization of Structured and Random Packings

Chao Wang[1,2], Micah Perry[1], Gary T. Rochelle[2] and Frank Seibert[1], (1)SRP, University of Texas at Austin, Austin, TX, (2)Chemical Engineering, The University of Texas at Austin, Austin, TX

Packing is widely used in distillation as well as post-combustion CO_2 capture because of its low pressure drop, good mass transfer efficiency, and ease of installation. In the CO_2 capture process, absorber and stripper performance are highly dependent on the effective mass transfer area of the packing (a_e). The stripper performance also depends on the liquid film mass transfer coefficient (k_L). Gas cooler and water wash performance depends on the gas film mass transfer coefficient (k_G). Our research is focused on consistent measurement and mechanistic model development to predict these important fundamental packing design parameters.

Five structured packings of various surface areas, corrugation angles and surface treatment and four high performance random packings of various nominal sizes have been studied using a 0.428 m diameter PVC column. The experimental results obtained in this work will be presented and compared with published data.

The purpose of this research is to work out generalized correlations for the prediction of the effective mass transfer area (a_e), liquid and gas film mass transfer coefficients for novel packings, which is highly needed in packed column designs. The generalized correlations will be based on consistent measurement in this work as well as published data.

(85d) INTALOX™ ULTRA™ Random Packing – Pushing the Boundaries

Izak Nieuwoudt, Koch-Glitsch, Wichita, KS and Pat Quotson, Koch-Glitsch, LP, Wichita, KS

INTALOX ULTRA packing is a new generation random packing that offers higher efficiency and/or capacity compared to earlier generation random packing families. Depending on the application, the user can opt for higher capacity without a loss of efficiency, or higher efficiency without a loss of capacity. INTALOX ULTRA random packing is currently used in several hundred applications, and this presentation provides case studies from several of those applications. In addition, data from F.R.I. testing of two sizes from the INTALOX ULTRA random packing family will be presented.

(85e) Raschig Super-Ring is Pushing the FRI Test Column to New Limits

Michael Schultes, Raschig Ring Department, RASCHIG GMBH, Ludwigshafen, Germany

Fractionation Research Inc. (FRI) is the most modern test institute in the world performing random packing tests under rectification conditions in an industrial scaled test column. FRI is testing random packings since more than 50 years under low, medium and high pressure conditions and has created one of the worlds largest data base for comparing packing efficiencies, pressure drops and capacities.

Beside the development of most modern packing generations over the past decades also tower internals like distributors, hold down plates and support plates have seen great improvements by design. High performance tower internals used 10 or 20 years ago in mass transfer columns are nowadays considered to be low performance devices especially in use with high performance packings.

Based on most recent FRI tests Raschig is able to disclose an impressive research program that underlines the impact of "old fashioned" and "new fashioned" high performance tower internals on Raschig Super-Ring's fluid dynamics and mass transfer characteristics. Last but not least the conclusions will demonstrate the importance of tower internal designs on packing performance in industrial columns.

Raschig Jaeger Technologies are very well known for troubleshooting as well as revamping cases of all kinds of packings including high performance third generation devices. Samples of such troubleshooting and revamp cases will be also presented.

(85f) Design of Packed Column Direct Contact Condensers for Vapor and Vapor Mixtures

Jan Mackowiak[1], Jerzy Mackowiak[1] and Andrzej Górak[2], (1)ENVIMAC Engineering GmbH, Oberhausen, Germany, (2)Department of Biochemical and Chemical Engineering, Laboratory of Fluid Separations, TU Dortmund University, Dortmund, Germany

In numerous industrial processes direct contact condensers (DCC) are used to condense vapor or vapor mixtures. In many cases, packed columns are used for this purpose, where the condensable vapor gets in direct contact with the liquid phase, which serves as a condensing agent. In this work miscible vapor-liquid systems are investigated.

For the modeling of condensation processes in packed columns, which are characterized by simultaneous heat and mass transfer, mostly empirical correlations are used, e.g. [1]. For the condensation of pure water vapor or vapor mixtures without non-condensable components, only few correlations are present in literature, valid for classical random packing types like Raschig rings or Intalox saddles. In case of the direct condensation of vapor and especially of vapor mixtures with non condensable components no experimentally validated correlations can be found for modern random packing types. Experimental data from relevant process in industrial scale in scarce, although such data would be very useful for the validation of models, which have been developed for direct contact condensation processes, e.g. [2].

In the present work results from experimental investigation of vapor condensation in pilot scale packed columns with diameters of 0,15 and 0,32 m and a packed height of up to 1,3 m are presented. In a first step, the direct contact condensation of pure water vapor using water has been investigated. Therefore, several different packing types and sizes made of stainless steel, ceramics and plastic have been used. Beside the classical Pall ring (1" and 2"), high performance stainless steel random packings types of the latest generation type McPac (1" and 2"), the ceramic R-Pac and ENVIPAC and DTNPAC made from plastic have been tested. Based on the experimental results, a dimensionless empirical correlation considering the packing type has been developed for the heat transfer driven process. It has been shown, that the strong relation between fluid dynamics and mass transfer acc. to the extend channel model [3, 4]can be extended to heat transfer processes as well.

In a second step, condensation of vapor out of saturated air and from vapor mixtures containing noncondensable carbon dioxide (CO2) and ammonia (NH3) has been studied. Due to the additional mass transfer resistance in vapor mixtures with non condensable components, the condensation rate decreases compared to the condensation of pure vapor. The aim of this experimental study was the generation of relevant data for the validation of a rate-based model for the condensation of multicomponent vapor mixtures in packed columns [2].

A comparison of experimental and calculated data shows good agreement for this complex heat and mass transfer process.

[1] BILLET, R., Verdampfung und ihre technischen Anwendungen, Verlag Chemie, Weinheim, 1981.

[2] MAÆKOWIAK, J.F., GÓRAK, A., KENIG, E.Y., Modelling of combined direct contact condensation and reactive absorption in packed columns, Chem. Eng. J. 2009, 149, 362-369.

[3] MAÆKOWIAK, J., Extended channel model for prediction of the pressure drop in single-phase flow in packed columns, Chem. Eng. Res. Des. 2008, 87, 123-134.

[4] MAÆKOWIAK, J., Model for the prediction of liquid phase mass transfer of random packed columns for gas–liquid systems, Chem. Eng. Res. Des. 2011, 89, 1308–1320.

(85g) UFM Valve: An Advanced Mini-Valve to Boost the Performances of the Fractionation Trays

Giuseppe Mosca, Technology, Sulzer Chemtech, Milano, Italy, **Mark W. Pilling**, Sulzer Chemtech USA, Tulsa, OK and Stefan Hirsch, Sulzer Chemtech, Tulsa, OK

Maximizing capacity of major equipment, such as distillation columns, is a key factor to meet the most challenging targets of every industrial plants i.e. Oil Refineries, Petrochemical Plants, LNG Plants, and Gas Fields.

In the last 20 years, a lot of effort and research work have been done to enhance the capacity and the efficiency of the mass transfer components, with the scope of pushing the distillation columns up to their hydraulic limit set by the vessel diameter.

Sulzer Chemtech has recently developed a new type of valve: UFM (Umbrella Floating Valve), which substantially improves the performances of the fractionation trays.

The purpose of this paper is to provide an overview of the characteristics of this new feature, its fields of application, and the achievable benefits in terms of column throughput, fractionation stages, and pressure drop.

MORE TOWER DAMAGES CAUSED BY WATER-INDUCED PRESSURE SURGE: UNPRECEDENTED SEQUENCES OF EVENTS

Ademaro Marchiori[1],
Ana Lidia Wild[3],
Aristides Yoshiaki Saito[1],
Arlan Lucas de Souza[2],
Carmen Mittmann[3],
Christian C. Anton[3],
Felipe Saldanha Duarte[3],
Sandro L.A. Pereira[3],
and Silvia Waintraub[2]

[1]PETROBRAS/HEADQUARTERS
[2]PETROBRAS/RESEARCH CENTER- CENPES
[3]PETROBRAS/REFAP – ALBERTO PASQUALINI REFINERY

Rio de Janeiro
Brazil

Prepared for Presentation at the AIChE 2013 Spring National Meeting,
San Antonio, April 28 – May 2, 2013

March/2013

Unpublished

Abstract

Three water-induced pressure surge cases in crude towers are presented. Although different and independent, all the cases have a common point: the water-induced pressure surge is associated to a pressure relief valve opening event.

In the first case study, a 126,000 bpd crude distillation unit had a significant decrease in one of the atmospheric tower side withdrawals after an emergency shutdown. The light diesel oil production was cut in less than half. A generalized tower temperature profile reduction was noticed: 27°F (15°C) temperature decrease in light diesel oil pan; 63°F (35°C) in heavy diesel oil pan and 18°F (10°C) in atmospheric residue withdrawal.

The second case study occurred at another crude distillation unit located at the same refinery with a capacity of 69,200 bpd. It was observed a deep reduction in the first side withdrawal (naphtha) yield that was compensated by the other side draw products yields. No loss in total diesel production occurred.

Compared to the second, the damages in the third case were higher and it was noticed a degradation of products to the bottoms residue. This problem happened at the same tower of the second case, one year later and it was caused by distinct conjugation of facts.

The paper presents the troubleshooting techniques applied to evaluate the different units problems (operational tests, gamma scans surveys, and thermograph readings), the root causes of these incidents and the mitigation actions.

Introduction

Crude Distillation Unit (CDU) is present in all refineries. It is the first unit that receives the crude, determines the maximum refinery throughput, minimizes residue production with a lower cost and separates the distillation products as LPG, Straight Run Naphtha, Kerosene and Diesel. A simplified sketch of an Atmospheric Crude Distillation Unit is shown in Figure 1.

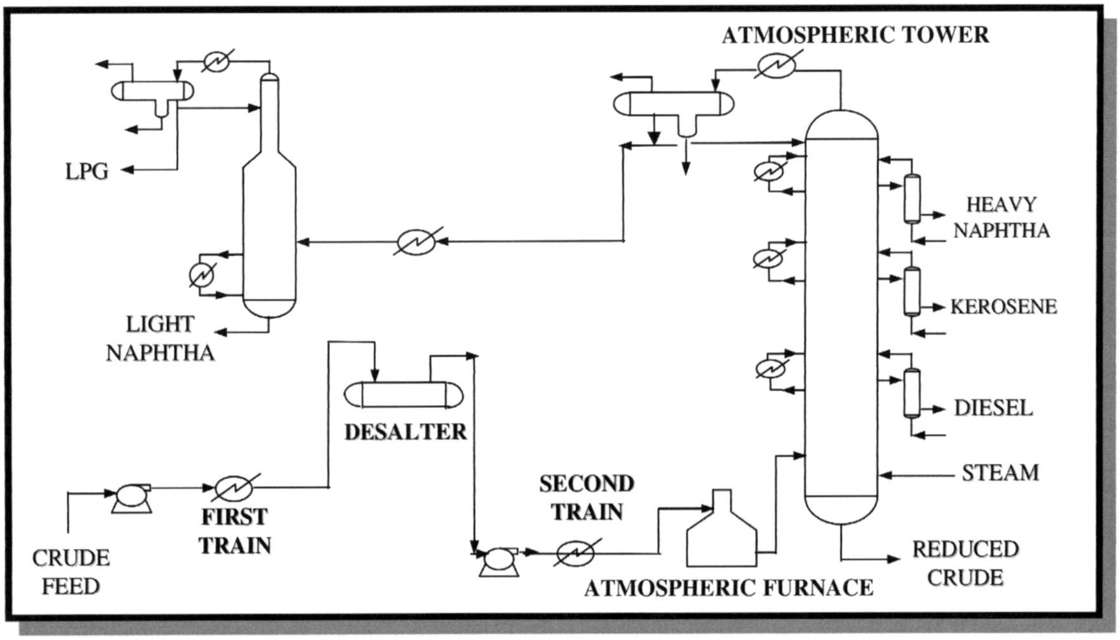

Figure 1 Atmospheric Crude Distillation Unit Sketch

As a CDU operates with very high feed input, a distillation tower malfunction implies in a deep profitability loss.

Despite of the huge improvement in distillation technology and knowledge, the number of tower malfunctions increases year after year. Tower internals damage has been the third most common tower malfunction, being most of the time caused by water-induced pressure (Kister, 2003).

Although other sources are also observed, the most common water-induced pressure surge is due to water carry over in stripping steam lines. Water accumulation in dead pockets is the fourth cause (Kister,2003). Kister (2006) published a special topic related to water-induced pressure surges, showing cases due to water in feed and slop, accumulated water in transfer line to tower and in heater passes, water accumulation in dead pockets, water pockets in pump or spare pump lines, undrained stripping steam lines, condensed steam or refluxed water reaching hot section and oil entering water-filled region.

Tower damages caused by water-induced pressure surges after pressure safety relief valve (PSV) opening were not easily found in literature and were not common to happen in Petrobras refineries. A great number of pressure relief valves are installed in a Crude Distillation Unit in order to avoid an overpressure that can occur due to many different causes, such as fire, electric power failure, water to condenser supply failure and valve inadvertent closure, among others (API 521, 2007).

There are two major common destinations of the liquid discharge of these pressure relief valves: atmospheric or preflash tower flash zone or a blowdown drum. The CDU units of the majority of Petrobras refineries have the discharge of liquid PSVs to the column flash zone. These PSVs liquid discharges should be composed only by hydrocarbons resulting in no harm to the tower. On the other hand, if water is carried over together with the liquid, a pressure surge can happen inside the column, caused by abrupt expansion of water when exposed to high temperatures. It is mandatory focusing to avoid water accumulation at all steps of the unit cycle life: basic design, detailed engineering, construction, inspection and all the time during operation, especially after shutdowns and start-ups.

In the present paper it will be described three case studies, all different and independent, but with a common point: the water-induced pressure surge was associated to a pressure relief valve opening event and caused a severe damage to the tower internals.

First Case Study: PSVs liquid discharge to a pumparound return

The simplified flowsheet of the Crude Distillation Unit with a nominal capacity of 126 KBPD is presented in Figure 2. The atmospheric column has four sidedraws (Heavy Naphtha, Kerosene, Light Diesel and Heavy Diesel) and the reduced crude is sent to a RFCC unit.

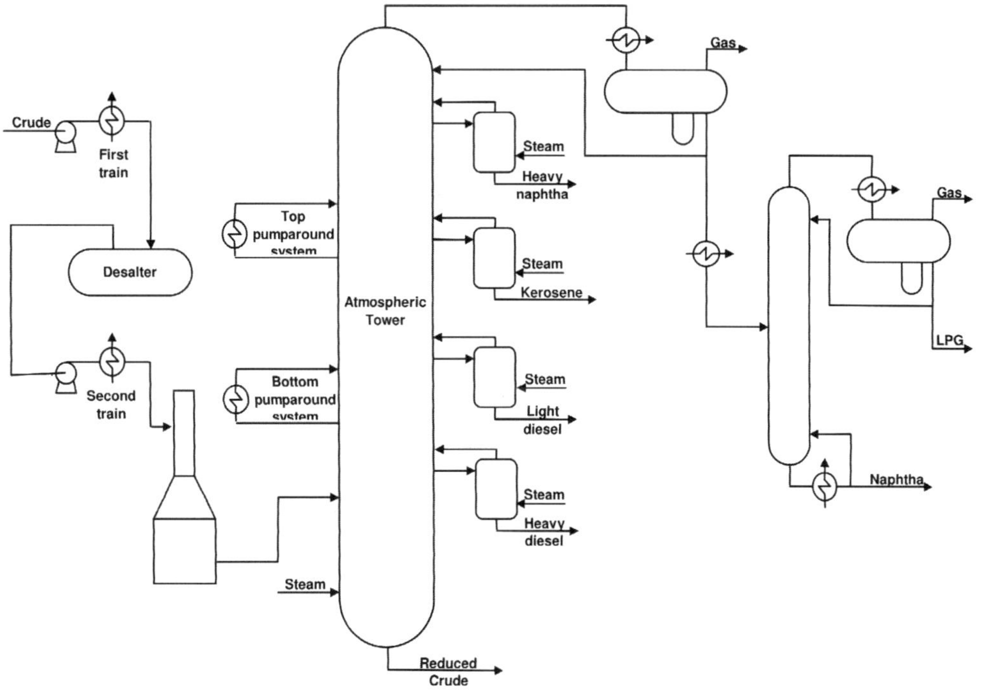

Figure 2 Fisrt Case Study – CDU Simplified Flowsheet

In January 2011, an electric power failure resulted in an emergency shutdown. The unit started-up after two days and it was observed that the light diesel oil production was cut in less than half. A generalized tower temperature profile reduction and a 1000 m^3/d loss in total diesel production was noticed as shown in Table 1.

VARIABLE	VARIATION
LIGHT DIESEL YIELD (m³/d)	- 2000
HEAVY DIESEL YIELD (m³/d)	+ 600
TOTAL DIESEL YIELD (m³/d)	-1000
LIGHT DIESEL TEMP. (°C/°F)	-15/-27
HEAVY DIESEL TEMP. (°C/°F)	-35/-63
REDUCED CRUDE TEMP. (°C/°F)	-10/-18

Table 1 Conditions after Emergency Shutdown

Thermograph readings and gamma scans surveys were done in order to look for possible damages inside the column.

LIGHT DIESEL NOZZLE	OUTLET LIGHT DIESEL VALVE OPENED	OUTLET LIGHT DIESEL VALVE CLOSED
	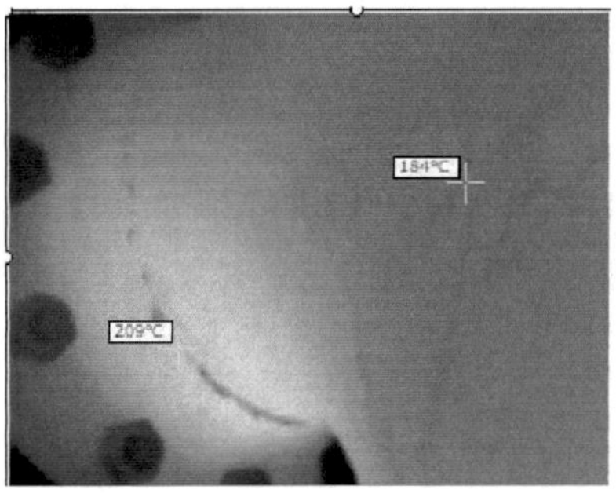	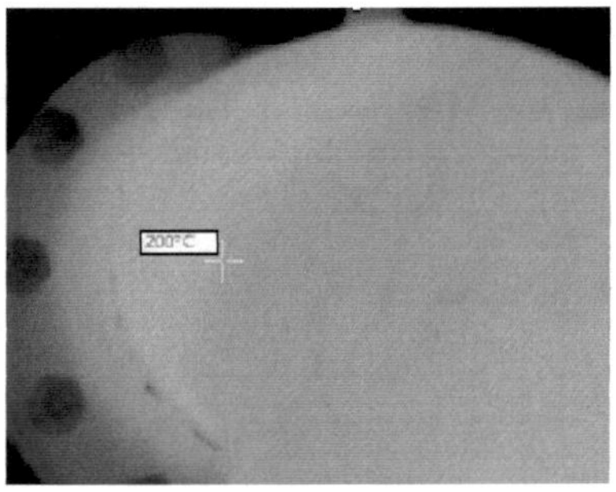
LIGHT DIESEL OULET LINE	TWO TEMPERATURES LIQUID AND VAPOR	ONE TEMPERATURE LIQUID

Figure 3 First Case Study – Thermograph Readings

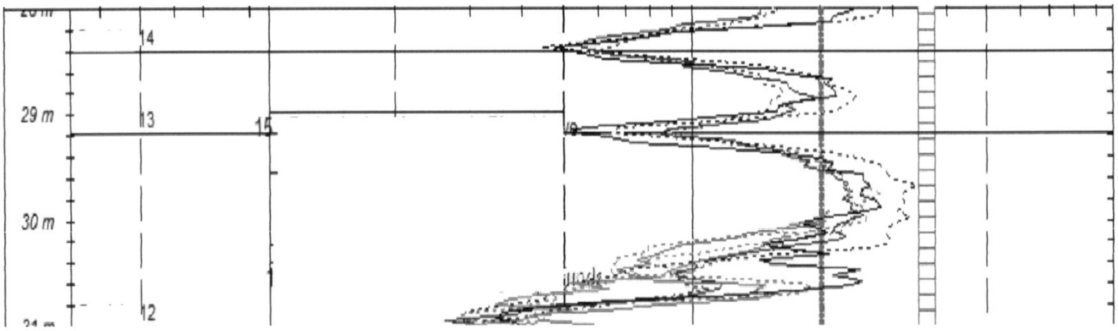

Figure 4 First Case Study – Crude Tower Gamma Scan (partial)

As it was possible to draw only half of the light diesel rate, thermografph readings were done close to its draw-off nozzle. The outlet light diesel valve was completely closed and as shown in

Figure 3, the collector channel was able to retain the liquid meaning that probably the light diesel tray itself (#13) should be harmed. The gamma scan analysis, partially presented in Figure 4, indicated that this tray liquid level was very low, a difference in liquid height between the north and south sides, and also pointed out that the region between the kerosene (# 21) and light diesel draw (#13) had problems.

The unit was shutdown in order to repair the internal mechanical tower damages. After inspection it was observed a lot of missing pannels, many opened trapdoors, some trays partially fallen and from observation it was concluded that an uplift from bottom to up had happened. The crude tower sketch is presented in Figure 5.

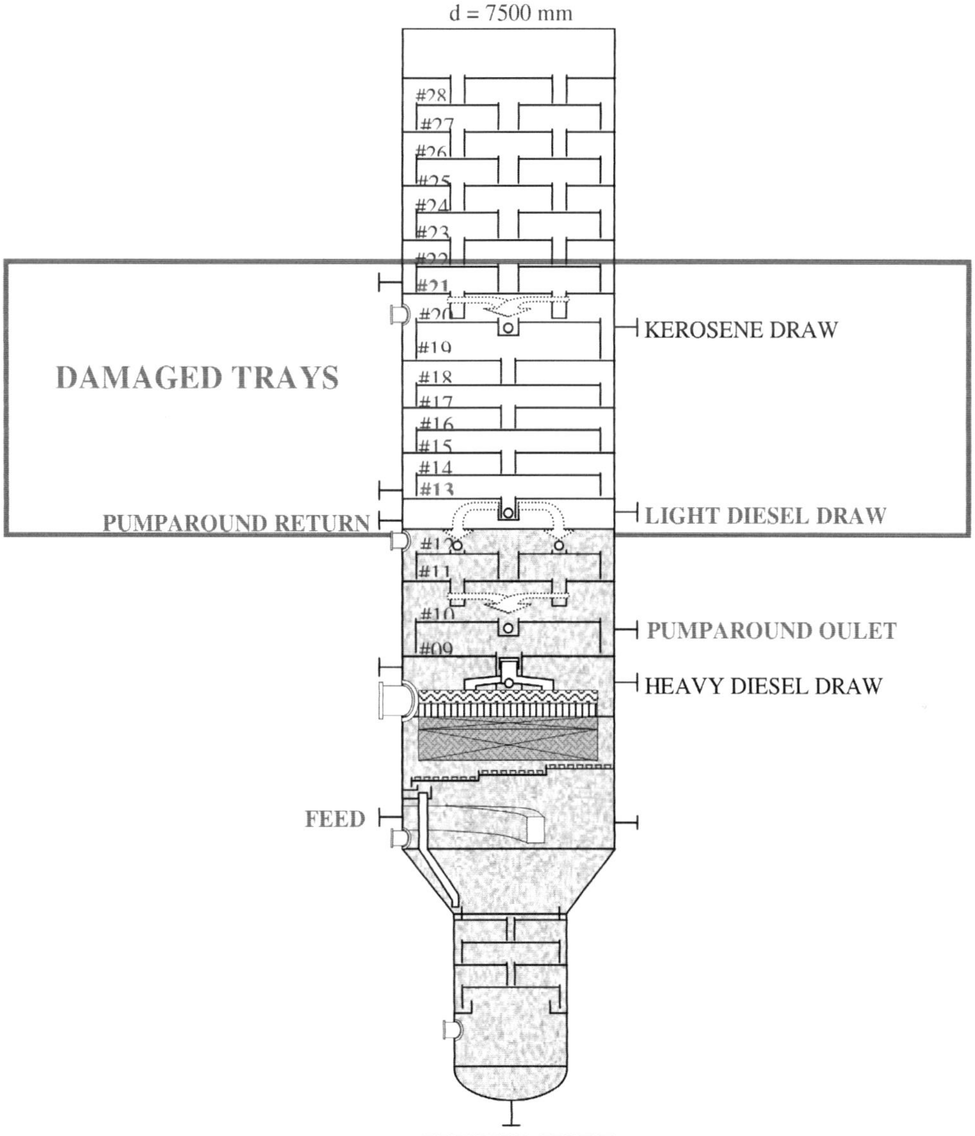

Figure 5 First Case Study - Atmospheric Crude Tower Sketch

TRAY # 21

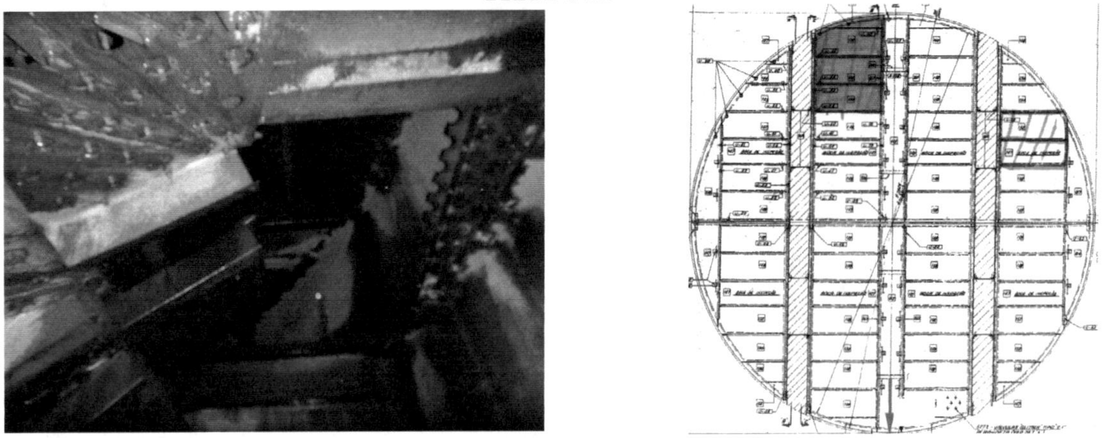

TRAY # 20

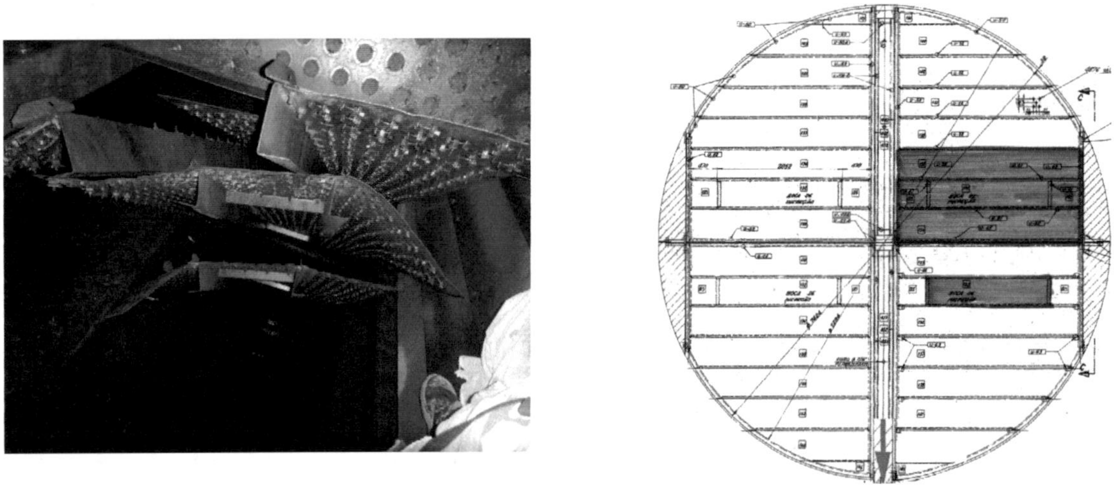

TRAY # 13 – LIGHT DIESEL DRAW

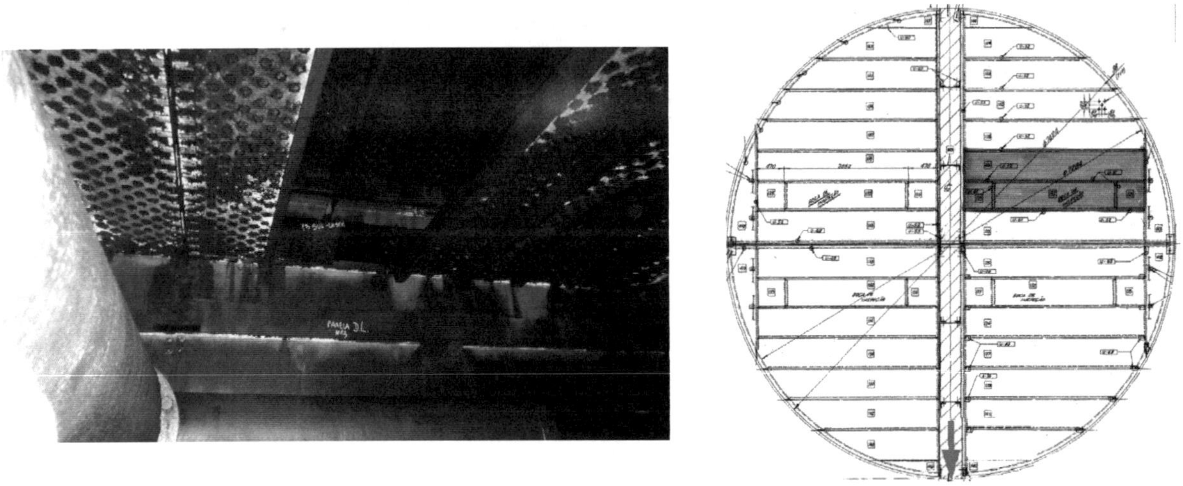

Figure 6 <u>First Case Study – Damaged Trays Pictures</u>

Some pictures of the damaged trays are shown in Figure 6. The solid marks represent missing panels and the dashed lines represent an upholstered region. The tower was fixed and after the new start-up the diesel yield returned to the expected values.

Investigation to find the causes of the damages showed that a water-induced pressure surge had happened due to water presence during the start-up after the emergency shutdown. Immediately after the heavy diesel pump was aligned it was observed an increase of 0.66 kgf/cm^2 in 30 seconds in top and flash zone tower pressures as it can be seen in Figure 7. It was observed by the operation team that the PSV of heavy diesel X crude exchanger had opened. The liquid discharge of this pressure relief valve is sent to the bottom pumparound system returning to the tower as showed in Figure 8. Unfortunately there are dead pockets that can accumulate water in this circuit and they were not drained because the shutdown was too short. This pumparound return is just below the damaged region of the crude tower and it was noted that the trays major harmed portions were concentrated in the southeast side.

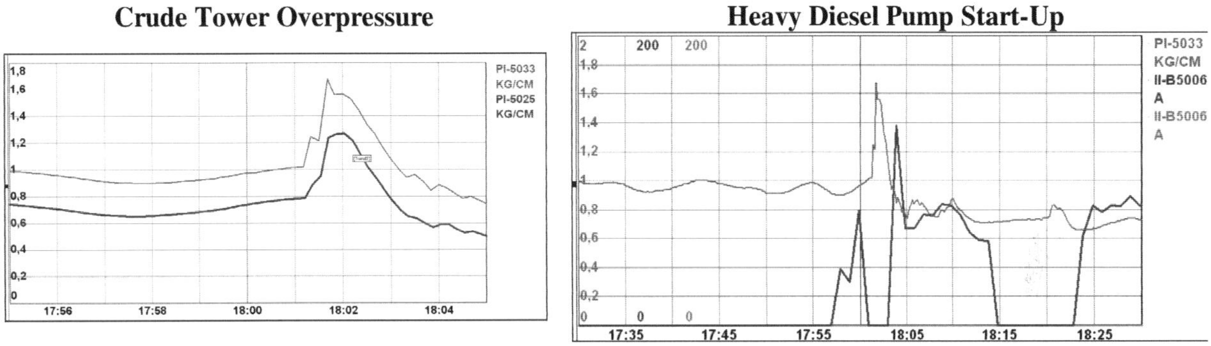

Figure 7 First Case Study – Overpressure

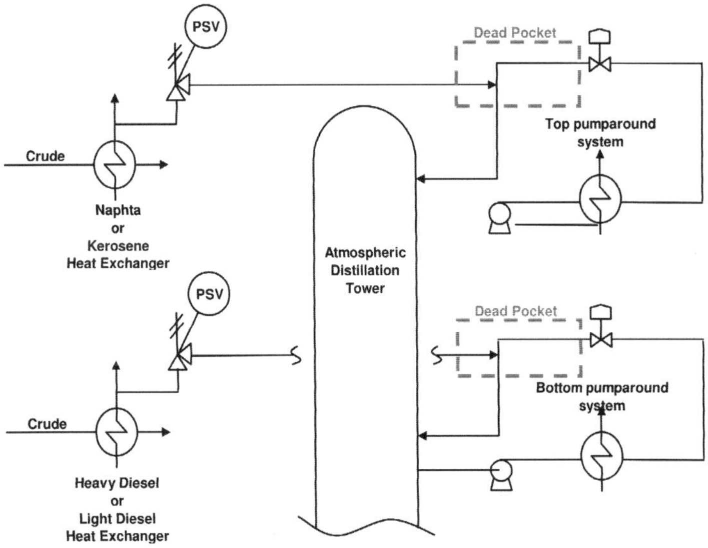

Figure 8 First Case Study – PSVs Discharged to Pumparound Returns

During the start-up operation the distillates pumps work with greater head values, close to their shut off, being common to happen opening of the PSVs that protect these systems. In this procedure the pumparounds pumps begin to operate later on after the tower is already heated. After returning from the emergency shutdown and at the same moment of the heavy diesel pump start-up, the PSV that protects the crude X HD exchanger opened, sending hydrocarbons to the pumparound system returning to the crude tower, carrying water together and causing the overpressure due to abrupt vaporization in high temperatures.

Ten months later another water-induced surge damaged the atmospheric tower of another crude unit located at the same refinery and this incident will be the next case study.

Second Case Study: PSVs liquid discharge to the atmospheric tower flash zone

The simplified flowsheet and control loops of the 69,200 BPD crude distillation unit are presented in Figure 9. There is no preflash drum nor preflash tower. The unit feed rate is controlled by FV-106 located after the booster pumps downstream of the desalters.

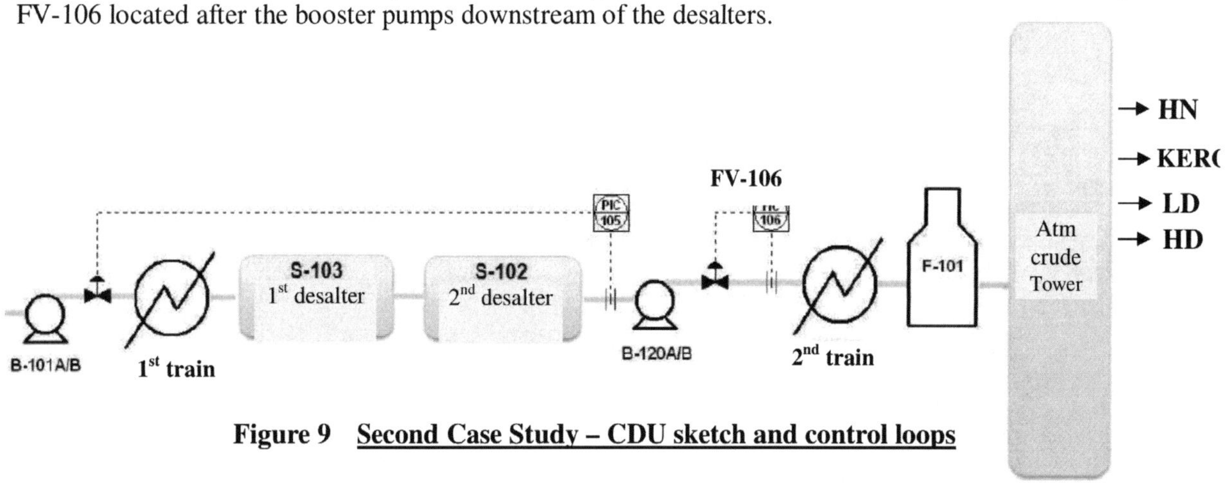

Figure 9 <u>**Second Case Study – CDU sketch and control loops**</u>

In November 2011 a mechanical problem happened to the feed control valve (FV-106) and it was abruptly closed. The charge pump remained sending crude to the unit, the system became over pressurized and the PSVs of both desalters opened, discharging to the crude tower flash zone. The crude tower flash zone pressure increased 1.3 kgf/cm^2 and the top pressure 1.1 kgf/cm^2 both in 1.5 min, as shown in Figure 10.

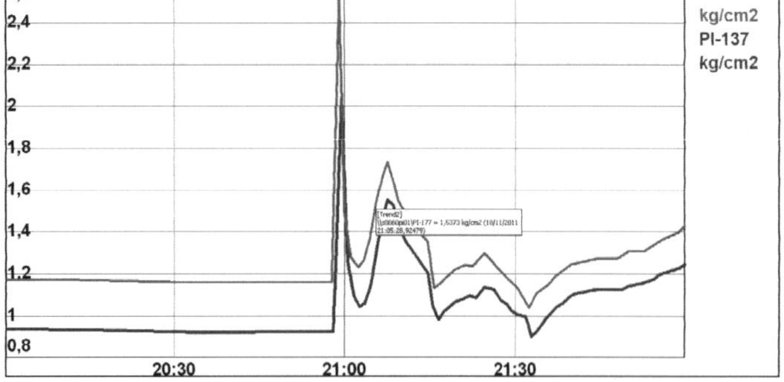

Figure 10 <u>**Second Case Study – Crude Tower Overpressure**</u>

After this event, it was observed a huge decrease in the first side-drawoff production (heavy naphtha), as demonstrated in Figure 11. This reduction was compensated by the lower sidedraws and no diesel loss to the bottoms residue was observed.

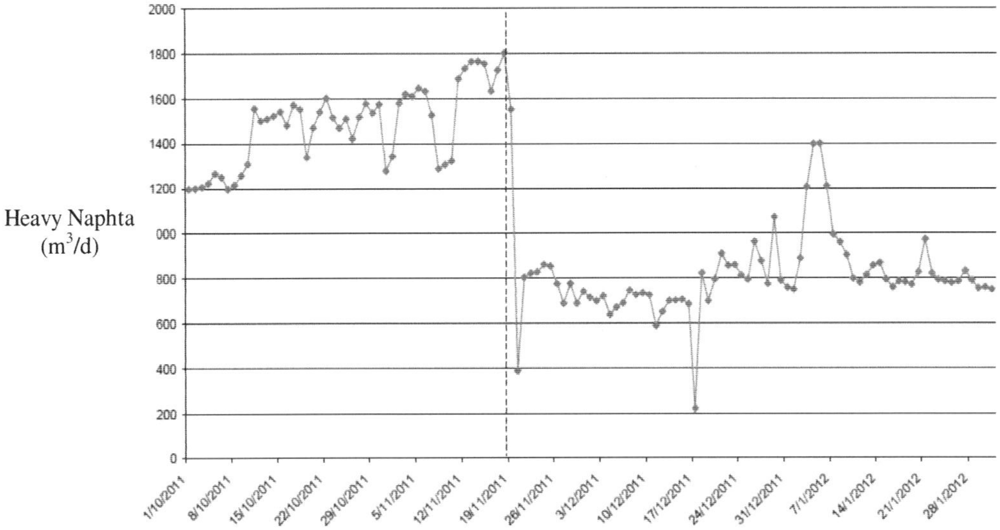

Figure 11 __**Second Case Study – Heavy Naphta Yield Reduction**__

A crude tower gamma scan was done in order to show the extent of the damages. It pointed out that only the trays located at the smaller diameter section were with problems.

One year later, in November 2012, it was the scheduled unit turnaround and the tower was inspected. Many trays were completely fallen (15 from 44), others partially damaged and from observation it was concluded that an uplift from bottom to up happened.

All the damaged trays were sieve type and were located at the tower upper section, in the smaller diameter region. The valve trays located at the larger diameter section were in place. These valve trays opened area are four times the opened area of the upper sieve trays.

The crude tower sketch is presented in Figure 12, where it can be seen in blue the valve trays, in red the sieve trays and in highlight the region that was damaged.

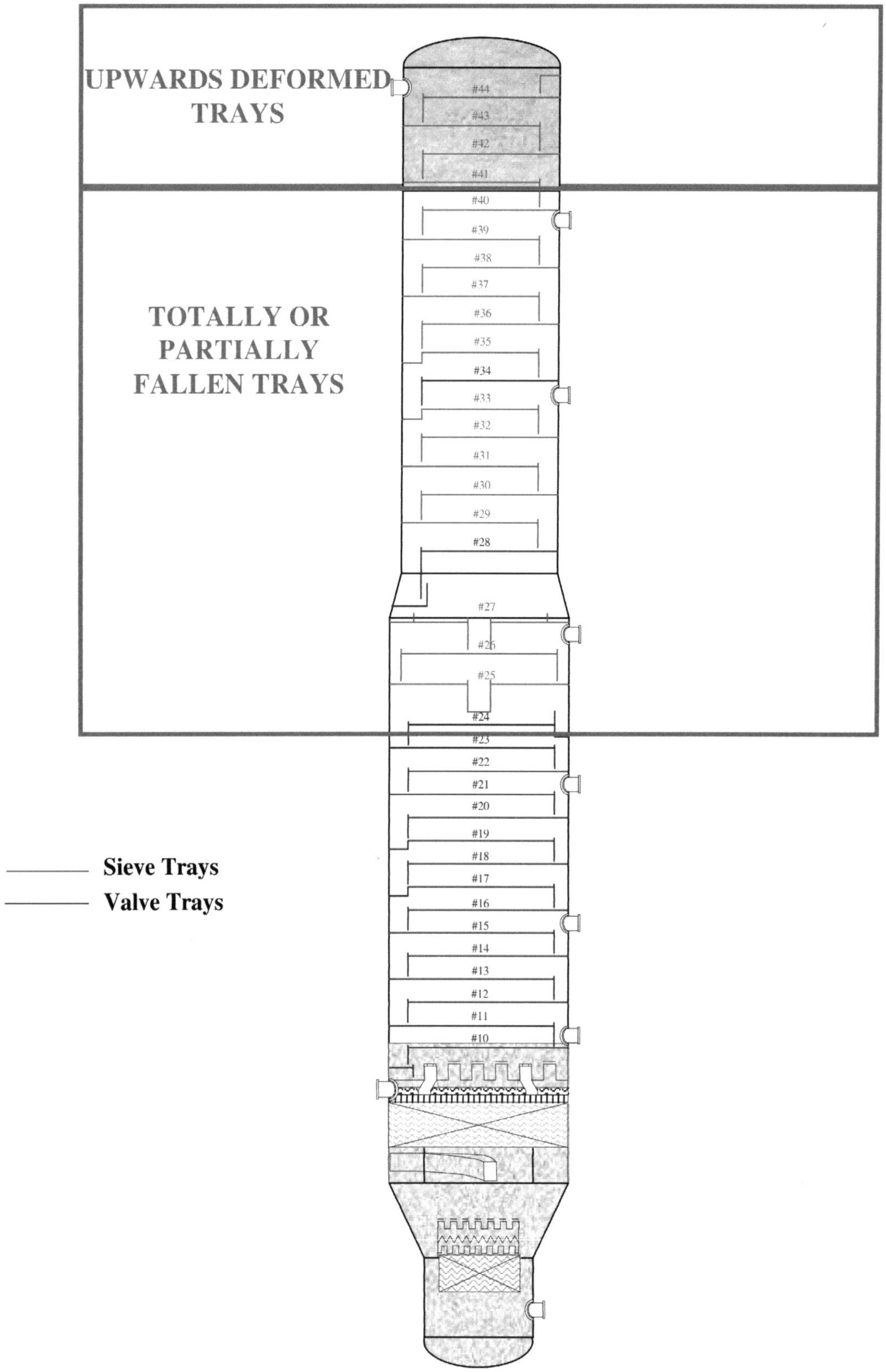

Figure 12 <u>**Second Case Study – Crude Tower Sketch**</u>

Although the pressure surge occurred at the flash zone, the packing and the valve trays located below the damaged trays and closer to the surge source were preserved. A possible explanation is based in the greater open area and tower diameter in this tower section.

Some pictures of the damage trays are showed in Figure 13.

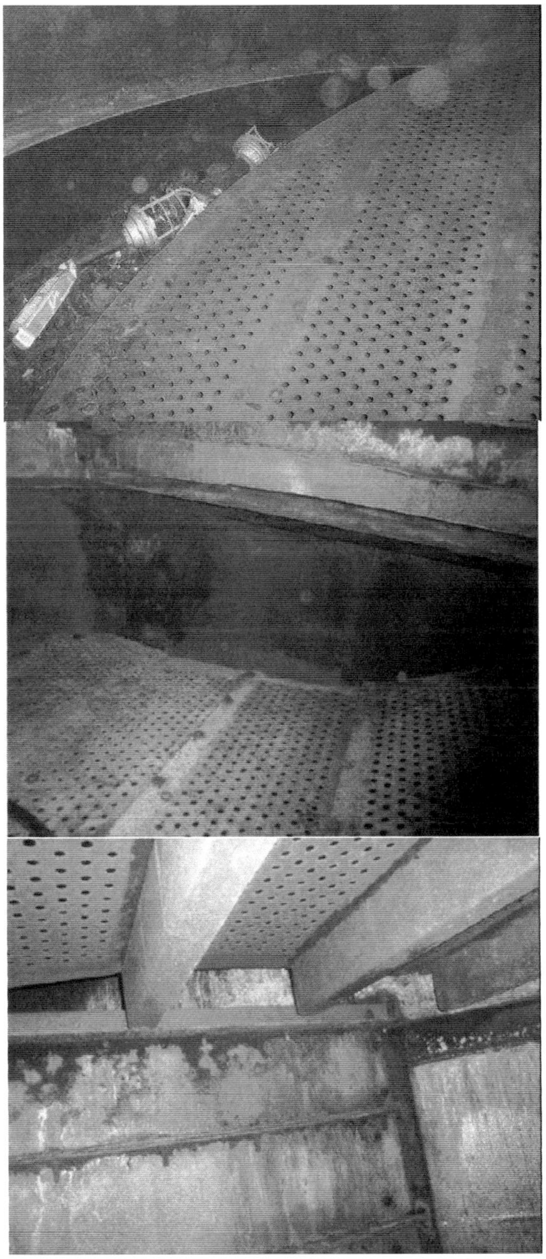

Tray # 29 fallen above tray #28

Tray # 40 fallen above tray #39

Tray # 42 with deformation upwards

Figure 13 **<u>Second Case Study – Damaged Trays Pictures</u>**

Investigation to find the causes of the incident showed a synchronism between the desalter safety relief valve opening, the atmospheric tower flash zone over pressure and an increase in water presence at the overhead drum. All of the observations led to a conclusion that the tray damages were caused by water-induced pressure surge.

The source of this water was attributed to a not drained dead pocket in the second desalter PSV discharge line as can be seen in Figure 14.

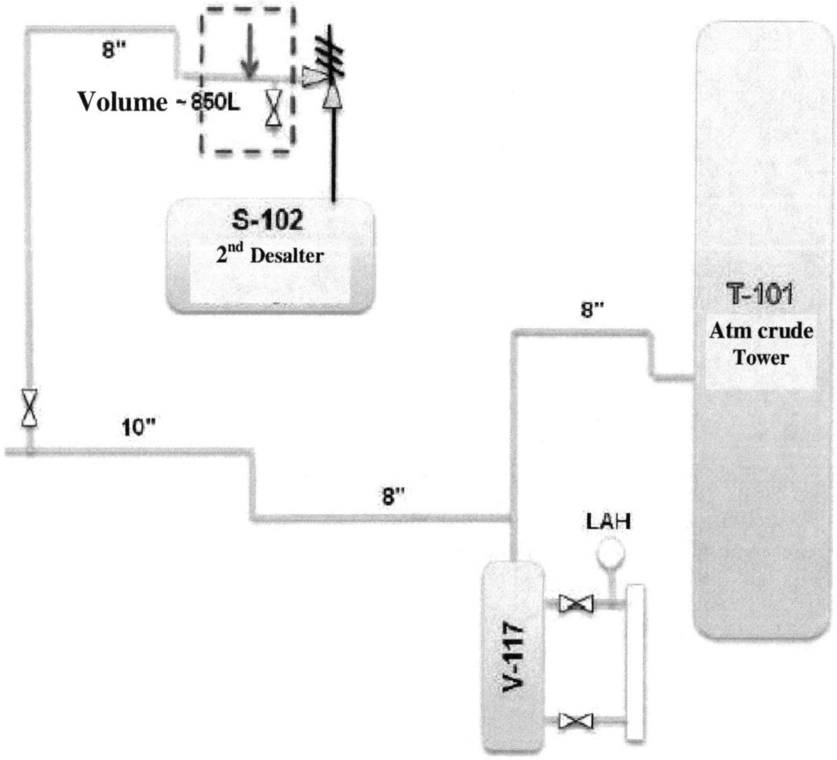

Figure 14 **Second Case Study – Water Accumulation in Dead Pocket**

The trays were fixed and the entire distillates yields returned to the normal values.

Unfortunately one month after the start-up another water-induced pressure surge happened at the same tower, which will be the third case study of this paper.

Third Case Study: PSVs liquid discharge to the atmospheric tower flash zone

In January 2013, after one month of operation, another incident happened to the same atmospheric tower. Due to a misunderstood interlock action, the feed charge booster pump B-120 A/B showed in Figure 9, was shutdown with continue operation of the feed charge pump B-101 A/B. Again the desalter safety relief valve opened and once more the atmospheric tower flash zone was pressurized. The flash zone pressure increased 0.9 kgf/cm^2 in 20 sec while the top pressure increased 0.85 kgf/cm^2 in 1 min, as can be seen in Figure 15.

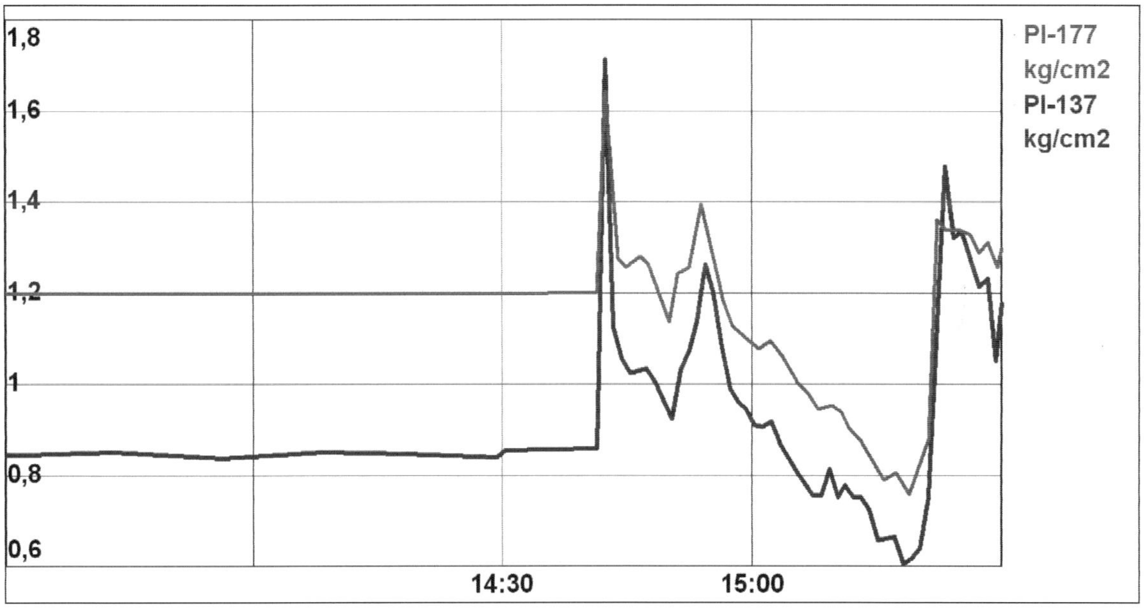

Figure 15 Third Case Study – Crude Tower Overpressure

After this event it was not possible to draw neither heavy naphtha nor kerosene. The lower draw-off pans were not able to compensate these losses and the bottom residue increased, meaning reduction of the total distillates yield, as indicated in Figure 16. A general decrease in the tower temperature profile was observed as shown in Figure 17.

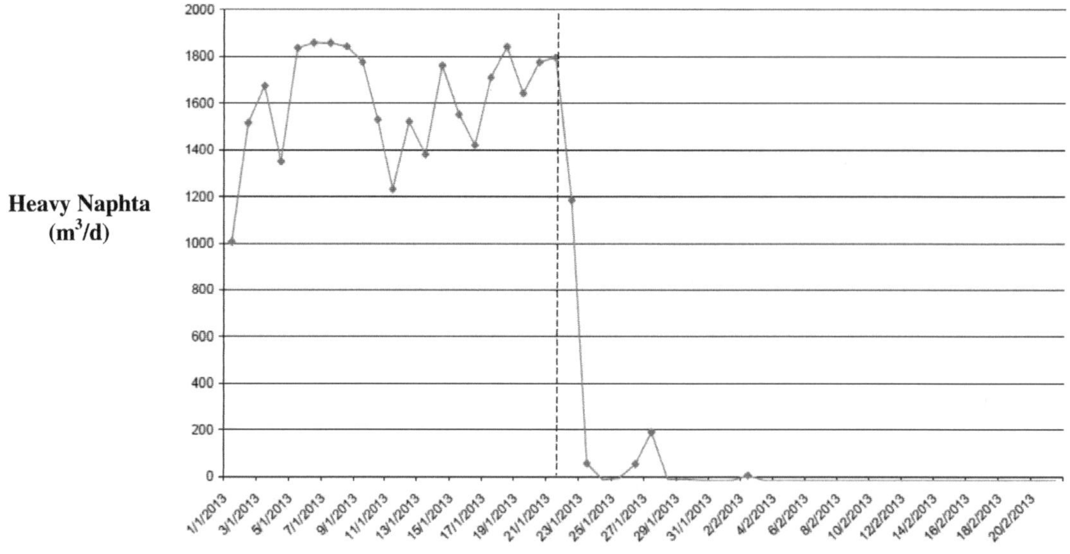

Figure 16 Third Case Study – Heavy Naphta Yield Reduction

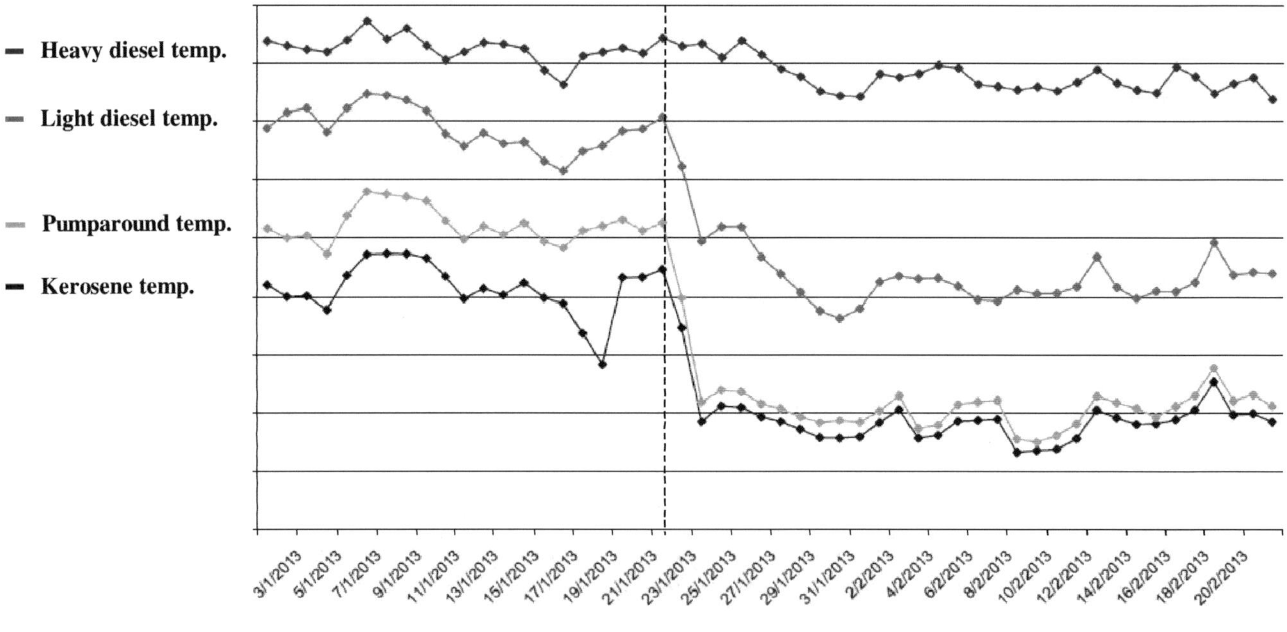

Figure 17 Third Case Study – Temperature Profile

The symptoms after this event were much worse than in case 2 and a new gamma scan (partially reproduced in Figure 18 in blue solid line) also indicated greater tower damage, showing a lot of trays partially damaged and trays from #40 to # 35 completely collapsed. Once again the damage trays were located at the tower smaller diameter section, from tray # 28 to tray # 43.

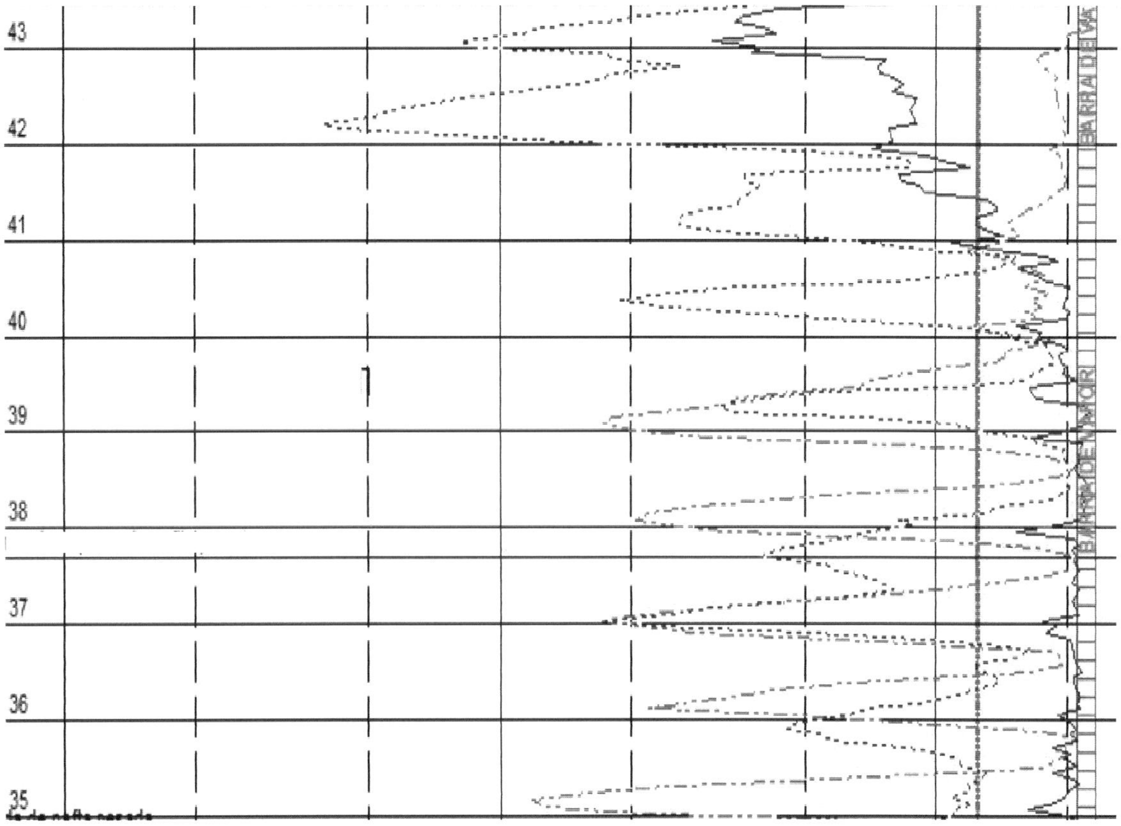

Figure 18 **<u>Third Case Study – Gamma Scan (partial)</u>**

Figure 19 shows schematically that many PSVs discharge liquid to the atmospheric tower. The heavy naphtha, kerosene and first pumparound systems have their PSV discharging to the first pumparound return, while for the light diesel, heavy diesel and second pumparound systems the discharges are sent to the second pumparound return, and the desalters (S-102 and S-103) and the preheating trains have their PSVs discharging to the tower flash zone.

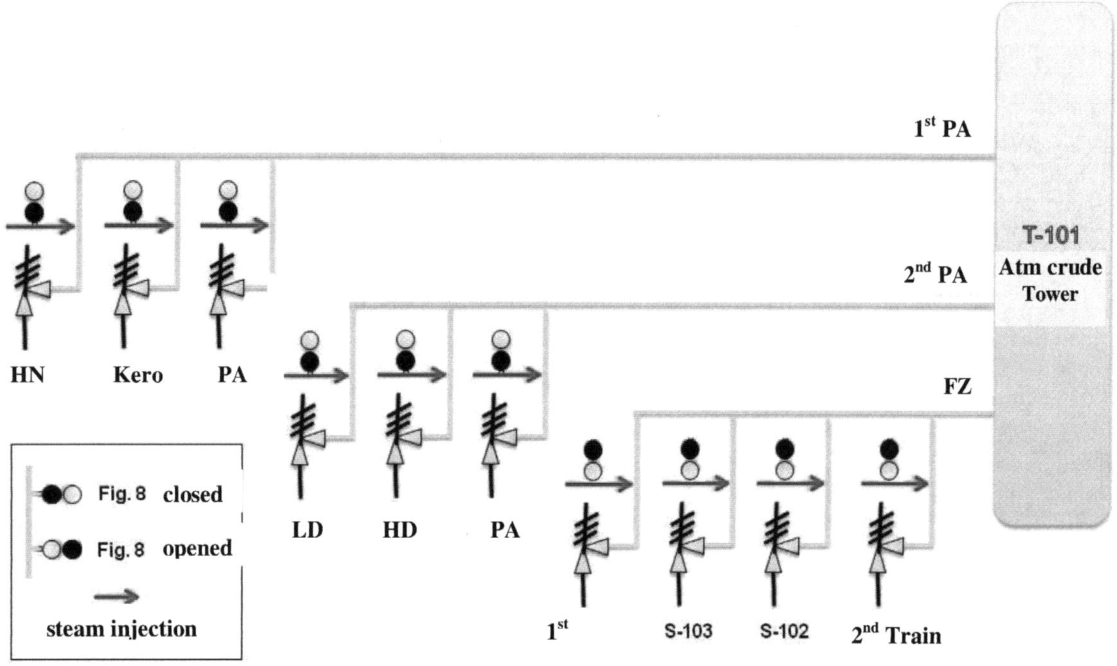

Figure 19 **Third Case Study – Safety and Relief Valves (PSV) discharges**

A multi-disciplinary team was designated to find the reasons of the tower damages. Once more the conclusion was water-induced pressure surge, but this time, the water source was found and was still in action during a visit to the operational site just two days after the incident.

During the site visit some bad practices were identified to be occurring simultaneously, which can be understood with Figures 19 and Figure 20:

- V-117 is a small drum used to indicate when any of these safety relief valves opens. Unexpectedly the level instrument connections were blocked and in fact there was no liquid indication;
- During the incident investigation it was drained a lot of clean water from V-117 for more than 30 min;
- This water source was caused by a non blocked steam injection to a valve used only with the purpose of PSV liberation . This valve is highlighted in dashed red line in Figure 20. Steam was flowing since the start-up and continued even after the tower damage event. Although this PSV has a double valve block, at that moment, only one valve was closed but allowing steam passage, and the second one was open due to a difficult access;
- The spectacle blind of this PSV steam injection system was not closed as it should be. As a matter of fact it can be seen from Figure 19 that all spectacle blinds located at the systems that discharge to the tower flash zone were open.

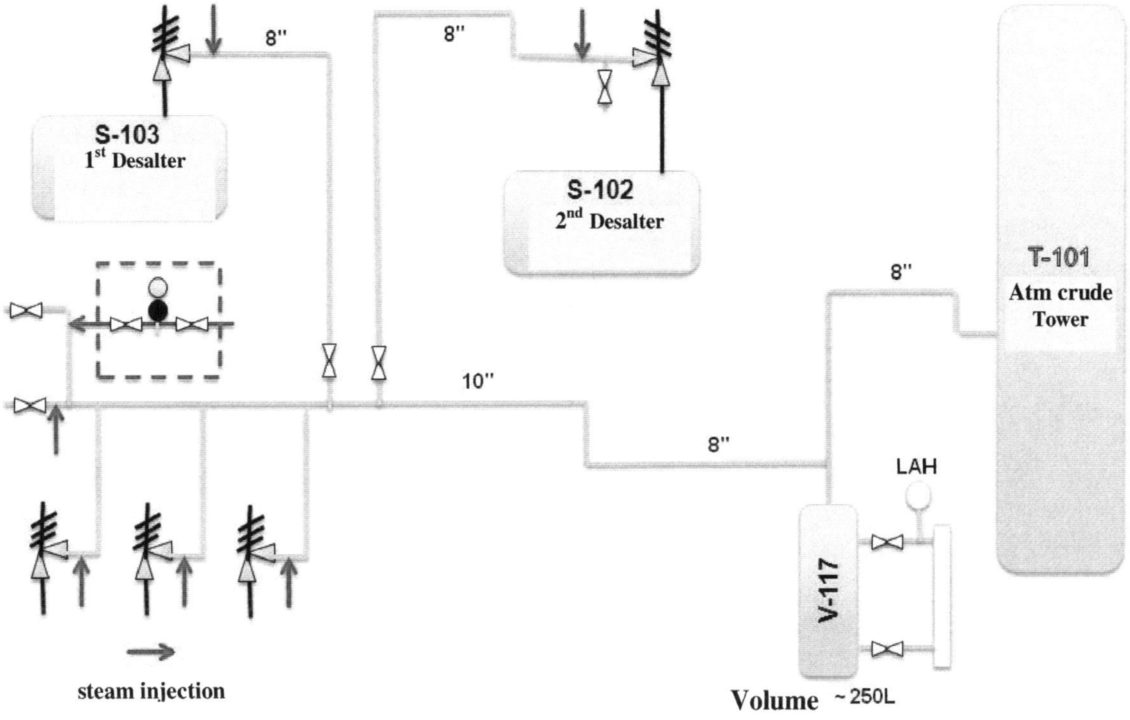

Figure 20 Third Case Study – V-117 LAH not aligned and steam injection not blocked

Conclusions

Towers malfunctions and as consequence losses in distillates yields and profitability, although undesirable, unfortunately are very common.

Three case studies of tower internals damage caused by water-induced pressure surge were presented.

All the cases have in common an atmospheric crude tower overpressure after a pressure relief valve opening event, discharging hydrocarbons and carrying water into the tower. Although all the cases are related to PSV opening and liquid discharging to the tower, the problem exists only due to the water presence. It should be mentioned that in this CDU many other times the desalter PSV opened discharging to the tower flash zone (in fact 16 times), but with no harm. In all of these events no increase in water accumulation was observed at the overhead drum.

It is mandatory to avoid water presence at the PSVs collectors that discharge liquid to towers. Focusing in eliminating water accumulation is necessary in all steps of the unit cycle life: basic design, detailed engineering, construction, inspection and all the time during operation, especially after shutdowns and start-ups. The liquid collector should be inclined with no dead pockets. There must be a reliable way of identifying liquid presence. Drainage is very important.

A more conservative design, but more expensive, is to have a blowdown drum collecting all the PSVs liquid discharges.

The case studies are summarized in Table 2.

	Case Study 1	Case Study 2	Case Study 3
Initial event	heavy diesel pump start-up after emergency shutdown	feed charge valve closure	feed charge booster pump shutdown
PSV location	crude X diesel heat exchanger	desalter	desalter
PSV discharge	pumparound atmospheric tower return line	atmospheric tower flash zone	atmospheric tower flash zone
Water source	possible undrained dead pocket	possible undrained dead pocket	unlocked PSV liberation steam valve and not aligned level indicator
Mitigation	avoid dead pockets during design and operational drainage routine	avoid dead pockets during design and operational drainage routine	operational training

Table 2 – Case Studies Summary

References

American Petroleum Institute, 2007, Pressure-relieving and Depressing Systems, ANSI/API Standard 521 5th Edition, Addendum May 2008.

Kister, H.Z., 2006, Distillation Troubleshooting (John Wiley & Sons, Inc., New Jersey, USA).

Kister, H.Z. 2003, What Caused Tower Malfunctions in the last 50 years ? Trans IChemE, Vol 81, Part A, January 2003.

(112b) Butyraldehyde Tower Capacity Limitations

Daniel R. Summers[1], Wagner Mônaco[2] and Marcos Andre Neiva[2], (1)Sulzer Chemtech USA, Inc., Tulsa, OK, (2)Elekeiroz, Camaçari - BA, Brazil

The Main Distillation tower in the only Oxo Alcohol Plant in South America was revamped in 2004. The tower experienced capacity limitations from the start. A study of the column capacity was enacted in 2005. New trays were provided in 2007. The capacity only increased marginally with the new trays. In 2008 another detailed capacity study was enacted including gamma scans of the tower. During this study, the top few trays showed unusual absorption on the Gamma Scans. The buildup of water in the tower was suspected. The spare pump line was drained and found to be totally full of water. However, upon further draining, an abundance of foam came out of the spare pump line. This paper will detail the lessons learned from this experience and the ultimate solution to this very interesting tower.

HIGH COST OF QUICK TURNAROUNDS AND ERRONEOUS PROCEDURES

Petr Lenfeld

Process Equipment Technology Group
GTC Technology Europe s.r.o
Brno, Czech Republic

Ian Buttridge

Process Equipment Technology Group
GTC Technology US LLC
Irving, Texas, USA

Prepared for Presentation

at the

Kister Distillation Symposium 2013

Topical Conference

at the

2013 Spring Meeting and 9th Global Congress on Process Safety

of the

American Institute of Chemical Engineers

San Antonio, Texas

April 28 – May 2, 2013

Introduction

A plant turnaround is always challenging. But with proper planning and reliable procedures in place, last minute time crunches and failures can be prevented. When haste combines with faulty methods, costly problems result.

Case Study #1: A sour water stripper trouble

A sour water stripper (SWS) is a small unit treating waste water from other units of the refinery. After treatment, the waste water is directed to the sewage system, waste water plant and ultimately to a nearby river. The plant has no spare SWS unit, so in case of failure, the whole refinery must be shut down as untreated water is prohibited from being discharged into the river.

A sour water storage tank may be used in an emergency with a capacity for five days of normal waste water accumulation. In the event that the SWS is down, it must be repaired and started within the five days, otherwise the whole refinery must be shut down as mentioned before.

In this instance, the SWS column was packed with metal random packing. Column performance was steeply decreasing, because of packing fouling. The column had to be cleaned immediately, before contaminants concentration in the treated water reached an unacceptable limit. A five-day turnaround was planned long enough to, shut-down SWS, unload metal rings, clean the rings and repack the column, and finally restart SWS unit. All refinery units were running at minimum capacity, directing all sour water to the five-day storage tank. It should be a routine cleaning operation.

Fire

It is well known that pyrophoric iron is formed or accumulated in SWS units (1). The standard safety precaution is continuous washing with water when unit is put out of operation to keep all internals wet and prevent pyrophoric iron self-ignition and column fire. In this particular unit operators had never run across pyrophoric iron before, therefore (and due to the winter conditions) this "wet procedure" was not applied. The SWS column was not continuously washed with water, but opened after steam-out under dry conditions. This short-cut caused a dangerous situation with serious plant operational consequences.

As one might expect when air entered the hot column pyrophoric iron self-ignited, resulting in a column fire. Despite manholes have been closed immediately and nitrogen fed into the column, the fire inside continued. Temperatures inside the column reached 2000°F - the column shell became warped and bent, and tower internals had been found melted into stainless steel icicles. The fire was extinguished in the last moment before total shell failure. The column fortunately did not fall onto the nearby piperack, but it had to be secured with a mobile crane to prevent its fall.

Figure 1: SWS column had to be secured with mobile crane

Figure 2: SWS column shell was warped and bent

Problem

Having no spare SWS unit, the five-day sour water tank slowly filled with water. Only five days were left to fix the problem or find another solution after which the refinery would have to be shut down. Plant shutdown in winter time was a nightmare for plant management due to the extreme time and cost involved.

Solution

8 AM **Fire**

10 AM **Column shell inspection**

It became clear the column can't be used anymore and must be replaced or shop-repaired at least. No chance to make repairs within five days – the shell shop-repair and plus damaged internals supply were estimated to require two weeks minimum.

General frustration

Emergency team setup for express column shell repair

12 AM **Got a solution**

A light-bulb in troubleshooter's head switched on.
Another column to process sour water!

How to make a spare SWS

Sour water stripper looks like a simple, "easy to find a replacement" distillation column. Not true! A sour water stripper is not a big complex column, but it is very specific in metallurgy and auxiliary equipment. The suitable spare column must have not only a sufficient capacity, appropriate design pressure, and material of construction (and the same its reboiler, condenser, reflux drum and reflux pump) – the crucial question is the column off-gas route! SWS off-gas is routed to a Sulfur Recovery Unit, and spare SWS must be connected to SRU as well. Considering all the factors, SWS is quite a unique unit. There is only one potential candidate for spare SWS meeting all requirements – the Amine unit regenerator.

Coincidentally, there was one idle amine regenerator at a nearby Hydrotreater unit, recently set on hot circulation to reduce sour water production.

At this moment two teams had been working in parallel – one team for SWS column repair and internals supply, and another team for spare SWS design. The GTC troubleshooter was member of both teams.

1) At 2 PM the Amine unit regenerator had been checked for sufficient capacity.
2) By the end of the day the temporary cross-connection piping was specified.
3) Within next two days the temporary lines had been designed, assembled, and pressure tested.
4) The amine regenerator control screens in DCS had been reprogrammed for new operating conditions.

Happy end

Spare SWS has been started successfully ten hours before the sour water tanks filled up. Plant shut-down had been avoided.

At this point there was enough time to make thorough shop-repair of damaged SWS column. The warped SWS column was cut into three pieces. The damaged part was replaced with a new cylindrical section. New internals were designed, manufactured and delivered by GTC in emergency schedule within 8 days.

SWS unit was successfully put back into operation after a two weeks' repair period.

Morals

1) SWS units are prone to fouling and pyrophoric formation. Special precautions shall be applied for SWS cleaning.
2) Trays should be preferred to packing for SWS service - trays are easier to clean and less prone to metal fire.
3) Standard five days capacity of sour water storage tank might be not sufficient in the case of serious trouble – a spare plan should be prepared to avoid costly plant shutdown.

Case Study #2: Mysterious LPG debutanizer

Problem description

A refinery was completely shut down for planned maintenance and cleaning. However, the turnaround was not well managed, causing the employees to rush their work and not follow some procedures properly. As a result LPG debutanizer after startup did not meet product quality regularly met before turnaround. Both top and bottom products were off specification and the column did not provide any fractionation despite operators maintaining the same reflux rate as before turnaround.

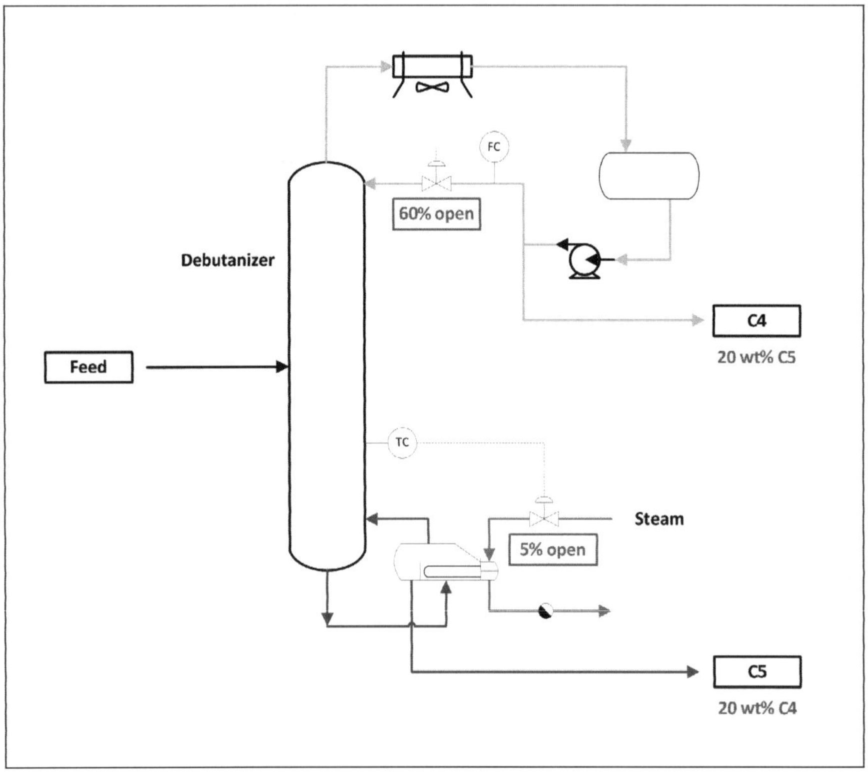

Figure 3: LPG debutanizer performance after turnaround

Problem analysis

Insufficient product separation had been first diagnosed as caused by extremely low tray efficiency. It was suspected the maintenance crew left open manways in the column trays. Later a close look at operating conditions showed the column was well refluxed, but the reboiler duty did not match. The reflux rate had been the same as before turnaround, but steam rate to reboiler was close to nothing.

The debutanizer column virtually violated an enthalpy balance. Instrument readings and control valve positions had been verified in the field, concluding both flowrate readings were right. Distillation columns can't run without any reboiling, but the mysterious debutanizer did.

Explanation of the mysterious behavior was elementary, as usual – reflux liquid was not coming down through the column, but it had been entrained back to reflux receiver and re-circulated.

Table 1: LPG debutanizer performance before and after turnaround

	Before turnaround	After turnaround
C_4 purity	0.5 wt% C_5	20 wt% C_5
C_5 purity	1 wt% C_4	20 wt% C_4
Reflux	valve 60% open	valve 60% open
Steam to reboiler	valve 50% open	**valve 5% open**

Finding the root cause

Heavy liquid entrainment is usually addressed as jet-flood, caused by extreme vapor load. In this case vapor rate could not cause any entrainment as the reboiler duty was almost none. One suspicion was that one of the tray downcomers was blocked with a piece of mineral wool insulation or other debris.

Plant management was not very pleased with this explanation – "The column has been just cleaned and inspected for perfect operating condition, but you claim the major malfunction is caused by some debris? Should the LPG plant be shut down again for an inspection and cleaning that was just performed?" Management needed a justification due to the expense. The troubleshooter proved that the problem was real and it could not be fixed without unit shutdown.

The short-term shutdown of the LPG unit was carried out, keeping the rest of the refinery running at a low capacity and temporarily flaring LPG as there was no spare LPG plant. The debutanizer column was open, after water washing and air purging. The inspector's first reaction was surprising – "Everything is fine, the column is clean!"

Mystery solved

The debutanizer column was equipped with two-pass trays, with side and central downcomers. On every other tray the central downcomer forms a wall in the column center, effectively blocking access from one side of the tray deck to the other. For this reason two-pass trays must have two manways, one per each segment, to ensure access to both column sections.

In case of LPG debutanizer the trays were of an old design, with separate I-beam support and bolted decks. The maintenance crew "did not waste time with opening both manways because trays had been always clean in the past". Only one manway has been opened in every tray for inspection and cleaning. On trays with central downcomers only one half of each tray had been cleaned as consequence. This wrong cleaning procedure has been applied for 20 years, since column first startup. The sections "behind downcomer" have never been inspected or cleaned. When both manways were opened, every other tray has been found covered with 1-inch layer of rust and dirt, with valves stuck and some downcomers blocked. Cleaning in the past was performed by water washing, so washed out dirt settled in non-accessed tray sections until they got completely blocked.

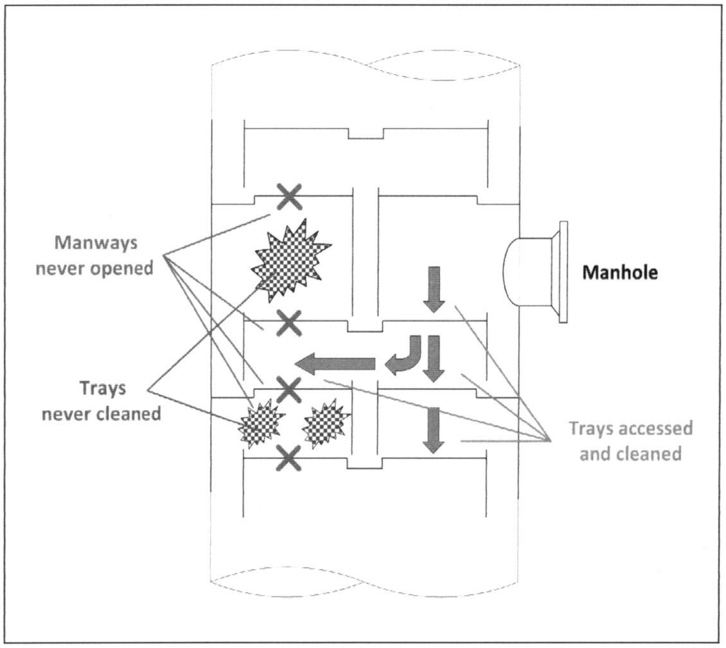

Figure 4: LPG debutanizer wrong cleaning procedure

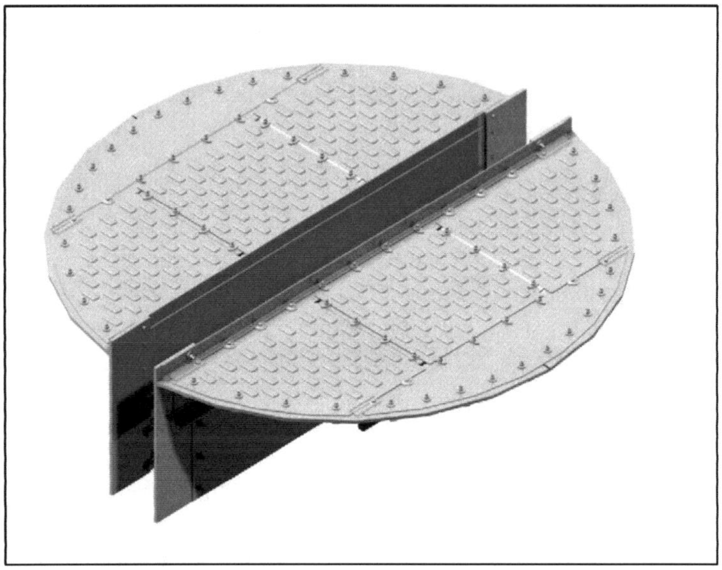

Figure 5: Modern design of two-pass valve tray with fast opening manways

Morals

1) Cleaning and inspection procedures shall not be compromised, despite turnaround time constraints.
2) "Sweeping dirt under the rug" is not the right cleaning procedure.
3) If the trays were of modern design, with fast opening manways and locks, the maintenance crew would not compromise inspection and cleaning procedure.

Case Study #3: Wrong level indication calls for column revamp

Problem description

A petrochemical plant solvent recovery column was revamped converting the existing 1-pass trays to 2-pass high performance trays to account for a large capacity increase and much higher liquid rates. After start-up of the unit unstable reflux level was observed making control of the column difficult.

The diagnosis by plant personnel was that the unstable reflux was due to flooding of the tower and entrainment carryover. The immediate solution proposed by plant personnel was to revamp the tower to packing.

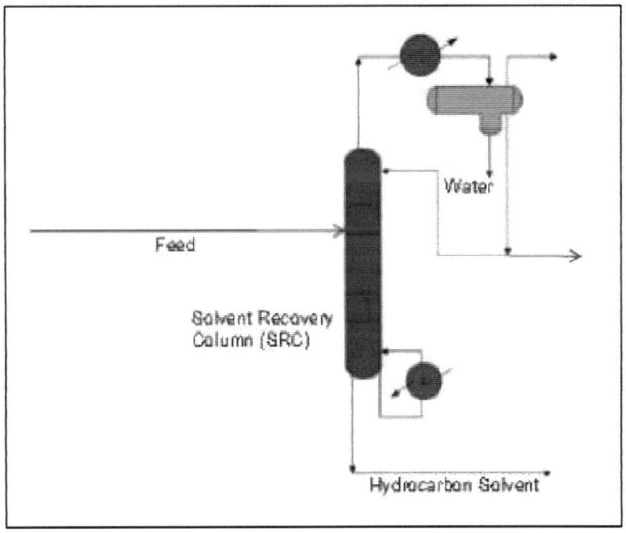

Figure 6: Tower configuration

Deciding on trays versus packing

The immediate reasoning for selecting packing was that a packing could be chosen that would appear to have more capacity than the high capacity trays currently in the tower, and that the nominal HETP (Height Equivalent per Theoretical Plate) nominally reported by vendors would be more than enough to achieve the number of theoretical stages required.

Calculated Flood Fraction of high performance trays = 79%

Calculated Flood Fraction of 125 m^2/m^3 X type structured packing = 70%

The nominal packing HETP for this type of packing was reported as 1.2m. This appeared to meet the separation requirements, as modeling had shown the revamped trays were generating 5 theoretical plates at 610mm tray spacing. The height available in the tower appeared more than enough to achieve the separation with the benefit of a lowered flood value.

It was at this point that fractionation specialists were called in and the dangers of applying a nominal HETP to different applications were noted. In this case of solvent separation where polar components are involved extrapolating HETP's used for low relative volatility non-polar hydrocarbons should not be considered. By cross consulting an operating data bank, it can be seen that for moderate pressure applications a structured packing surface area of 200 to 250 m^2/m^3 would be required to achieve the same or better efficiency than the trays, regardless of nominal values reported.

At the specific liquid flux evaluated at these conditions the capacity of even a high capacity structured packing at this specific surface area would be much less than the high performance trays currently installed in the tower. Finally, the revamp scope and downtime required to complete a revamp from the tray to packing solution would be costly.

Root cause evaluation

Examining the design of the trays themselves resulted in no immediate cause for concern, and indeed rating methods resulted in a predicted flood value of 79%. Entrainment and flood were not expected at these conditions.

An examination of the raw data and the actual symptoms would need to be conducted. The main operating issues came about as extremely unstable overhead level and the inability to control the overhead pressure. As the overhead pressure increased the operators reacted by reducing steam to the tower bottoms or decreasing reflux to the tower. The overhead reflux drum level would dramatically rise to high levels and level hunting began. If overhead pressure decreased below the set point the reflux was increased or steam was increased. Overall it was difficult to reach a steady state operating point.

The unstable operation was assumed to be caused by flooding in the tower. When evaluating the tower for other flooding signs (such as higher pressure drop) it was found that no flooding symptoms where observed. The bigger problem was diagnosing what was the root cause, and what was a symptom of the root cause. This would only be found after examining the tower under more stable conditions.

Reduced feed charge operation

At a feed charge rate of 50% of the revamp conditions a clearer picture was observed. At these conditions, what became noticeable was that the bottom level hunting preceded the overhead reflux hunting. In this case the bottom level would rise resulting in increased steam or decreased reflux, which would cause instability in the overhead level. The bottom level would dramatically fall as well reacting to the step changes. At these conditions the overhead pressure was relatively stable.

Field observation

Incorrect field measurements in particular level measurements are not uncommon issues faced by troubleshooters. The best way to confirm that these are the issue is walking to the unit and a visual check. *In this case it became immediately obvious when liquid was observed draining into the top level nozzle at the bottom of the column.*

Tower internals revised orientation and process control nozzles

In general tower internals designers should evaluate the new internals of a tower for the arrangement relative to process control nozzles. In the case of simple revamps where the number of passes remain the same this is relatively trivial. In the case where the tower is being converted from trays to packing, or in the case where the number of passes changes some extra care should be considered. In this particular case the internal equipment designer was provided with information to evaluate all process and control nozzles that were located inside the new trays, however no information was shown for the bottom level nozzles relative to the reboiler return. Figure 7 compares the prior revamp bottoms arrangement to the new arrangement.

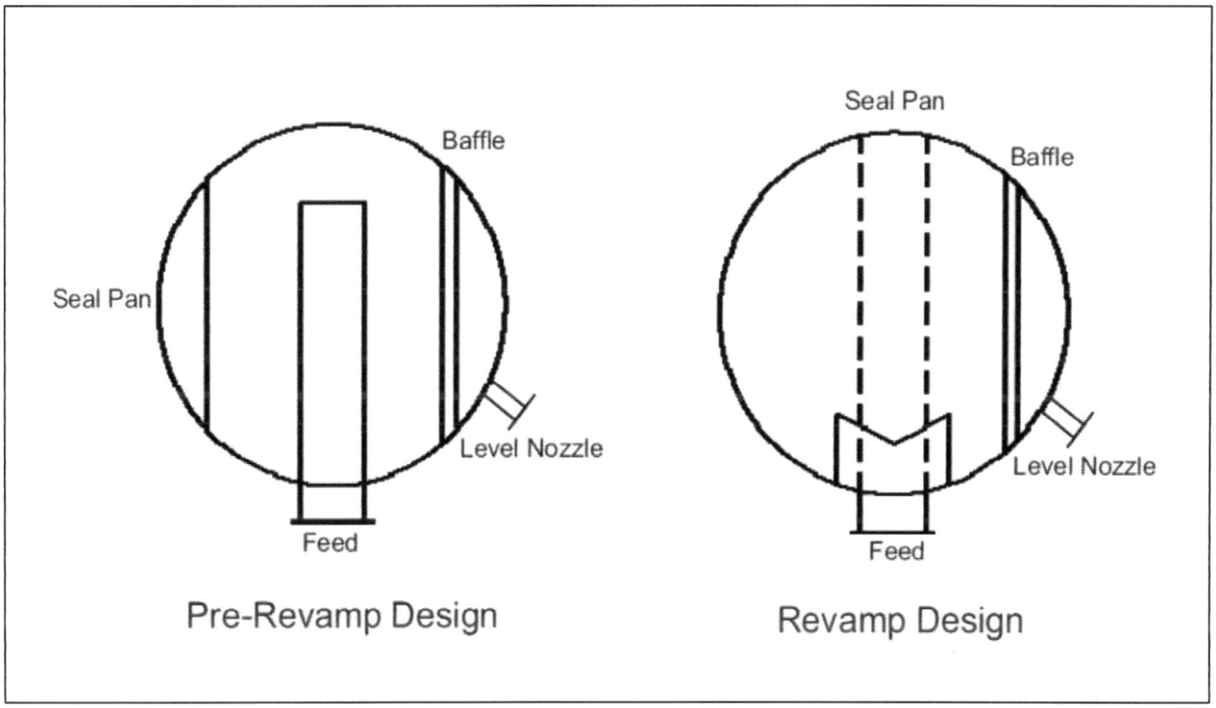

Figure 7: Pre-revamp and post revamp configurations

The revised arrangement directed the two phase reboiler return directly into the top level nozzle, which resulted in liquid draining into the level. This resulted in control issues. A seemingly high level in the product draw saw operation increase the reboiler steam, which in turn would follow with a rise in reflux flow.

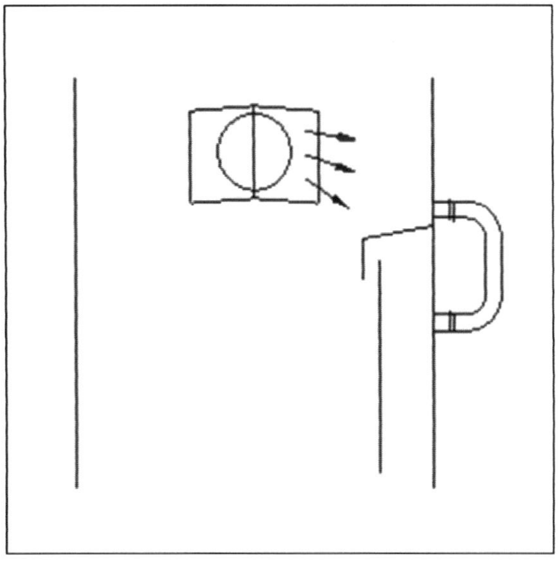

Figure 8: Liquid entrained into top level nozzle

Modifications

Adding new level taps at an elevation below the top of the reboiler baffle arrangement was probably the most straight forward option, however adding new nozzles may require PWHT or at least additional approval and time. In order to further minimize the downtime a simple solution of installation a half pipe to seal off the top level nozzle from the reboiler return was used. This was simply extended through the baffle. The result was much more stable tower operation at the full range of feed rates. The sketch in Figure 9 displays the solution.

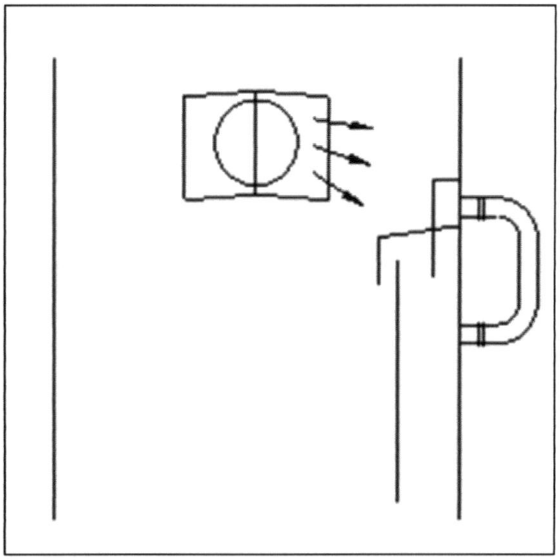

Figure 9: Level nozzle field modification

Morals

1) An evaluation of tower internals selections in regard to process nozzles should be considered not only for mechanical interference, but also in regards to a detailed analysis in regards to fluid flow and process control. An extra review step with all parties involved can further eliminate future issues.
2) Careful troubleshooting and diagnosis by specialists in the particular engineering discipline is a key to correct diagnosis and preventing further costly revamps that do not address the root cause issue.

References

1. Mary Kay O'Connor Process Safety Center, "Best Practices in Prevention and Suppression of Metal Packing Fires", Texas 2003

(112d) Ruptured Rupture Disc

A. Y. Ogundeji, T. C. Thurber and Mike Resetarits, Fractionation Research, Inc., Stillwater, OK

The FRI distillation unit in Stillwater, OK, contains 11 relief valves and 2 rupture discs. One of those rupture discs protects the steam system that drives the FRI reboilers. On 4 June 2012, FRI purposely blew that rupture disc. Video cameras were placed at four outside locations and at one location inside the control room. Birds seemed to know when the disc was about to rupture. The event was loud. As a result of this study, FRI gained additional confidence in the discs that had been used successfully for many years.

Enclosed vessels, the ones that experience internal pressures in excess of ambient pressures, must be protected from over-pressurizations. Those vessels can be protected with relief valves or with rupture discs. When those valves/discs blow open, the resultant rapid depressurizations are shocking, frightening and loud. When those valves/discs fail to open, or when they were not engineered into the processing system in the first place, the resultant rapid vessel depressurizations - - - are far worse - - - and people sometimes get hurt.

(112f) Tower Scanning – Now a Quantitative Tool, Part B

Lowell Pless, Tracerco, Pasadena, TX

Tower gamma scanning and neutron backscatter techniques are well established in the process industries as qualitative tools to help troubleshoot separation towers, both trayed and packed towers. In this presentation it will be shown that through careful and innovative application of gamma scanning and neutron backscatter techniques, quantitative information about separation towers can be extracted.

Part B: For packed towers some form of a grid-scan, a series of multiple vertical scans through packing, has been used to detect liquid mal-distribution through packing. A new technique will be introduced whereby gamma scan data is used to determine relative process densities through packing. Relative process densities can be converted into liquid holdup fractions, either of which can be used as a quantitative measure of liquid mal-distribution. This concept will be applied to several case studies to illustrate how quantifying gamma scan data can provide more useful information than by traditional qualitative analysis.

Influence of Surface Tension on the Performance of Mist Eliminators – Part II

Wolfgang Heikamp Rhodius GmbH
Prof. Dr. Michael Schultes Raschig GmbH
Attilio Praderio Conoco Phillips

AIChE Conference 28th April – 3rd May 2013 San Antonio

My special thanks to:

Attilio Praderio Conoco Phillips, for the great ideas,

Prof. Michael Schultes Raschig, for the support and trust

Mr. Armin Pitzer for the work in the lab

Content

- Introduction
- Review last results, open points
- Test rig
- Results
- Conclusions

Introduction

Typical mist eliminator systems are:

cyclone system
knitted wire mesh
random packing
vanes

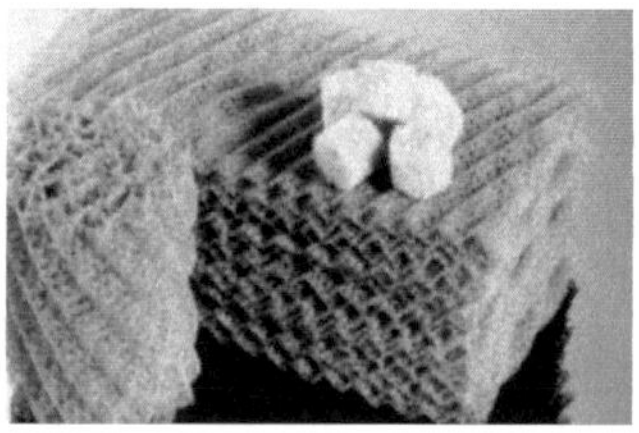

Introduction

RHODIUS
Safety and Environmental Solutions
RASCHIG JAEGER TECHNOLOGIES
ConocoPhillips

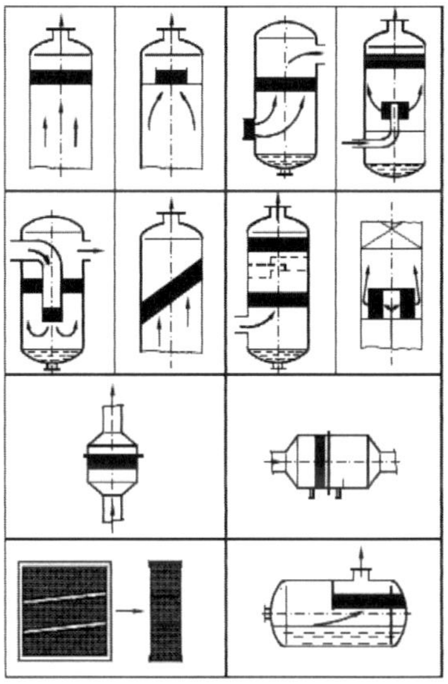

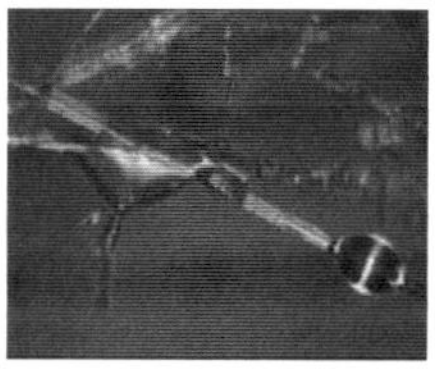

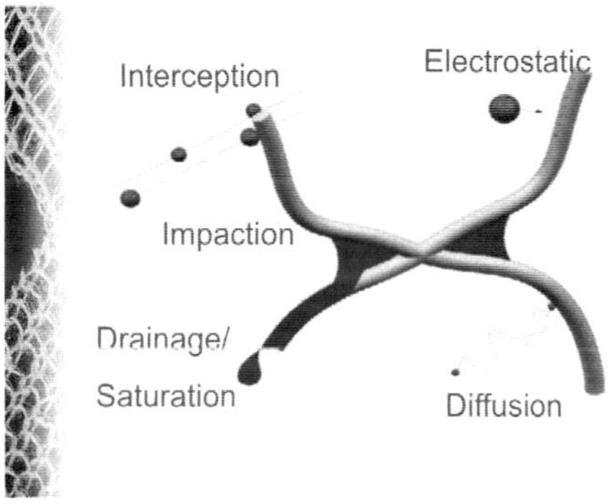

Introduction

RHODIUS
Safety and Environmental Solutions
RASCHIG JAEGER TECHNOLOGIES
ConocoPhillips

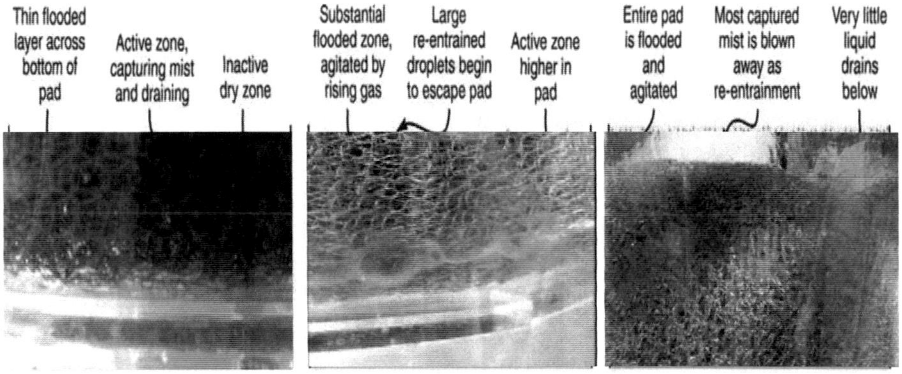

Point A:
Optimum velocity, light liquid load

Point B:
Higher velocity and load: re-entrainment begins

Point C:
Excessive velocity and load: flooding, full re-entrainment

www.amistco.com

Introduction

Calculation mode A:

Souders – Brown: $U = K \sqrt{(\rho_L - \rho_G) / \rho_G}$

typical K values:	0.3	ft/s:	minimum entrainment
(p. e.)	0.2	ft/s:	moderate pressure
	0.45	ft/s:	high pressure
	0.2	ft/s:	de – entrainment steam
			R. J. Robichaux
	0,3	ft/s	div. autors

K factor is an empirical value and change with pressure, application

Introduction

Calculation mode B:

acc. Bürkholz: semi empirical model

-> dimensionless groups based on experiments

dimensionless groups: identify the independent parameter

$$\eta_F = f\ (D, d, H, v, \Delta p, \varrho_T, \varrho, \mu, g)$$

related the parameter to the basics unit [m, s, etc.]

Stokes-, Weber-, Reynolds-, ...

=> F (x) = A(x)^a + B(x)^b + C(x)^c + ...

Introduction

RHODIUS Safety and Environmental Solutions
RASCHIG JAEGER TECHNOLOGIES
ConocoPhillips

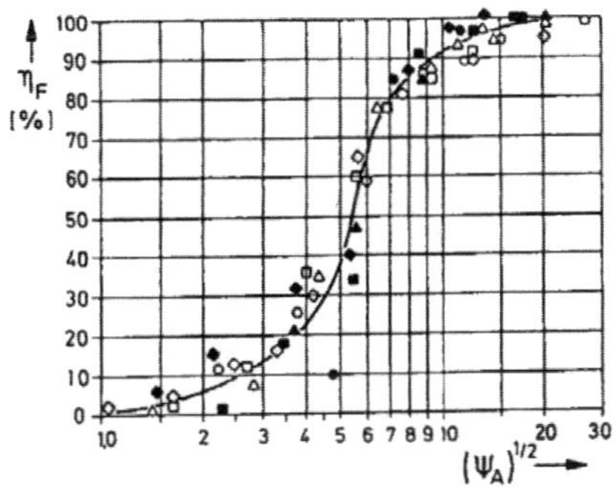

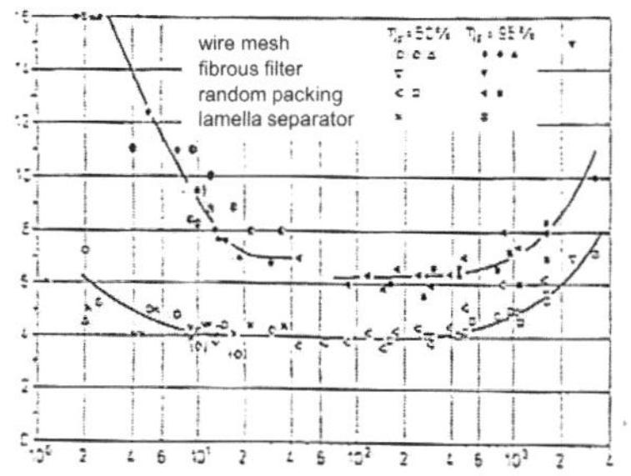

$$(\psi_A)^{1/2} \equiv \frac{1}{2}\cdot(\varrho_T^{1/2}\varrho^{-1/6}\mu^{-2/3})\cdot(\Delta p^{1/3}D*^{-1/3}d)$$

Rhodius selected method B, but
both methods have not included the surface tension

Introduction

RHODIUS Safety and Environmental Solutions
RASCHIG JAEGER TECHNOLOGIES
ConocoPhillips

Important applications with low surface tension:

- high pressure distillation of light Hydrocarbons

Low surface tension cause smaller droplets in the mist, the smaller droplets have lower separation efficiency.

The max. velocity in the KWM – Separator is lower, because the mist eliminator is starting to flood earlier.

➡ **Lost of performance and liquid**

Introduction

RHODIUS Safety and Environmental Solutions · RASCHIG JAEGER Technologies · ConocoPhillips

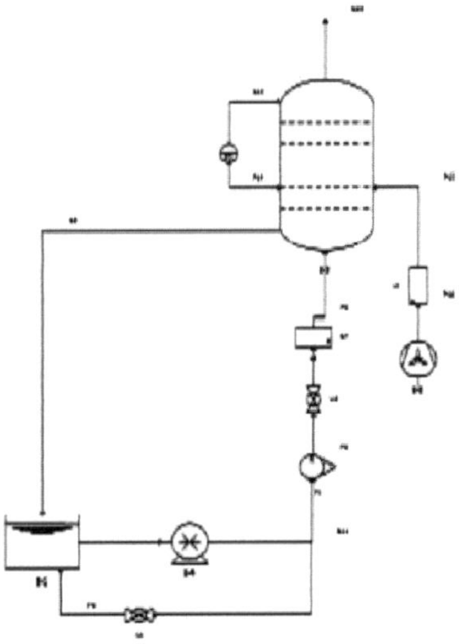

Review

RHODIUS Safety and Environmental Solutions · RASCHIG JAEGER Technologies · ConocoPhillips

graphical method

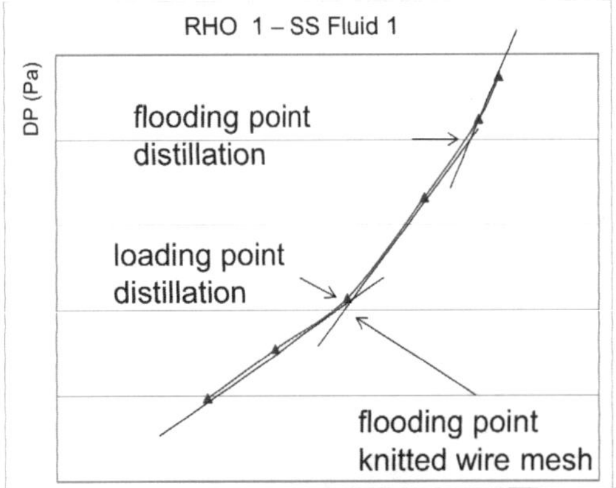

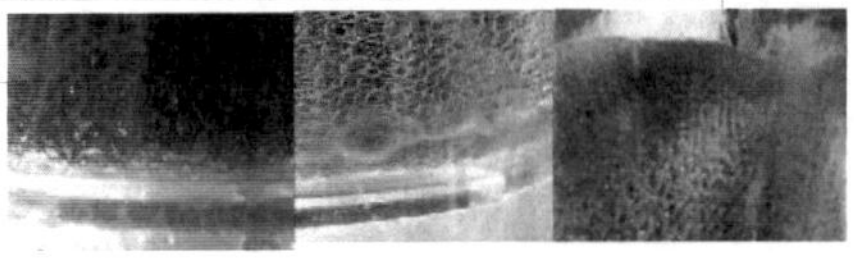

visual method

first drops coming out

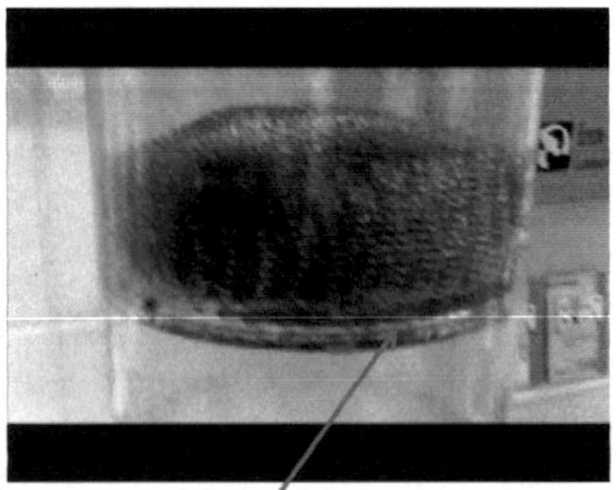

liquid level is rising

Review

no flooding

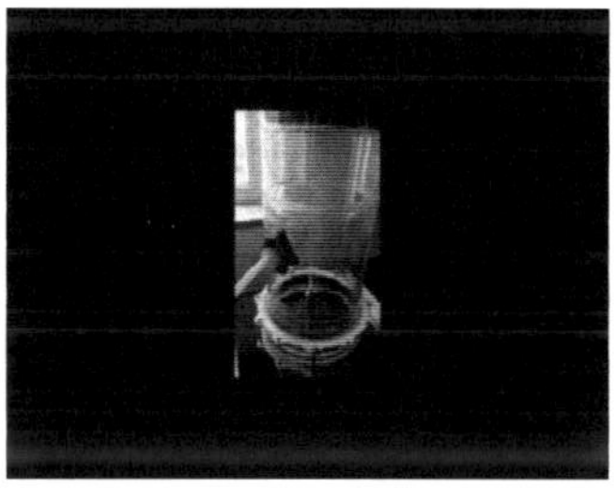

flooding

Review

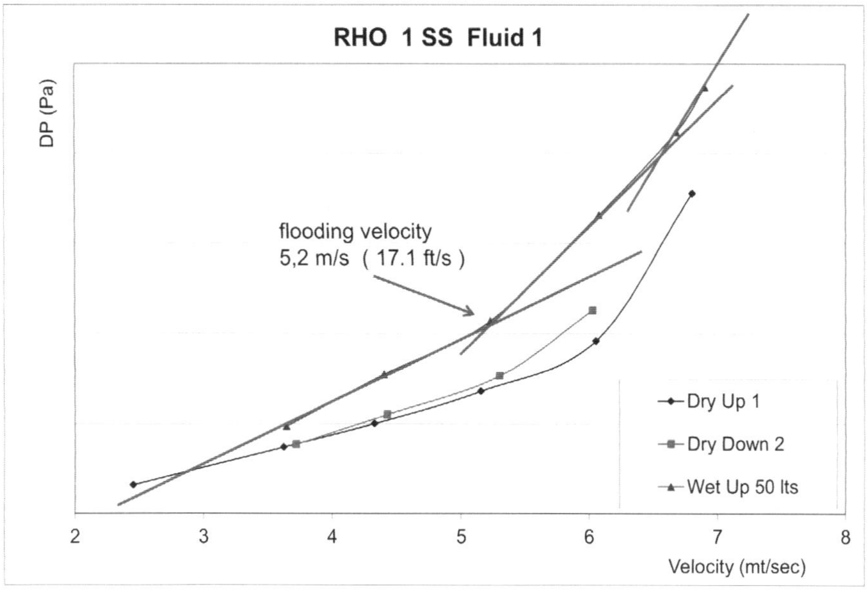

Review

Measurement Flooding Point

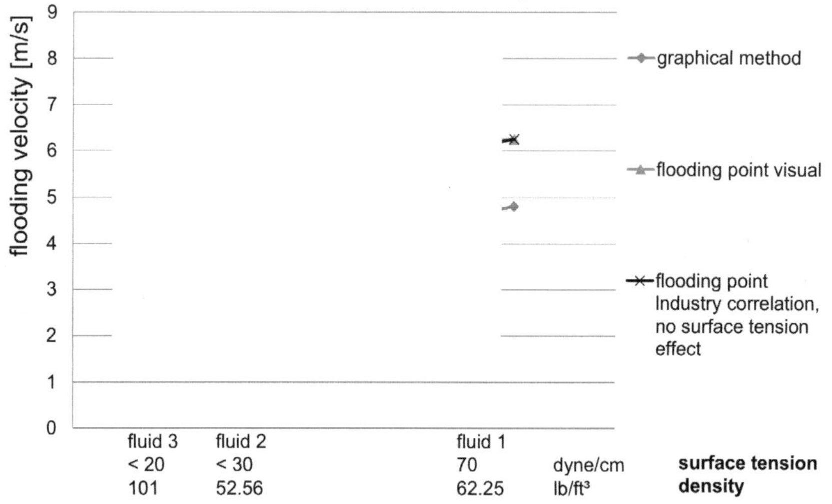

Review

Flooding velocity acc. density

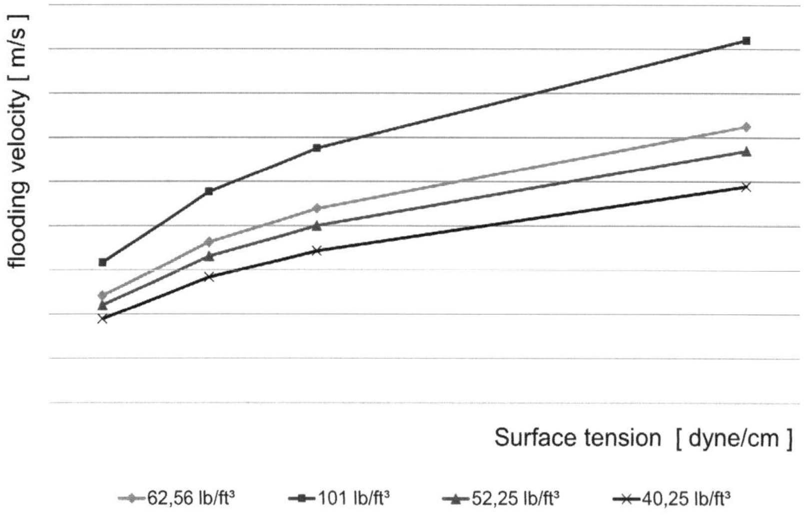

Review

RHODIUS Safety and Environmental Solutions
RASCHIG JAEGER TECHNOLOGIES
ConocoPhillips

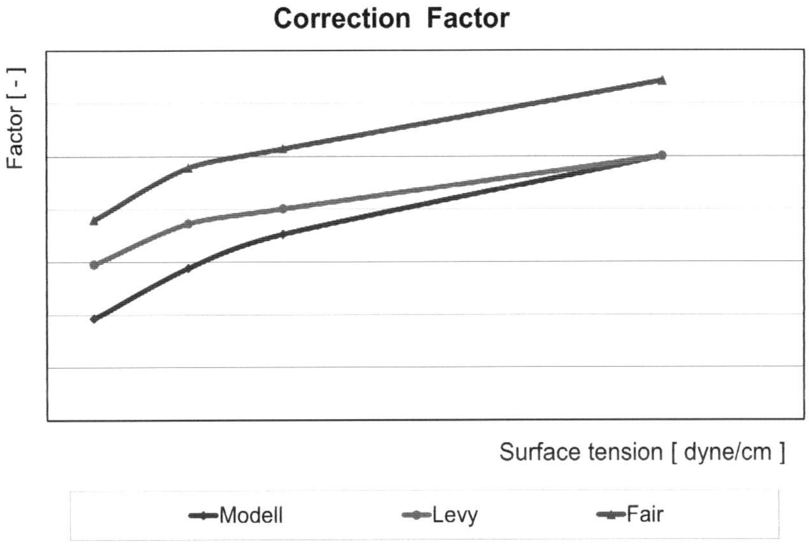

RHODIUS Safety and Environmental Solutions
RASCHIG JAEGER TECHNOLOGIES
ConocoPhillips

Open Points

18

Open Points

- Influence of density
  different fluids with various density similarly surface tension
  different fluids with same density and different surface tension
- Wettability
  different material, stainless steel, plastic
- Co Knit Style

Test rig

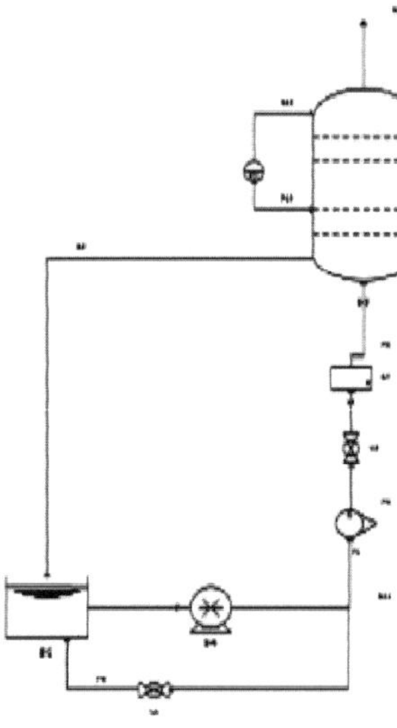

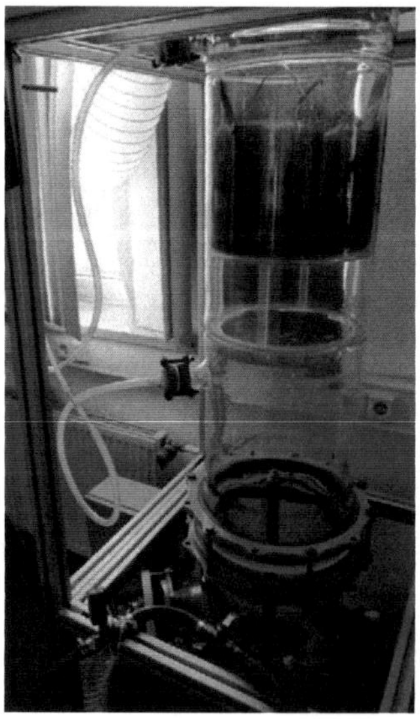

Test rig

RHODIUS Safety and Environmental Solutions
RASCHIG JAEGER TECHNOLOGIES
ConocoPhillips

First run is with dry air,
increase velocity and reducing velocity.

=> dry pressure drop

Open liquid pump and justify the flow rate.

=> wetted pressure drop

Observation if flooding occur, if not increase gas flow rate.

=> determine flood point

drop size distribution at spray nozzle below mist eliminator

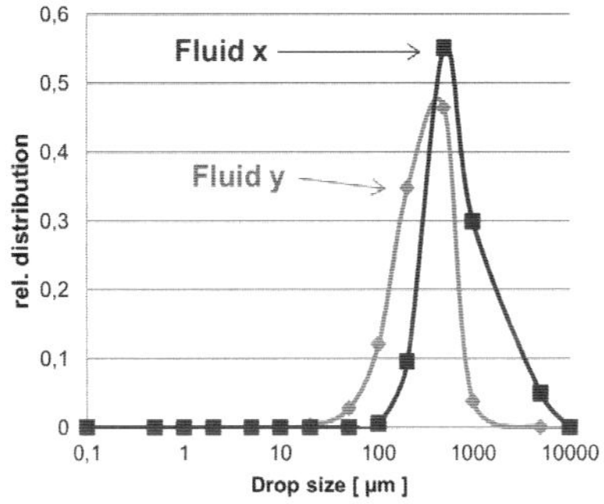

Dropsize distribution measured with Spraytec

Test rig

RHODIUS Safety and Environmental Solutions
RASCHIG JAEGER TECHNOLOGIES
ConocoPhillips

Substance data I

	Fluid 1	Fluid 2	Fluid 3	
Density	998	842	1622	kg/m³
	62.25	52.56	101.26	lb/ft³
Viscosity	1	1,2	1,249	mPas
	1	1.2	1.249	cP
Surface tension	70	< 30	< 20	mN/m
	70	< 30	< 20	dyne/cm
Vapor pressure	31,7	59	22,55	mbar
	0.46	0.82	0.33	PSI

Test rig

Substance data II

	Fluid 4	Fluid 5	Fluid 6	
Density	1110	970	994	kg/m³
	69.5	60.6	62.1	lb/ft³
Viscosity	20	1,2	1,42	mPas
	20	1.2	1.42	cP
Surface tension	50	30	40	mN/m
	50	30	40	dyne/cm
Vapor pressure	5	3,1	3,2	mbar
	1	0.07	0,07	PSI

New Results

24

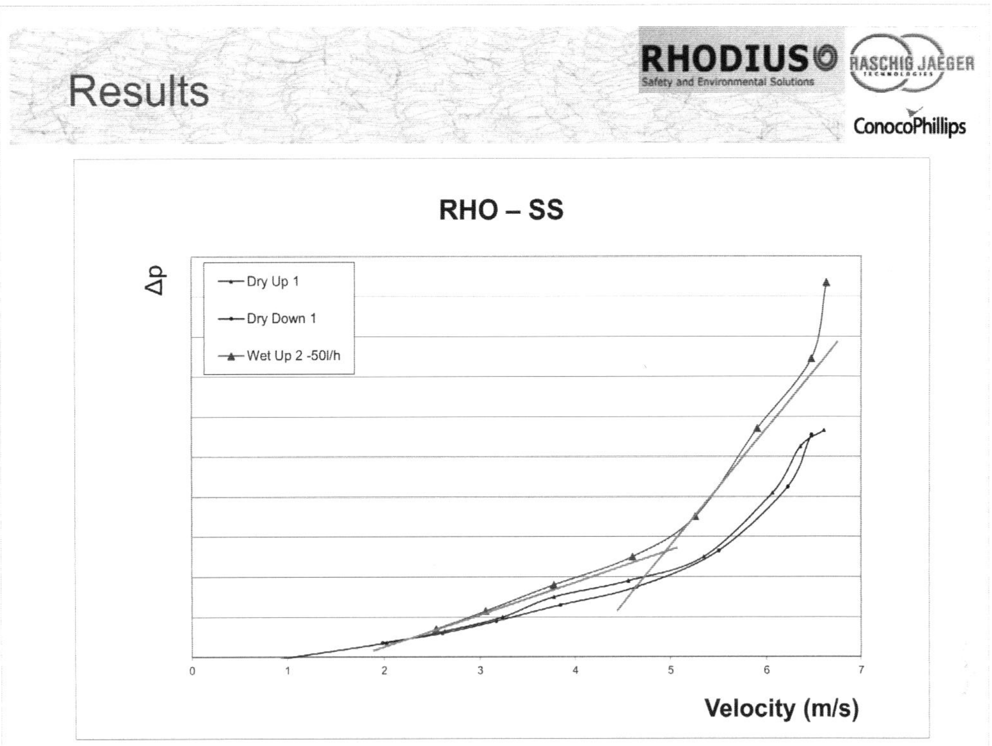
Results
RHODIUS
Safety and Environmental Solutions
RASCHIG JAEGER
TECHNOLOGIES
ConocoPhillips
RHO – SS
Δp
Dry Up 1
Dry Down 1
Wet Up 2 -50l/h
0
1
2
3
4
5
6
7
Velocity (m/s)

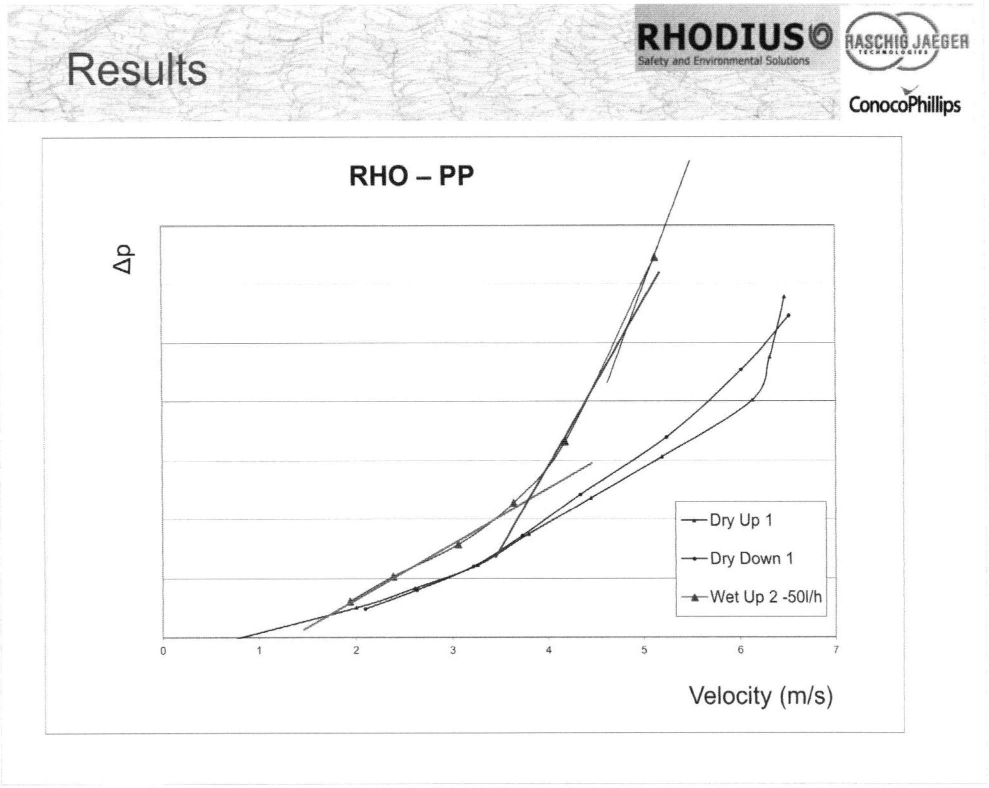
Results
RHODIUS
Safety and Environmental Solutions
RASCHIG JAEGER
TECHNOLOGIES
ConocoPhillips
RHO – PP
Δp
Dry Up 1
Dry Down 1
Wet Up 2 -50l/h
0
1
2
3
4
5
6
7
Velocity (m/s)

Results

RHODIUS Safety and Environmental Solutions
RASCHIG JAEGER TECHNOLOGIES
ConocoPhillips

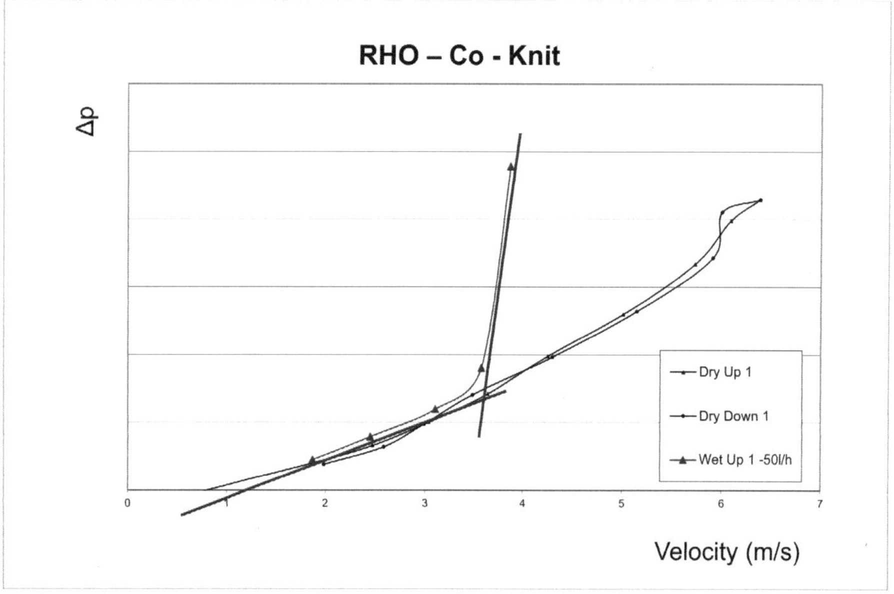

Results

RHODIUS Safety and Environmental Solutions
RASCHIG JAEGER TECHNOLOGIES
ConocoPhillips

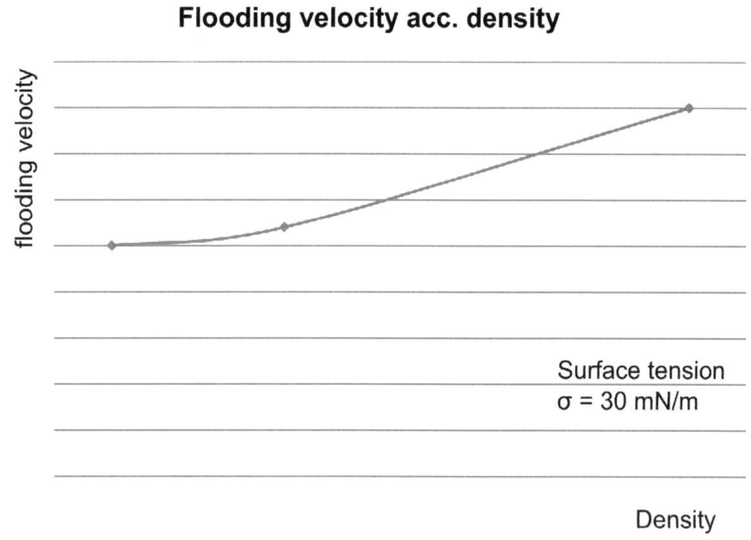

Results

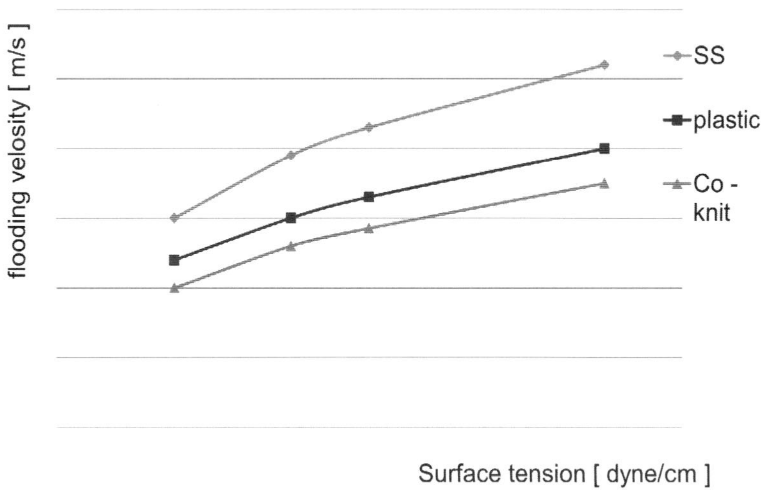

Conclusions

Test rig is able to reproduce the results

Different liquids to validate the influence of density and material

The graphical method with the selected point is a good approach

Lower surface tension reduce flooding velocity

The influence of density is shown

Influence of material (wettability) is shown

Surface tension exponent for plastic is equal to SS,
Co knit material have a different behaviour

Thank you
for
your
attention

Questions?

(130a) Design Guidelines for Cryogenic Distillation

Karl Kolmetz, Linde Process Plants, Tulsa, OK

Cryogenic distillation is highly energy intensive process. Small distillation improvements can result in a large return on investment. The separation energy for cryogenic distillation in the production of natural gas can be greater than fifty percent of the plant's total energy requirements. The separation energy for cryogenic distillation in ethylene production can be greater than twenty five percent of the plant's total energy requirements.

The removal of the light components such as hydrogen, methane, ethane and carbon monoxide in cold boxes, demethanizer and deethanizer distillation columns can be optimized to reduce the ten year life cycle cost of the plant. The ten year life cycle cost would be the equipment capital charges and the energy charges during the ten-year equipment life. Additional non-quantized benefits would be environmental and operational. This guideline will discuss simulation techniques, tray efficiencies factors, tray hydraulics and case studies of cryogenic distillation.

AIChE PAPER 130b

ADVANCEMENTS IN C2 RECOVERY DISTILLATION

Paul T. Scott, P.E.
Senior Technical Advisor- Olefins
KBR, Inc.
Houston, Texas

Yong Li Ma
Plant Manager
Lanzhou Petrochemicals
Lanzhou, P.R.C.

Michael J. Tallman
Manager, Catalytic Olefins Technology
KBR, Inc.
Houston, Texas

Prepared for Presentation at the 2013 Spring National Meeting
San Antonio, Texas, April 29 - May 2, 2013

KBR Internal Reference Paper 2373

ADVANCEMENTS IN C2 RECOVERY DISTILLATION

Paul T. Scott, P.E.
Senior Technical Advisor- Olefins
KBR, Inc.
Houston, Texas

Ma Yong Li
Plant Manager
Lanzhou Petrochemicals
Lanzhou, P.R.C.

Michael J. Tallman
Manager, Catalytic Olefins Technology
KBR, Inc.
Houston, Texas

Abstract: In today's competitive market, continuous improvements in process technology are essential. The ethylene recovery section of an olefins plant is one of the most capital and energy intensive areas of the recovery section, so technology improvements in this area can lead to significant competitive advantages.

In this paper, KBR will present two unique and patented new distillation processes for the recovery of ethylene:

- Advanced Deethanizer – a patented and commercialized distillation process in which a portion of the ethylene product is produced directly from the Deethanizer, applicable in either grass roots or revamp scenarios. The paper will also include the description of a further refinement in the process, also patented and commercialized, in which several equipment items are deleted for lower cost
- Divided Wall Column (DWC) – a brief description of DWC concepts along with an application for ethylene recovery and purification will be provided

ADVANCEMENTS IN C2 RECOVERY DISTILLATION

Paul T. Scott, P.E., KBR
Ma Yong Li, LPC
Michael J. Tallman, KBR

Introduction

Outside of the cracking furnaces the majority of the energy consumption in the ethylene process is incurred by the major centrifugal compressor drives. Of these systems ethylene refrigerant is provided at the lowest temperature levels normally required by the process and is therefore the most expensive (energy intensive). Technological improvements in this area can lead to significant reductions in overall plant energy consumption and, potentially, capital cost.

In this paper, KBR will present unique and patented new distillation processes for the recovery of ethylene:

- Advanced Deethanizer – a patented and commercialized distillation process in which a portion of the ethylene product is produced directly from the Deethanizer, applicable in either grass roots or revamp scenarios. The paper will also include the description of a further refinement in the process, also patented and commercialized, in which several equipment items are deleted for lower cost
- Divided Wall Column (DWC) – a brief description of DWC concepts along with an application for ethylene recovery and purification will be provided

The ADVANCED DEETHANIZER, a KBR patented and commercialized distillation process, is described and advantages for this unique process flow scheme are reviewed. A comparison is made to alternative designs including both conventional high pressure and low pressure heat pump systems. Further refinements to the flow scheme are explored. Finally, a brief description of divided wall distillation concepts is provided along with a discussion of its use in ethylene product recovery and purification.

Evolution of Flow Sheet

KBR has been an innovator with regard to ethylene plant process flow sheet development for decades. KBR is the result of a merger between the two former competitors, the M. W. Kellogg Company (MWK) and Brown & Root Engineering (B&R), in 1998. A variety of process flow schemes have been offered and successfully commercialized over the years between the two companies.

High Pressure C2 Recovery

One such scheme entails the traditional Demethanizer first/ back end Acetylene Hydrogenation process flow scheme. The C2 Product Recovery area for this scheme is depicted in Figure 1 and high pressure Deethanizer and C2 Splitter towers were used. The C2 Refrigeration system was closed loop and not integrated with the C2 Splitter. The Deethanizer tower pressure was set at 29.6 bar and the C2 Splitter tower 22.8 bar, with both towers condensing against a somewhat elevated propylene refrigerant level (-29°C /1.2 bar). The C2 Splitter tower was effectively a propylene heat pump as mid-level compressor extraction vapor was used to drive the tower reboiler with the condensed propylene liquid from the reboiler feeding directly to the C2 Splitter overhead condenser.

This allowed the propylene refrigeration compressor to be only two stages (no atmospheric propylene level), reducing the equipment count and energy consumption for that important system. On subsequent versions of this flow scheme, the system was further integrated by arranging Deethanizer overhead to condense against C2 Splitter inter-tray liquids as well as cracked gas in Demethanizer chilling train.

The scheme included C2 acetylene hydrogenation inter-spaced between the two towers. Ethylene product was taken as a liquid sidedraw off of tray 10 and pumped either to above critical pressure for delivery to customers via pipeline or to lower pressure storage. Boiloff vapors from pressurized storage were recovered back to the C2 Splitter tower upstream of the condenser.

This system worked well and was reasonably energy efficient for this time period due to the heat integration. However, without a propylene refrigeration level at atmospheric pressure the relatively high operating pressures led to:

- Increased fouling tendencies in the Deethanizer reboiler
- Some shift of load into C2 refrigeration
- Increased number of stages in both towers (60-65 trays was typical for the Deethanizer and 135-145 trays typical for the C2 Splitter to make polymer grade product)

Having acetylene hydrogenation between the two towers requires the process gas to be reheated then re-cooled (inefficient with regard to energy), and inert material is inserted back into the process flow scheme, requiring unreacted H_2/CO/CH_4 to be removed via a 10 trays pasteurization section on top of the C2 Splitter. Green oil removal and guard dryers were also recommended/ required to handle secondary reaction products from hydrogenation, particularly when the catalyst was fouled and operating at elevated temperatures. The hydrogenation reactor was spared to handle periodic off-line *in situ* regeneration.

Nevertheless, this configuration or some variation thereof was widely used and successfully operated in numerous units sold in the late 1980s and early 1990s.

Two designs were completed during this time period with intervening acetylene extraction used instead of hydrogenation. This eliminates the need for the Green Oil Wash tower and the pasteurization section on top of C2 Splitter. It also allows opportunity to take ethylene product as a vapor directly from the overhead of C2 Splitter, depending on the pressure requirements, with substantial energy savings.

Figure 1.
CONVENTIONAL
CLOSED LOOP
REFRIGERATION

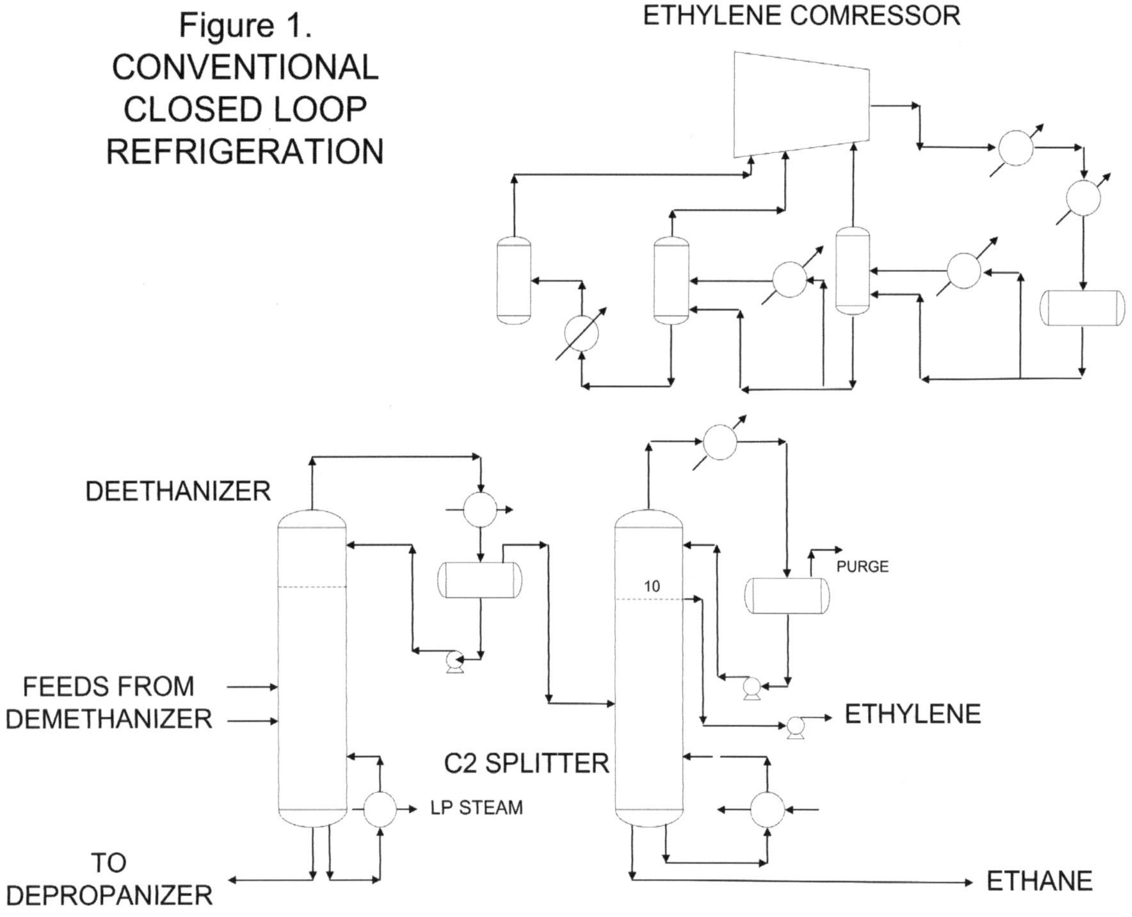

Front-End Acetylene Converter and Heat Pump C2 Splitter

An alternative design was pioneered by Brown & Root which comprised a patented "front-end" acetylene hydrogenation design coupled with low pressure heat-pumped towers and open-loop refrigeration in the cold section of the process. This scheme has many advantages relative to the high pressure design described above and was first introduced in the late 1960's and was developed through the years up to the time of Brown & Root's merger with MWK.

Figure 2 shows the base configuration for the C2 Recovery area with a high pressure Deethanizer coupled to a heat-pumped C2 Splitter. In this scheme the Deethanizer tower operates as a conventional high pressure tower condensing against propylene refrigerant. However, the C2 Splitter operates at low pressure and is integrated into the C2 Refrigeration system. High purity product is taken from the accumulator drum on the discharge side of the compressor.

The C2 Splitter overhead operates at the ERC 2nd stage suction pressure. 2nd stage discharge vapor is condensed in the C2 Splitter reboiler supplying heat to reboil the column. Condensed reboiler liquid becomes the reflux for the same tower (heat-pump configuration). The total reflux requirement is satisfied with some supplemental liquid from the accumulator drum.

In this scheme the ethylene refrigeration system is open-loop. Ethylene product is drawn from the ethylene accumulator and is pumped to the battery limits as a liquid or as a dense phase vapor after re-heating.

One minor variation on this scheme had the Deethanizer tower operating at medium pressure (13.4-15.5 bar). This required atmospheric level propylene refrigerant in the condenser but eliminated fouling of the reboiler and the need for a spare exchanger. Quench water was used for heating instead of low pressure steam.

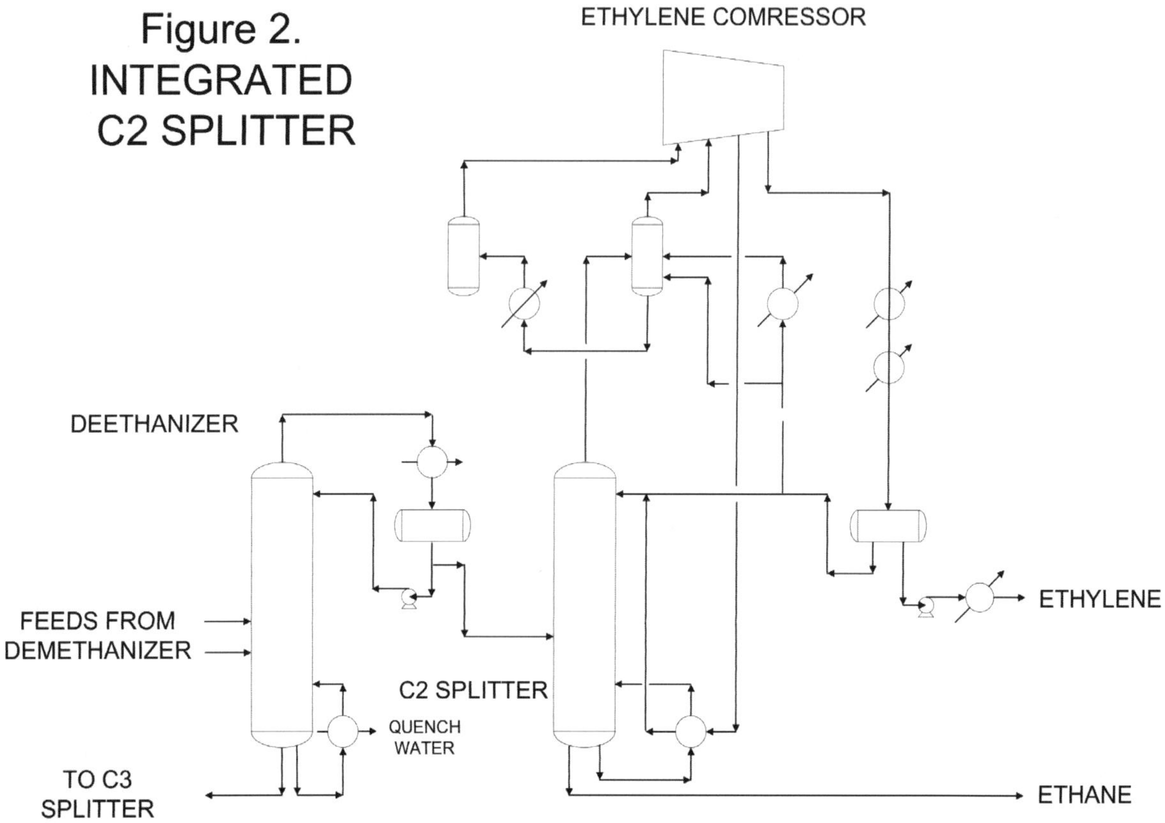

There are many clear advantages of this approach. Some are attributable directly to Front-end Acetylene Hydrogenation:

- Reactor space velocity is maintained independent of the furnace throughput - stable reactor operation and minimum flaring during startup
- Dilution effect of the hydrogenated reflux minimizes the temperature rise over the reactor
- Very high hydrogen partial pressure (in excess of reaction stoichiometry)

The net effects of the above are elimination of green oil formation and a high selectivity of the reaction toward ethylene. Consequences of these effects are:

- No green oil removal system required
- No spare reactor is required
- No regeneration system is required, lowering cost and eliminating associated operating and safety issues

- Long catalyst life
- No pasteurization section for C_2 splitter
- No high purity hydrogen production is required
- No CO injection is required
- All hydrogenated MAPD is selectively converted to propylene - increases propylene product for the case where Front end Depropanizer is used
- High and constant ethylene gain across the reactor

Integration with the refrigeration compressor makes it possible to design the C2 Splitter at low pressure but condense reflux at high pressure without the use of low temperature level of refrigerant. The advantages attributable to the use of the low pressure heat-pumped towers integrated with the open-loop refrigeration system include:

- Total refrigeration compressor power reduced by 6% relative to conventional high pressure C2 systems
- Low pressure enables better fractionation in the tower as a result of higher relative volatilities for the key components, lowering reflux requirement and reducing the number of stages compared to running at high pressure
- Reduced total equipment count: the C2 Splitter tower reflux drum function is replaced by the existing C2 Refrigeration compressor accumulator drum, the C2 Splitter reboiler and condenser are now the same service and the tower reflux (cold service) pumps are eliminated
- Additional significant capital cost savings result from reduced tower height and wall thickness (total installed weight), more than offsetting the increased cost per ton for lower temperature rating of the purchased steel – this is of extreme importance for today's very large capacity grassroots ethylene plants
- Operating inventories of high purity product are lower at the lower operating pressure with the smaller equipment sizes and reduced number of equipment items. This means quicker recovery to on-specification after upsets and is considered safer to operate
- Reboiler temperature is reduced sufficiently to totally eliminate reboiler fouling in most applications, eliminating the spare reboiler while reducing initial capital cost and on-going maintenance costs

Advanced Deethanizer

Figure 3 shows the process flow scheme of an Advanced Deethanizer with heat-pumped C2 Splitter. The Advanced Deethanizer concept is to add rectifying trays to the top of the Deethanizer, allowing a portion of the polymer grade ethylene product to be produced directly from the Deethanizer, reducing the throughput of the C2 Splitter and Ethylene Refrigeration Compressor. The idea is effectively a "Distributive Distillation" concept and was patented by KBR (US Patent #5,678,424) with the first commercial installation at Dow (Terneuzen, the Netherlands) in 2001. It has since been installed successfully in three additional locations. Adding trays to the top of the Deethanizer does not change the internal reflux requirements or utilities for the overall system. It does allow however for up to 30% of the total ethylene product to bypass C2 Splitter and C2 Refrigeration, resulting in energy and capital savings. Total compressor power requirement is reduced by 7% relative to the heat pump scheme described above (and 13% relative to the high pressure system). Installed capital costs for the system are reduced at the same time due to the smaller size for C2 splitter and C2 Refrigeration Compressor.

In this scheme, the Deethanizer separates the feeds to the column into an overhead high purity ethylene product, a side draw mix C2 stream, and a bottoms C3+ stream that can be recycled to the cracking furnaces. The column overhead vapor, polymer grade ethylene is totally condensed and is sent directly to the C2 Refrigeration compressor accumulator drum. The mixed C2's side product feeds the C2 Splitter for final ethylene purification.

While the Advanced Deethanizer has the obvious benefits as described above for grass roots applications, it is also extremely valuable in debottleneck applications. Since approximately 30% of the total ethylene can be by-passed around the C2 Splitter a 30% expansion of the plant capacity, a typical target in many debottlenecking projects, can be achieved without modifying or replacing the C2 Splitter or the C2 Refrigeration Compressor.

Note that the Advanced Deethanizer concept cannot be applied without the inclusion of the Front end Acetylene Hydrogenation. Front end Acetylene Hydrogenation eliminates the presence of both C2 acetylene and inerts (H2, CO, CH4) in the Deethanizer / C2 Splitter tower system which would otherwise contaminate the high purity Deethanizer overhead ethylene product. The Advanced Deethanizer is, however, applicable to those plants which recover acetylene as a product by extraction. In one revamp project involving a 30% expansion of such a plant an Advanced Deethanizer was installed upstream of a conventional high pressure C2 Splitter, avoiding a replacement or parallel C2 Splitter. Although the C2 Refrigeration compressor was replaced in this project due to other added refrigeration loads, the change was small enough that the new compressor could utilize the same "footprint" avoiding modifications to the compressor deck, related piping, and structure; thus the costs of the expansion project were significantly reduced.

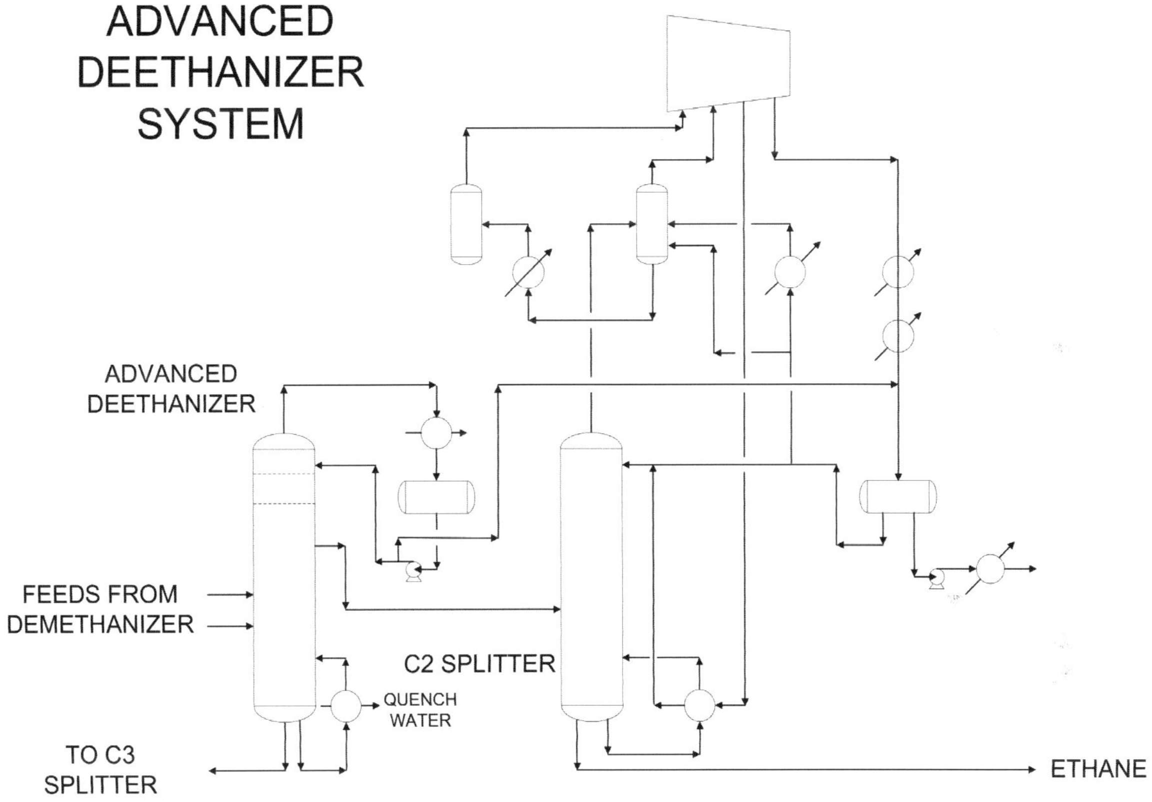

Figure 3. ADVANCED DEETHANIZER SYSTEM

Further Refinements – The Improved Advanced Deethanizer

Figure 4 shows a further refinement to the process flow scheme. In this scheme, the C2 Refrigeration compressor discharge system includes capacity for the reflux requirement for the Deethanizer resulting in further capital cost savings through elimination of several equipment services and reduction of the overall equipment count. Namely, the Deethanizer condenser, reflux drum and reflux pumps are eliminated. Reflux for Deethanizer tower rectifying section is supplied by the ethylene product pump. The Deethanizer gross overhead vapor, in essence high purity ethylene product, joins the C2 Refrigeration compressor discharge upstream of the condenser.

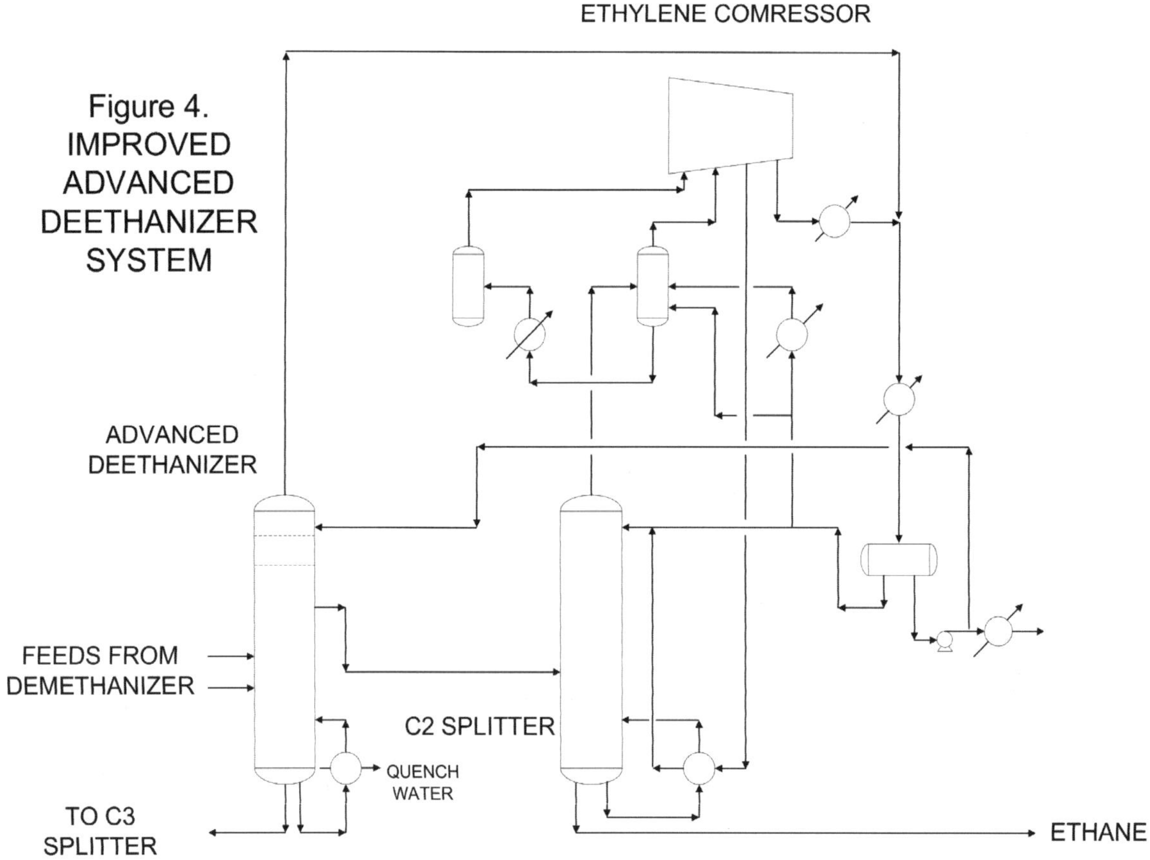

Figure 4. IMPROVED ADVANCED DEETHANIZER SYSTEM

For many projects where ethylene product is consumed locally the pressure, delivery requirements are on the order of 50 bar as ethylene vapor. Conventional product delivery systems are designed to condense the ethylene vapor from the final discharge stage of the C2

Refrigeration compressor (for heat-pump systems), then add a high head pump plus one or more exchangers to re-vaporize/ re-heat the vapor to the required delivery conditions. In such situations it is possible to add one additional stage to the C2 Refrigeration compressor, the discharge of which is medium pressure heated vapor available for export without further processing. Compared to conventional product delivery, this scheme has significant power savings and reduction in overall capital cost through

- Replacement of one set of high head cryogenic pumps with a low head design (also results in higher reliability/lower maintenance), plus
- Elimination of the product vaporizers/ re-heat exchangers.

This scheme is depicted in Figure 5. This scheme was used in the original design for the most recent ethylene plant in Lanzhou, China, including additional process heat integration to reduce energy consumption. Lanzhou's long-time plant manager notes that the scheme has operated very well since its start-up and that Lanzhou is very satisfied with the design. There have not been any problems or concerns with being able to consistently meet ethylene product specifications with this design.

Figure 5.
ADD EXTRA STAGE OF COMPRESSION

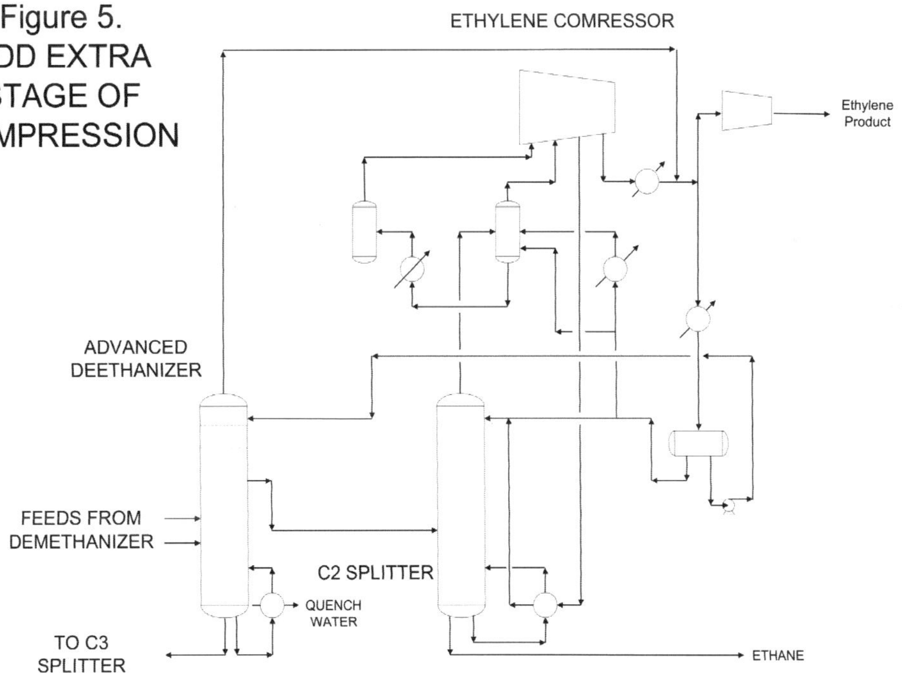

A final variant of the design consists of the Deethanizer tower operating at a reduced pressure that is the same as or slightly above the C2 Splitter tower pressure and inter-linking the two tower. The improved relative volatilities at the lower operating pressure reduce the size and number of stages for the Deethanizer and thus the installed cost. The Deethanizer condenser and reflux drum are still not necessary. However, the low head Deethanizer Reflux pumps are re-located from the C2 Refrigerant Accumulator drum to take suction from an intermediate draw-off pan in the C2 Splitter, thereby utilizing C2 Splitter inter-tray liquid to reflux the Deethanizer. The Deethanizer tower bottom is reboiled against mid-level propylene refrigerant vapor, reducing overall refrigeration requirements. This process flow scheme has also been successfully commercialized and is depicted in Figure 6.

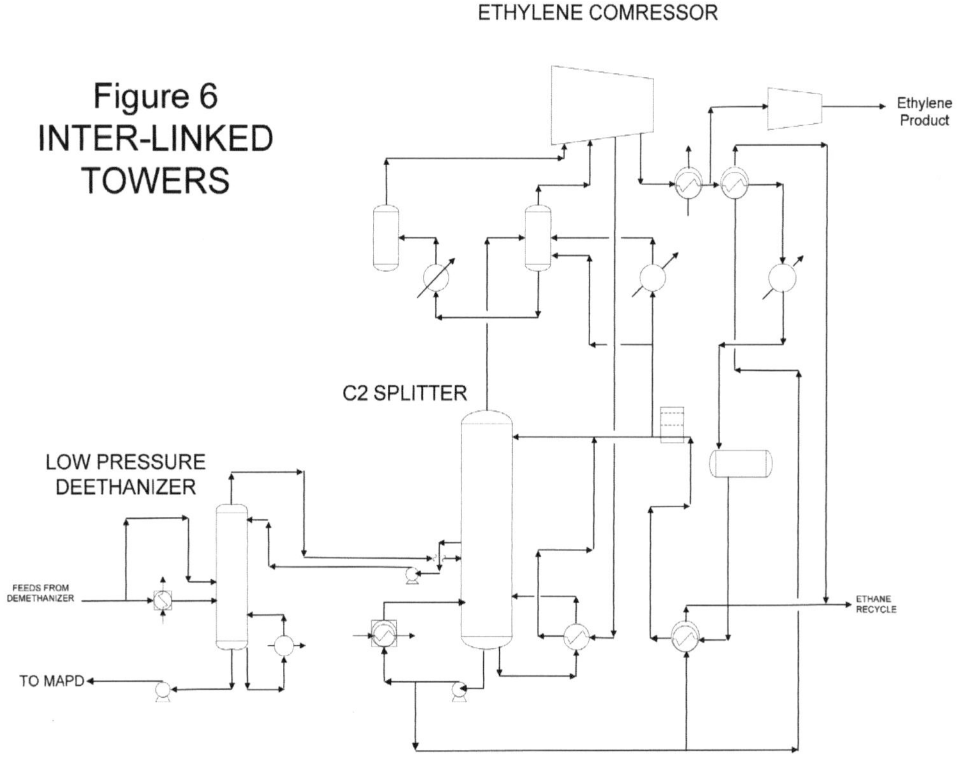

Figure 6
INTER-LINKED
TOWERS

Divided Wall Column

Divided Wall Columns (DWC) or, as they are sometimes referred, Tri-Product Columns, have been in use in the chemical industry for several years. DWC can be used whenever multi-component mixtures have to be separated into pure fractions. The separation of a multi-component mixture into three product streams in conventional fractionation requires at least two columns operating sequentially. With DWC, the fractionation is accomplished in only one consolidated tower.

A vertical wall is introduced into the middle of the column, creating a feed and draw-off section. The intervening wall is gas and liquid sealed. It permits the separation of the low and high boiling fractions in the feed section. The medium boiling fraction is then concentrated in the draw-off part of the column. This setup eliminates the second column, along with its reboiler and condenser. Divided Wall Columns are an alternative to multi-column systems and can reduce investment and operating costs.

In Figure 7 and DWC arrangement for the C2 Recovery area is shown, in which a 26-stage low pressure deethanizer is inserted within a 69-stage C2 Splitter. The tower is refluxed from the C2 Splitter, similar to the scheme described in Figure 6. The Deethanizer distillate flows directly to the C2 Splitter as an internal vapor. The Deethanizer is reboiled against Acetylene Converter effluent, reducing mid-level PR. Chilling and condensing of the compressor side draw is the same as the base flow scheme. The net ethylene circulation is subcooled, combined with the condensed side draw, and further subcooled. Ethane product is recycled to the cracking furnaces after refrigeration recovery.

Combined power requirement for this case is estimated to be slightly higher than for the High Pressure Advanced Deethanizer, as shown in Table 1. However, the total installed cost is the lowest of all the cases described.

Figure 7
Divided Wall
Column

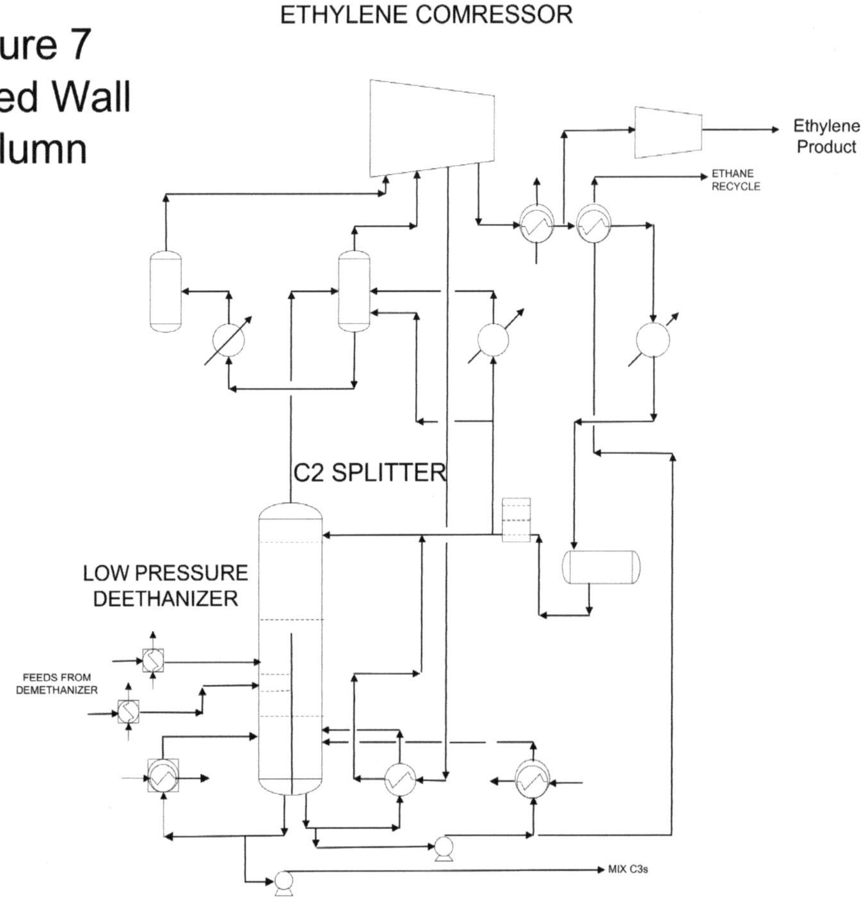

Table 1
Comparison Of Refrigeration Power
(Basis: 600 kta Ethylene Plant)

	Scheme 1	Scheme 2	Scheme 3	Scheme 4
	Conventional HP Deethanizer	Heat Pumped C2 Splitter	Advanced Deethanizer	Divided Wall Column
C2 REFRIG COMP., BHP, frame size	-BASE-/ 38M6-5	114%/ 38M6-5	71%/ 32M6-5	114%/38M6-5
C3 REFRIG COMP., BHP, frame size	-BASE-/ 56M	85%/ 56M	94%/ 56M	84%/ 56M
TOTAL REF. COMP., BHP	-BASE-	94%	87%	93%

Summary

The Advanced Deethanizer process flow scheme represents a patented, innovative design that, along with the Improved Advanced Deethanizer, reduces both energy and capital costs compared with conventional offerings. The use of an Advanced Deethanizer scheme is particularly attractive for revamp applications up to 130% of original design capacity.

The Improved Advanced Deethanizer scheme has been commercially operated with great success, most recently in the Lanzhou Ethylene Plant in China without any operating concerns.

KBR continues to innovate in this important area of the ethylene process flow scheme with several examples given in this paper including the inter-linked low pressure Deethanizer/C2 Splitter system, C2 vapor product delivery, and use of Divided Wall Columns.

(130c) C2 Splitter Debottlenecking and Operation

Daniel R. Summers, Sulzer Chemtech USA, Inc., Tulsa, OK

C2 Splitters are the "Money Towers" of the ethylene plant. With recent interest in lower price ethane for cracking in ethylene plants in North America, there is heightened interest in getting as much capacity out of these "Money Towers" as possible. Two recent revamps of C2 Splitters will be presented along with operating data showing the capacity and performance of the new High Capacity trays employed.

(130d) High Performance Trays and Tubes for Reduced Utility and Energy Consumption

Brian Nowak, UOP LLC, Tonawanda, NY

In today's environment of rising energy costs, reduced carbon footprints, low product margins, and high feedstock prices, modern-day petrochemical producers are maximizing production from equipment assets while minimizing operating costs. The use of high performance fractionation trays and heat exchanger tubes in existing equipment are an economical means of reducing utility consumption and improving operating profitability. This is particularly true in energy-intensive fractionations having low relative volatility between the components such as C2 and C3 splitters, demethanizers, deethanizers, and depropanizers. Energy savings can be realized by:

- Lower column operating pressures to achieve desired separation
- Reduced condenser and reboiler duties
- Close heat exchanger temperature approaches and mean temperature differences (MTD's)
- Reduced condensing pressure and temperature in ethylene and propylene refrigerant condensers
- Compressor horsepower savings in refrigeration systems
- Minimized cooling water consumption

The revamp of existing columns and exchangers to incorporate UOP MD™ trays, High Flux™ tubes, and High Cond™ tubes have been shown to have a 2-year payback. This paper will present case studies which demonstrate the reliable operation and reduced operating costs that can be achieved with high performance trays and tubes.

AIChE Paper

EVALUATION OF FOULING PRECURSORS RESIDENCE TIME IN DISTILLATION TOWER

André Bernard
Process Design Engineer
NOVA Chemicals (Canada) Ltd.

Prepared for Presentation at the 2013 Spring National Meeting
San Antonio, TX, April 28th – May 2nd, 2013

AIChE Paper

EVALUATION OF FOULING PRECURSORS RESIDENCE TIME IN DISTILLATION TOWER

André Bernard
Process Design Engineer
NOVA Chemicals (Canada) Ltd.

Abstract

Fouling mechanisms in distillation towers are often evaluated by conducting batch experiments or running hot liquid process simulation (HLPS) in laboratories. In most cases however, it is difficult to relate the laboratory results to actual operation as the residence time of fouling precursors is often unknown in a distillation tower. Evaluating the residence time of fouling precursors in distillation towers is a difficult task as it involves vapor/liquid equilibriums (VLE) and internal recycles, which may differ from one component to another. Tracer test on towers are known methods that are effective for evaluating hydraulic distributions. However, they provide little insights with regards to specific components distribution. This paper discusses a method for estimating the residence time distribution (RTD) of fouling precursors in distillation towers. The method is based on simulating a tracer test experiment using a dynamic model of the system. The tracer is configured as a pseudo-component that has a boiling point, specific gravity and molecular weight identical to a specific fouling precursor. By doing so, the tracer follows similar VLE and internal recycle rates as its assigned fouling precursor and can then be used for estimating the residence time of such component.

Introduction

Fouling of distillation towers is a problem that affects capacity and run lengths in chemical plants. Beside financial losses, unplanned shut down requiring vessel entries adds exposure to employees from a safety standpoint. A significant amount of work has been done and published on reaction mechanism and components involved in fouling services. The majority of the work has been conducted in laboratories and provides valuable information regarding conditions promoting undesired reactions. Extension of this work has led to the development of reaction inhibitors and/ or dispersants. The effectiveness of those chemicals is often first tested in laboratory batch experiments or hot liquid process simulation (HLPS). The dosage in the plant relies on laboratory results or is based on experience from similar operating systems. Residence time of fouling precursors and inhibitors and/ or dispersants in distillation towers is however a big unknown and makes it difficult to correlate the work done in laboratory to the actual tower operation. The performance of a tower known to foul is often monitored through frequent gamma scans. This technique is extremely valuable for identifying operation anomalies but provides little or no information with regards to residence time and the extent of fouling reactions. Radiotracer techniques, while not as widely used, provide good information on distribution and residence time. In some cases, the characteristics of the tracer may emulate the flow path and residence time of given fouling precursors. However, depending on the system it can in some cases be challenging to position the detectors in appropriate locations. The test is also done on a live system for a given set of operating parameters. It would therefore be difficult to extrapolate to a different mode of operation. For instance, investigating a fouling excursion experience on a tower would normally involve an evaluation of historical operating data. A combination of plant data, process simulation and gamma scans (if available) will provide good information on compositions, temperature profile and tower operation. Residence time of reactants in the system remain however unknown. Knowing that chemical reactions are driven by compositions, temperature and residence time; quantifying those three parameters would provide a good understanding on the extent of reaction. Similarly, understanding the residence time impact along with composition and temperature profiles in evaluating operating mode expected by future direction in plant operation would provide a more accurate upfront evaluation on expected conditions promoting fouling. Building on the methodology used with radiotracer techniques and the concept of residence time distribution (RTD) discussed in reaction engineering text books; a method was developed for evaluating residence time of given components in a distillation tower. The method is based on simulating a tracer test on a distillation tower using a dynamic model. The simulation environment gives full flexibility with regards to the tracer properties and tower operating conditions. This paper discusses the method itself and the different factors affecting the residence time of components within a distillation tower. Examples on how it could be applied for evaluating different operating conditions on Primary Fractionators and Debutanizers in ethylene plants is also briefly discussed.

Hydraulic Residence Time vs Components Residence Time

Differentiating hydraulic and components residence time is a concept that is often poorly understood. For most, the definition of residence time is the ratio between volume and flow rates. This definition applies a on single phase system with ideal flow patterns and no recycles streams such as plug flow systems and completely mix vessel often referred as completely stirred tank reactor (CSTR). As illustrated on Figure 1a, without recycle flow (Q_R = 0); the residence time of each component contained within the feed stream is equivalent to the ratio of the vessel volume over the feed rate Q or V/Q. In this case, hydraulic and components residence time are equivalent. By introducing a recycle flow, the hydraulic residence time becomes equal to the ratio of the vessel volume over the total flow (Q + Q_R). For a similar inlet flow Q, the total flow to the vessel increases due to the recycle flow Q_R, which decreases the vessel hydraulic residence time while the time spent by the fluid in the system remains unchanged and equivalent to V/Q.

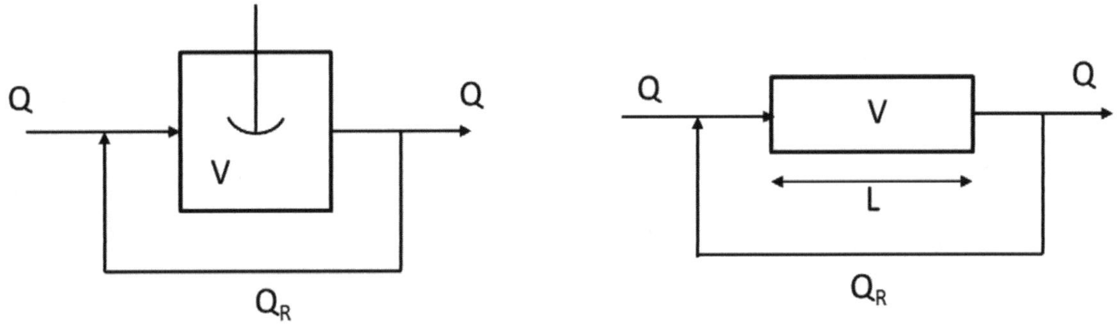

CSTR **Plug Flow**

Figure 1a: CSTR & Plug Flow Hydraulics

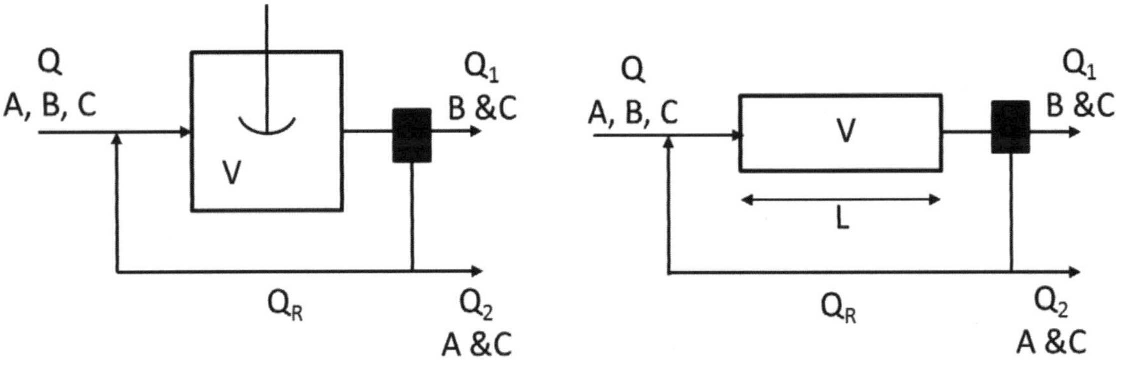

CSTR **Plug Flow**

Figure 1b: Impact of Recycle and Separation on Residence Time

The systems illustrated on Figure 1a are relatively simple; however introducing a form of separation at the recycle point makes the evaluation of components residence time less straightforward. In the systems illustrated on Figure 1b, the hydraulic residence time will remain equivalent to the ratio of the vessel volume V over the summation of the feed rate Q and recycle rate Q_R. However, the residence time of component A, B and C will differ and depend on the extent of separation and recycle rates. Distillation tower hydraulics can be compared to a complex network of CSTR and plug flow type hydraulics combine with internal and external recycles. The vapor liquid equilibrium involved at each stage of separation also add to the complexity of the system. Therefore each component present in the feed stream will likely have different residence time in the tower. By knowing the vapor and liquid traffic in a tower, a fair estimate of the liquid and vapor hydraulic residence time on a tray and downcomer can be attempted. A similar evaluation would however be more challenging on a pack tower as it requires evaluating the liquid and/ or vapor holdup at each stage. A number of correlations have been published in literature regarding holdup prediction. Those evaluations can become fairly complex and still won't provide information regarding the residence time of specific components. Given the level of complexity a practical way of evaluating residence time is through the use of non-reactive tracers. This practice has been well documented in Reaction Engineering for evaluating the average residence time in chemical reactors with non-ideal flow pattern. Radio tracers are known to be used in distillation system for evaluating distribution patterns and in some cases residence time. The evaluation relies on injecting an inert non-reactive tracer in the feed stream and collecting it in the outlet of the process unit.

Tracer Experiments, Residence Time Distribution (RTD)

Element of fluids will take different routes and may take different length of time to travel through a process unit. The distribution of these times is called the exit age or the residence time distribution (RTD) of the fluid. The most popular methods for evaluating the residence time distribution is based on a tracer pulse experiment. Other tracer inputs such as a step or periodic inputs can also be used; this discussion is however limited to pulse tracer inputs. This involves an instantaneous or rapid injection of a given mass of tracer in the system. The concentration of tracer in the outlet of the process unit is then measured over time. This analysis can be performed analytically on simple system such as those illustrated on Figure 1a without recycle flow. To illustrate this, let's assume that a mass of tracer B is dropped in a vessel at time t=0. For the CSTR vessel illustrated on Figure 1a the material balance can be described by the following equation:

$$Q(C_{Bin} - C_{Bout}) = V\frac{dC_B}{dt} \quad (1)$$

Where:

Q = Flow rate
C_{Bin} = Concentration of B in the inlet
C_{Bout} = Concentration of B in the outlet

V = Vessel volume
t = Time

C_{Bin} in the inlet streams equals zero as B is injected in the vessel at time t=0. Since we assumed a homogeneous and perfectly mixed liquid in the vessel, C_{Bout} equals C_B in the vessel at all time. Equation 1 then becomes:

$$V \int \frac{dC_B}{C_B} = -Q \int dt \tag{2}$$

Equation 2 can be transformed into:

$$V \ln C_B = -Q\, t + K_1 \tag{3}$$

Constant K_1 is evaluated from known condition at time t=0, where C_B equals the total mass of B (M_B) divided by the vessel volume (V).

$$K_1 = V \ln C_{B0} = V \ln \frac{M_B}{V} \tag{4}$$

Equation 3 can also be modified to express C_B as a function of time.

$$C_B = e^{\left(\frac{-Q\,t + K_1}{V}\right)} \tag{5}$$

Equation 1 still applies to the Plug Flow vessel illustrated on Figure 1a. However unlike the CSTR vessel, C_{Bout} does not equal C_B at all time. It is assumed that the Plug Flow vessel behaves like a pipe with no axial dispersion. As shown below, the mass of B is assumed introduced at the inlet in an infinitesimal layer across the pipe diameter. The layer containing the tracer travels across the pipe at the same velocity as the fluid being fed. The entire mass of tracer will then exit the vessel at a time equivalent to the pipe length divided by the fluid velocity.

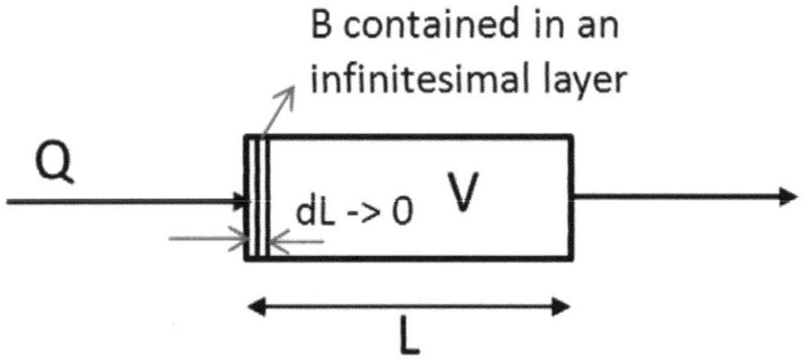

Figure 2: Plug Flow Hydraulic Characterization

Using equation 5 and the above plug flow analyses, the distribution of B can be plotted over time as shown on Figure 3. As illustrated, for identical vessel volume and flow rates; the distribution of fluid elements is quite different between a perfectly mixed and plug flow vessel. In the perfectly mixed vessel, Fluid element of B exit the vessel immediately after injection. The amount of B exiting the vessel then decreases over time until complete depletion. The time spent by different element of B is often referred as the age of the fluid elements. In the plug flow hydraulic, the entire fluid elements travel along the length of the vessel with the entire mass of B exiting the vessel at the same time.

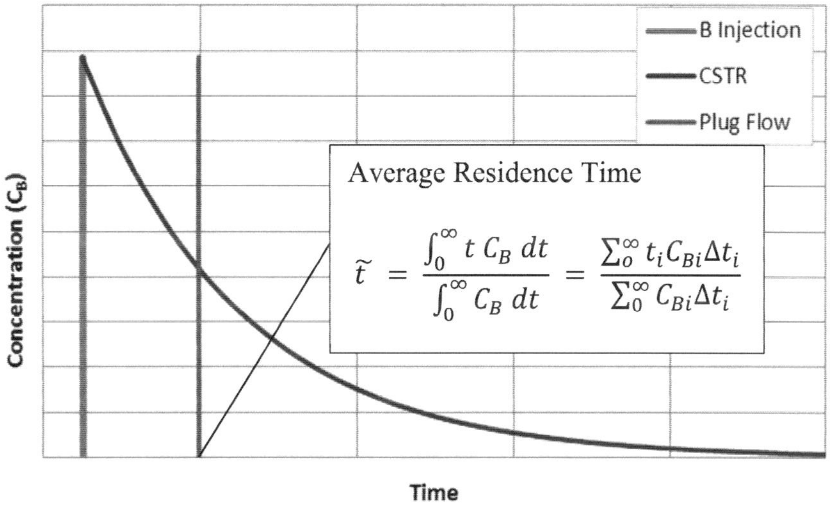

Figure 3: Distribution of B over Time

The mass of B that has left the vessel can be calculated from the above distribution at any time by integrating the distribution curve between time 0 and time t and multiplying by the flow rate. As t reach infinity, M_B becomes equal to the initial mass of B injected in the system. For practicality, the tracer experiment can be concluded at a time for which equation 6 equals the initial mass of B (or tracer) with reasonable tolerance.

$$M_B = Q \int_0^t C_B\, dt = Q \sum_0^t C_B\, \Delta t \tag{6}$$

The average residence time of the distribution can be obtained by the following expression. Again, for practicality the integration or summation should be performed up to a time where the mass of tracer recovered (as described by equation 6) is equivalent to the total mass of tracer injected within reasonable tolerance.

$$\tilde{t} = \frac{\int_0^\infty t\, C_B\, dt}{\int_0^\infty C_B\, dt} = \frac{\sum_0^\infty t_i C_{Bi} \Delta t_i}{\sum_0^\infty C_{Bi} \Delta t_i} \tag{7}$$

In the completely mixed system evaluated above, the residence time of fluid element B will on average equal the hydraulic residence time. With the plug flow hydraulic the entire mass of B will leave the system at a time equivalent to the hydraulic residence time of the vessel.

Residence Time Distribution and Process Simulations

In practice, tracer experiments are conducted on live systems. It is effective in providing the average residence time of the tracer and its distribution. However, there are circumstances for which a live unit might not be available. New design, predicting impacts of processing new feedstocks, evaluating historical data are examples. In the case of a simple vessel, such as those analyzed above; a mathematical model can easily be developed. However, developing such models for more complicated process units such as distillation towers would involve time consuming and tedious work. As an alternative, conducting tracer experiments in a dynamic simulation is an option believed to provide a reasonable estimate. The methodology described here is based on using [1]Aspen Plus® Dynamics for predicting the residence time distribution.

For practicality and to make analysis of a system having multiple outlets easier, the residence time distribution will be expressed in term of components mass flow. The concentration of tracer is equivalent to the tracer mass flow (m) divided by the total volumetric flow rate (Q). Equation 8 can then be obtained after making this substitution in equation 7.

$$\tilde{t} = \frac{\int_0^{\infty} t\, m\, dt}{\int_0^{\infty} m\, dt} = \frac{\sum_0^{\infty} t_i m_i \Delta t_i}{\sum_0^{\infty} m_i \Delta t_i} \tag{8}$$

Simple tracer experiments were conducted on Aspen Plus® Dynamics Flash2, Pipe and RadFrac blocks for understanding the capability in reproducing hydraulic flow patterns.

Aspen Plus® Dynamics Flash2 Block

The Flash2 block assumed a homogeneous vapor/ liquid mixture in equilibrium. Because of the homogeneity within each phase, the age of fluid element contained in the block would follow a residence time distribution similar to a completely mixed vessel (CSTR). This assumption was validated by simulating a tracer experiment on a Flash2 block with conditions similar to the CSTR distribution described by equation 5. The Flash2 block distribution expressed in mass flow of tracer; is illustrated on Figure 4 and compared to the CSTR distribution after conversion from concentration to mass flow. The results displayed on Figure 4 confirm that a Flash2 block provides a perfectly mixed hydraulic behavior.

[1] Aspen Plus® Dynamics is a registered trademark of Aspen Technology, Inc

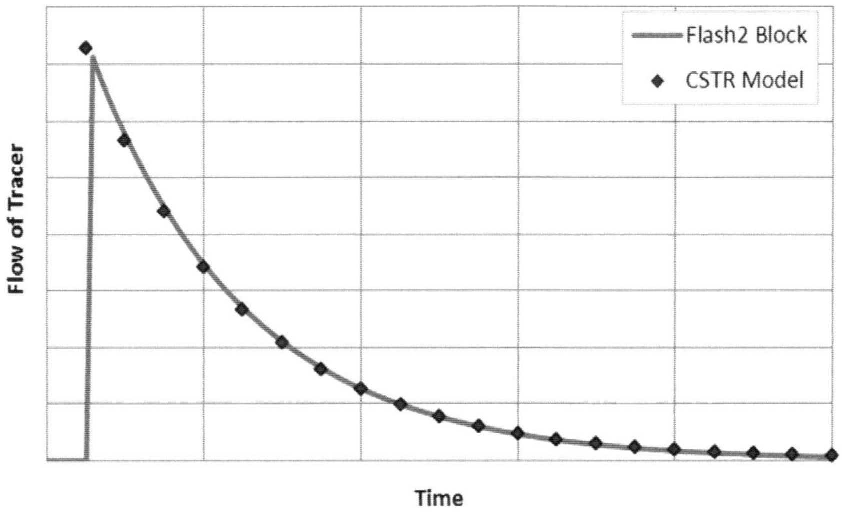

Figure 4: Flash2 Block Residence Time Distribution

Aspen Plus® Dynamics Pipe Block

In turbulent flow regime, the flow through a pipe should follow a plug flow type hydraulics. It is therefore expected that a pipe block in Aspen Plus® Dynamics would emulate the same behavior. As for the Flash2 block, a tracer experiment was simulated on a pipe block. The simulation did not provide the expected results. As shown on Figure 5, the response to the simulated tracer experiment provided a distribution curve typical of flow through a pipe but with significant axial dispersion. The pipe block in Aspen Plus® Dynamics uses the integrate solution method by default. By changing the solution method to "Const-dp-dl" and by enabling the pipe time delay option, a response matching expectation of plug flow type hydraulic is obtained.

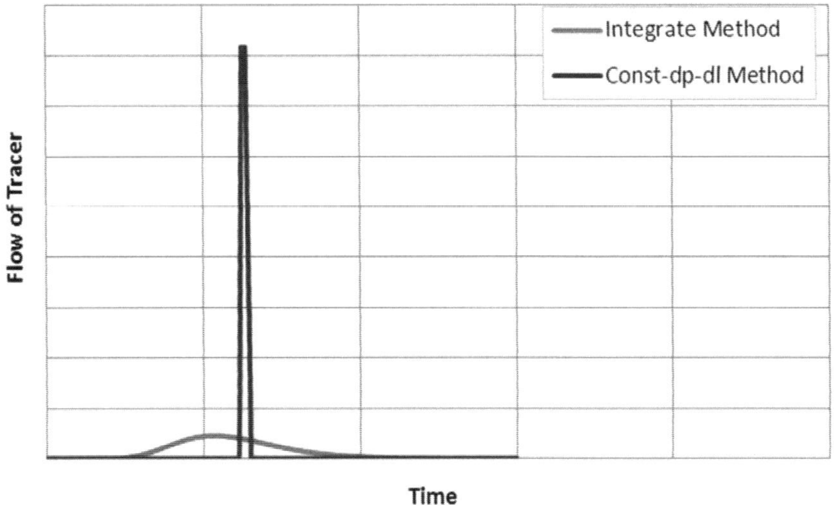

Figure 5: Pipe Block Residence Time Distribution

Aspen Plus® Dynamics RadFrac Block

Distillation towers are modeled with the RadFrac block in Aspen Plus® Dynamics. Residence time of fluid will depend on the dynamic configuration of the sump and reflux drum geometry along with stage hydraulics. The sump and reflux drum are assumed to have perfectly mixed distributions similar to the Flash2 block. The stage hydraulics can be specified as a simple tray or simple packing. A rigorous method is also available but required to input the tray rating option. As will be discussed later, the simple tray and packing options are believed adequate for calculating the average residence time. For simple tray, the Francis weir equation is used to relate liquid flow rate to the tray liquid holdup. By default the RadFrac block will use a ratio of weir length to column diameter equivalent to a one pass tray (L_w/D = 0.7267). The active area as a percentage of the tray area is also set by default at 90%. Those 2 values can be adjusted to reflect other tray configurations. For packing, the liquid flowrate and hold up are correlated through a packing constant calculated from a specified initial volume fraction (default value set at 0.05).

Tray Hydraulic Residence Time

Tray hydraulic residence time and tower internal circulation driven by vapor liquid equilibrium are the two main components driving the average residence time of a specific component on a tray. A simulation of a hypothetical debutanizer has been built for weighting the contributions of those two parameters. The simulation is based on a 7 ft diameter tower equipped with two pass sieve trays. It is configured with 30 equilibrium stages (including condenser and reboiler) with feed located at stage 15. Table 1 summarizes the tray geometry assumed for the rectifying and stripping section.

Table 1: Debutanizer Assumed Tray Geometry

	Rectifying	Stripping
Tower Diameter (ft)	7	7
Number of Passes	2	2
Active Area (ft^2)	32.423	30.33
Hole Area (ft^2)	3.242	3.033
Side Downcomer Width (in)	8	8.5
Center Downcomer Width (in)	4	7
Side Downcomer Area (ft^2)	3.729	4.076
Center Downcomer Area (ft^2)	2.332	4.079
Side Weir Length (in)	98.6	101.3
Center Weir Length (in)	167.8	167.4
Weir Height (in)	2	2
Downcomer Clearance (in)	1.5	1.5

The tray hydraulic residence time has been estimated at the top and bottom of the tower from the clear liquid height (h_c), the downcomer liquid height (h_{dc}) and the clear liquid flow rate. The clear liquid height on the tray was calculated by summing up the weir height and the liquid head over the weir, which was obtained from a corrected Francis weir formula. The downcomer clear liquid height was calculated by summing up the tray clear liquid height, the tray pressure drop and head losses under the downcomer. The total tray pressure drop and head losses under the downcomer were obtained from a tray rating program. The process conditions for each section and results are summarized on Table 2.

Table 2: Tray Hydraulic Residence Time Estimation

	Rectifying (Top)	Stripping (Bottom)
Vapor Flow Rate (lb/hr)	176,805	146,680
Vapor density (lb/ft^3)	0.7215	0.8157
Liquid Flow Rate (lb/hr)	99,660	202,400
Liquid Density (lb/ft^3)	34.96	38.11
Total Tray Pressure Drop (in of Liq.)	2.8	2.54
Losses Under Downcomer (in of Liq.)	0.35	1.14
Clear Liquid Height on the tray (in)	3.13	3.68
Clear Liquid Height in the Downcomer (in)	6.28	7.36
Tray Liquid Residence Time (sec)	10.7	6.3
Downcomer Liquid Residence Time (sec)	2.5	1.7
Total Tray Liquid Residence Time (sec)	13.2	8.0

As intuitively expected, the hydraulic residence time on a tray is in the order of magnitude of seconds. The ability of the RadFrac block in reproducing those residence times was verified with a tracer experiment conducted on a dynamic model of the debutanizer. The model was configured to reflect the tray geometry described on Table 1 using the simple tray option. A pseudo component was configured with a high boiling point and molecular weight such that it would remain in the liquid phase and would not participate in any vapor liquid exchange. By doing so, the tracer will only follow the liquid path and its average residence time will equal the tray hydraulic residence time. The tracer experiment was conducted at locations similar to those reported on Table 2. The tracer residence time distribution is illustrated on Figure 6. A first observation from the plot is that the distribution follows the characteristics of a perfectly mixed volume. The average residence time from the tracer experiments are also in the order of magnitude of seconds. However, because of the calculation methodology in Aspen Plus® Dynamics, they differ from those calculated on Table 2. Parameters affecting liquid holdup and liquid residence time as calculated on Table 2 and predicted by Aspen Plus® Dynamics are listed and compared on Table 3. To compensate for the difference between the number of theoretical stages and actual number of trays, Aspen Plus® Dynamics defines two different heights of clear liquid. One which represents the liquid level on a real tray and one who represents the liquid level on a model stage (equilibrium stage).

Figure 6: Tray Liquid Residence Time Distribution

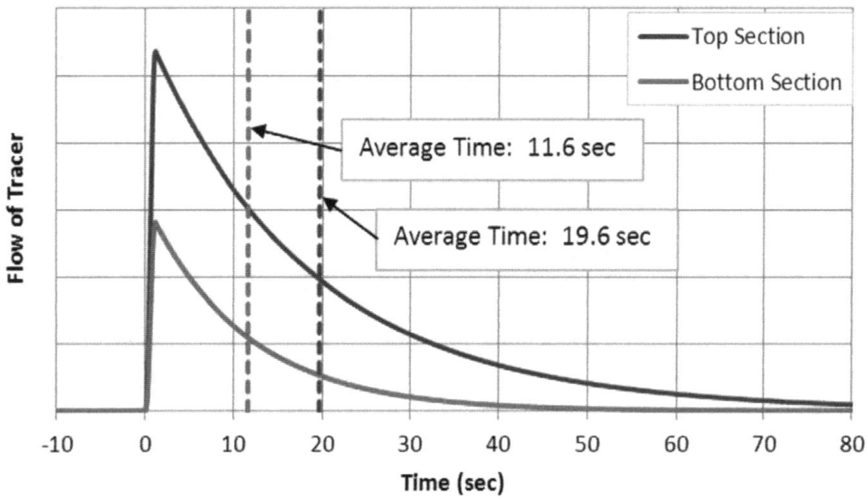

The liquid level reflecting a real tray is used for flow calculation, while the model stage liquid level is used for holdup calculation. Since the total number of theoretical stages is normally different than the actual number of trays, using the real tray liquid level would lead to discrepancies in holdup between the model and the actual tower. Therefore, the residence times calculated from Aspen Plus® Dynamics variables in Table 3 and the average residence times evaluated from the tracer distribution curves on Figure 6 are associated to a single equilibrium stage. Multiplying the global efficiency (number of stages/total tray count) would provide the actual tray hydraulic residence time. This value is provided on Table 3 and is referred as "Hydraulic Residence Time Corrected for Stage Efficiency". As shown, after making the correction for efficiency, the tray residence time calculated by Aspen Plus® Dynamics aligns with the value calculated from a rating program (Table 2). Note that with the simple tray option, the downcomer residence time is omitted in Aspen Plus® Dynamics; which makes the total residence time short by a few seconds.

The average residence times calculated by the tracer experiments distribution curves illustrated on Figure 6 are within 0.5 to 1 seconds of what has been calculated directly from Aspen Plus® Dynamics variables. Given the short residence time, this difference is assumed reasonable and demonstrates the ability of estimating residence time distribution with the RadFrac block in Aspen Plus® Dynamics.

Table 3: Tray Hydraulic Residence Time Calculation Comparison

	Rectifying (Top)		Stripping (Bottom)	
	RadFrac	Table 2	RadFrac	Table 2
Liquid Flow Rate (ft^3/hr)	2,864	2,852	5,363	5,310
Clear Liquid Height on the tray (in) / RadFrac Liquid Level on Real Tray (in)	3.13	3.13	3.67	3.68
RadFrac Liquid Level on Model Stage (in)	5.49	N/A	6.47	N/A
Clear Liquid Height in the Downcomer (in)	N/A	6.28	N/A	7.36
Active Area (ft^2)	32.423	32.423	30.33	30.33
Liquid Volume (ft^3)				
Tray	14.839	8.45	16.35	9.29
Downcomer	N/A	1.95	N/A	2.50
Total	14.839	10.4	16.35	11.79
Hydraulic Residence Time (sec)				
Tray	18.7	10.7	11.0	6.3
Downcomer	N/A	2.5	N/A	1.7
Total	18.7	13.2	11.0	8.0
Hydraulic Residence Time (sec) Corrected for Stage Efficiency				
Tray	10.7	10.7	6.3	6.3
Downcomer	N/A	2.5	N/A	1.7
Total	10.7	13.2	6.3	8.0

Impact of Vapor Liquid Equilibrium on Residence Time

So far the discussion has been limited to the evaluation of hydraulic residence time. Vapor/ liquid equilibrium and internal circulation driven by reflux and reboil ratios are other factors affecting the age of a component within a specific tray and the entire tower. A tracer experiment was simulated on the debutanizer model for quantifying this impact. A pseudo component having a C5 aromatic boiling point, molecular weight and specific gravity was configured and used as tracer in the model. The tracer was injected as a pulse function on the feed stage. The tracer was sampled from the tower bottom product stream. The tower sump level was brought to a minimum to minimize the sump holdup contribution on the average residence time estimation. Figure 7 illustrates the residence time distribution curve. Using equation 8, the average residence time of the tracer in the tower was estimated at 10 minutes and 36 seconds.

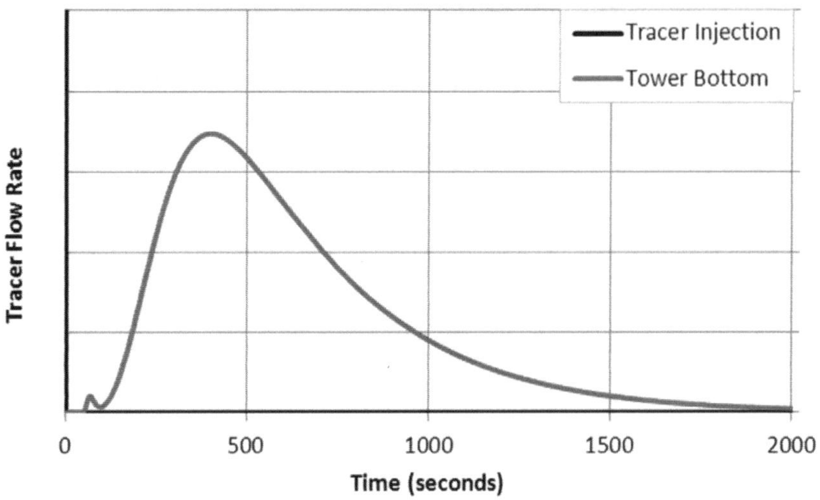

Figure 7: Debutanizer C5 Residence Time Distribution

The average residence time of stages located below the feed point were estimated by trending the amount of tracer truly leaving the equilibrium stage. Since 100% of the mass of tracer exits the system in the bottom product stream, any element of tracer leaving stage N as a vapor will eventually return as a liquid from the stages above. Because of vapor/ liquid equilibrium, some of the tracer exiting stage N as liquid will recycle back as vapor after being stripped in stages below stage N. As illustrated on the schematic below the actual mass of tracer truly exiting stage N is equivalent to the difference between the tracer liquid rate leaving stage N and the tracer vapor rate entering stage N.

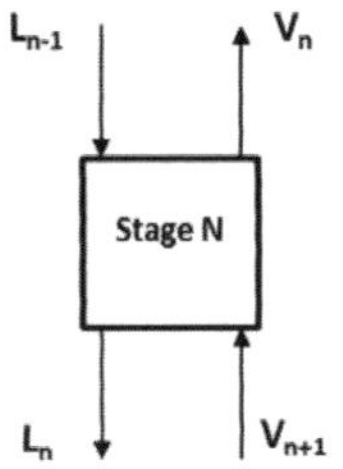

Tracer Leaving Stage N = $L_n x_n - V_{n+1} y_{n+1}$

Using this calculation, residence time distributions for a number of stages were plotted as illustrated on Figure 8a. Since the mass of tracer must travel its way down the tower, a lag time is expected before any noticeable amount of tracer gets detected in a particular stage. The further away from the injection point (feed stage) the longer the lag time. Figure 8b illustrates lag times for the different stages evaluated. The average residence time calculation for each stage must be corrected for this lag time.

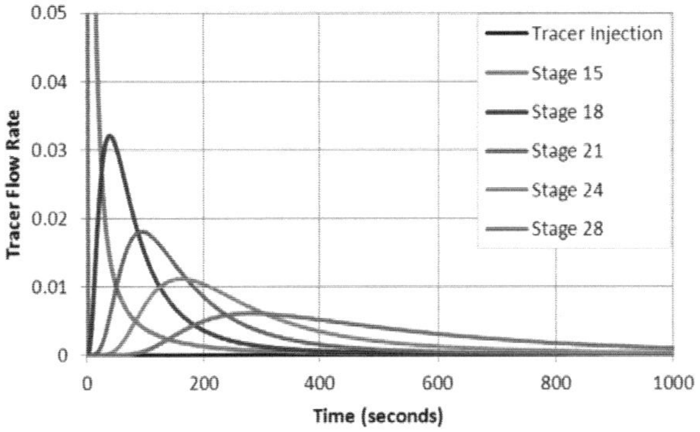

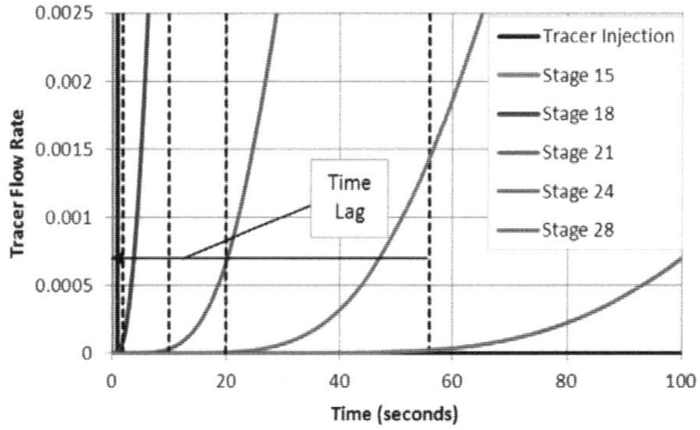

Figure 8a: Stages C5 Residence Time Distribution

Figure 8b: Stages C5 Residence Time Distribution Time Lag

Table 4 provides the stage average residence time of tracer as calculated from the residence time distribution curve. The tracer average residence time is then corrected for the lag time and compared against the stage hydraulic residence time calculated from Aspen Plus® Dynamics RadFrac block variables. As shown, the hydraulic residence time contributes 2 to 23% of the tracer average residence time. The hydraulic residence time contribution is less as stages approach the tower bottom. This is explained by the internal recirculation of the tracer which is more pronounced near the bottom.

The internal recirculation is driven by the boilup ratio and volatility of the component. The lighter fraction of components recover in the bottom of the tower will have higher circulation rate then heavier components. At the limit heavy component residence time will approach the stage hydraulic residence time. To illustrate this, the tracer experiment was repeated on the debutanizer with the tracer characteristics configured to emulate components similar to styrene.

Table 4: Debutanizer Stripping Section Stage Residence Time Summary

	Average Residence Time from RTD Curve (seconds)	Lag Time (Seconds)	Average Residence Time of Tracer (Seconds)	RadFrac Stage Hydraulic Residence Time (seconds)
Stage 15	50.5	0	50.5	11.7
Stage 18	112.2	2	110.2	11.7
Stage 21	195.9	10	185.9	11.5
Stage 24	322.6	20	302.6	11
Stage 28	538.6	56	482.6	11

In this second experiment, the average residence time of the tracer in the tower dropped from 10 minutes, 36 seconds to 3 minutes, 22 seconds. As shown on Table 5, the stage hydraulic residence time contribution increases with the heavier tracer. At the feed stage, the stage hydraulic residence time contribute to 90% of the residence time of components similar to styrene. However, as the tracer moves down the tower; the vapor/ liquid equilibrium component quickly becomes the dominant factor.

Table 5: Tracer Residence Time Comparison

	C5 Tracer		Styrene Type Tracer	
	Average Residence Time of Tracer (Seconds)	Hydraulic/Average Residence time %	Average Residence Time of Tracer (Seconds)	Hydraulic/Average Residence time %
Stage 15	50.5	23.2	12.9	90.6
Stage 18	110.2	10.6	48.1	24.3
Stage 21	185.9	6.2	77.8	14.8
Stage 24	302.6	3.6	102.2	10.8
Stage 28	482.6	2.3	129.9	8.5

The above discussion demonstrates that each component contained in a stream feeding a distillation tower will have different residence time in the tower. That residence time will depend on the volatility of the component. Even though a component's tray residence time may approach the tray hydraulic residence time in certain location of the tower, the vapor/ liquid equilibrium element quickly becomes the dominating factor as the component migrates towards its natural exit point in the tower (bottom or overhead product stream).

Tracer Injection and Sampling

Injection of the tracer occurs over a given period of time. Ideally, for a pulse input, this time must be kept to a minimum. The importance of the time span is directly related to the dynamic of the system being studied. In the case of the Debutanizer, tray hydraulic residence times were evaluated within a 5 to 15 second range. The tracer injection was done by ramping up the mass rate, holding the injection rate and, finally, ramping the rate back down to zero. This procedure was done over a 1 second period. As stated earlier, the stage hydraulic residence time calculated from the tracer was found to be 0.5 to 1 second above what was calculated from Aspen Plus® Dynamics RadFrac block variables. The over estimation may be caused by the time use for injecting the tracer. However, given the vapor/ liquid equilibrium contribution on the component residence time, this error is relatively negligible.

Sampling of the tracer must be done at specified time intervals. The time interval selected may influence the accuracy of the test if not properly done. Integrating the mass of tracer injected and recovered over time provides the total mass of tracer injected and recovered. Ratio of the tracer recovered over injected provides the fraction of tracer recovered. A ratio approaching 1 with reasonable tolerance is an indication of a successful test. Inability to reach a ratio of 1 is caused either by an incomplete test (tracer still flowing in the system) or by a sampling time that is too large.

Examples

Primary Fractionator and Quench Oil Circulation

Because of feedstock market conditions, many plants originally designed for liquid cracking are converting their operation for processing larger volumes of ethane feed. The significant reduction of fuel oil make from cracking furnaces on ethane feed makes this conversion challenging for the Primary Fractionator and Quench Oil Circulation system. To understand the impact on residence time of fuel oil components in the quench oil streams, a tracer experiment was simulated on a dynamic model of a quench system typical of liquid crackers. The system modeled, illustrated on Figure 9, includes quench oil fittings at the cracking furnaces and quench oil mid reflux on the Primary Fractionator. The volume of this system is not limited to the Primary Fractionator and must also account for other process units and quench oil piping. A combination of RadFrac and Pipe blocks was used to replicate as accurately as possible the flow pattern and hold up of the fluid. Depending on boiling range, fuel oil exits the system at the bottom of the PFO and PGO strippers.

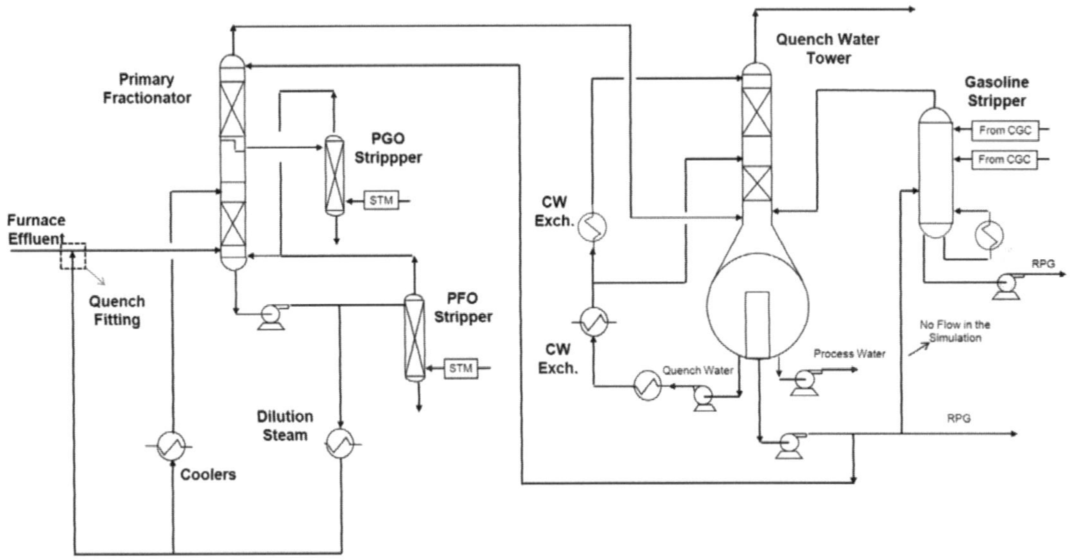

Figure 9: Ethylene Plant Liquid Cracker Typical Quench System

A first test conducted on a mix feed of naphtha/ propane/ butane was simulated by injecting a tracer configured with characteristics equivalent to heavy fuel oil components (boiling point 450°C). The tracer was injected in the heater effluent stream and sampled from the PFO and PGO Stripper bottoms. Insignificant amounts of tracer were found in the PGO stripper bottom as most exited the system from the PFO stripper bottom. The first test provided an average residence time of 5.1 days.

The quench system was then re-simulated with a feed reflecting ethane/ propane cracking. A similar tracer test was conducted, which provided a tracer average residence time of 68.9 days. This test concludes that the average residence time expected for heavy fuel oil would increase from 5.1 to 68.9 days when transitioning the plant feed slate from naphtha/ propane/ butane to ethane/ propane. A stable and reliable quench system normally operates with oil residence time of less than 6 -7 days. The dramatic residence time increase calculated on ethane/ propane cracking suggests that without external compensation (compatible flux oil), the quench system would not be operable with those operating conditions.

Debutanizer False Load

Tower operation below turndown in C3 and C4 areas is another consequence of ethylene plant transitioning towards higher volume of ethane feed. The typical low cost remedy is to false load the towers by operating at higher reflux and reboil ratios. The impact on residence time was evaluated by simulating the debutanizer at three different reflux ratios (1.1, 1.5 and 2.0). The tracer configured as a typical C5 component was found to reside in the tower at time relatively similar for the three different reflux ratios evaluated. It is however important to realize that, while average residence time of components remains unchanged, composition and temperature profile will change.

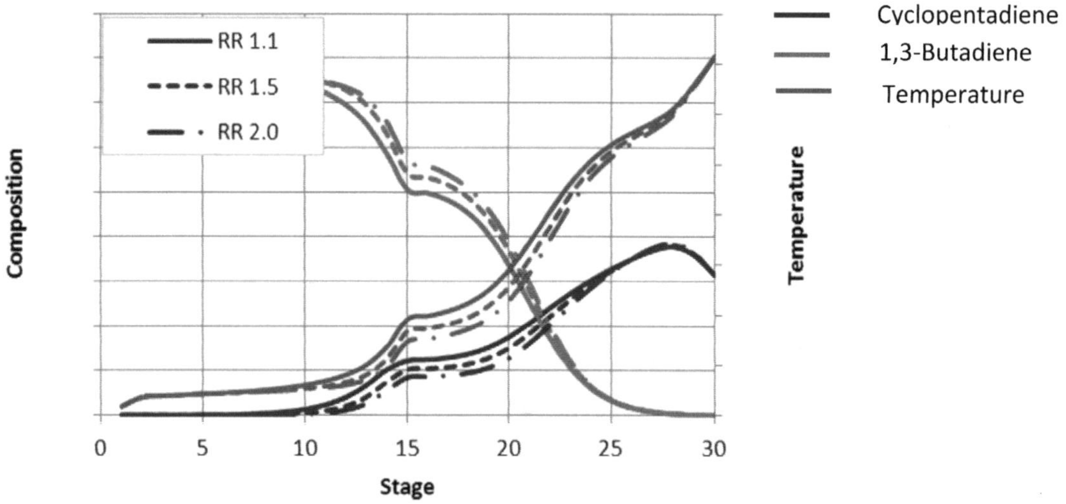

Figure 10: Composition & Temperature Profile

As shown on Figure 10, the composition of cyclopentadiene decreases in the stripping section of the tower as the reflux ratio increase, while the composition of 1,3-butadiene increases. Consequently, a cooler temperature profile is observed at higher reflux ratio as the C4 components composition in the stripping section increases.

Conclusion

Using a dynamic process simulator such as Aspen Plus® Dynamics is an effective tool for simulating tracer test residence time distribution. It also offers the advantage of configuring tracers that will emulate vapor/ liquid equilibrium behavior of specific components. As demonstrated, vapor/ liquid equilibrium is a major contributor with regards to how long a specific component will reside in a distillation tower. Because, the stage or tray hydraulic component is a minor contributor, the simple tray or packing option in Aspen Plus® Dynamics is sufficient for accurately evaluating residence time of components.

Perfectly mix or plug flow hydraulic flow patterns can be created in Aspen Plus® Dynamics by using Flash2 and Pipe blocks. Those are useful when integrating the model with associated systems such as the Primary Fractionator example. It must be understood however that the flow pattern affects the shape of the distribution curve but not the residence time. For instance, plug flow and perfectly mixed vessel of similar volume with similar flow rate will have identical average residence time despite the fact that their distribution curves are different. The RadFrac block provides a perfectly mixed stage, which is suspected to be close to reality with trays having short path length. With long path length, the liquid flow from the inlet of the tray to the downcomer inlet may get closer to a plug flow hydraulic with axial dispersion. This may affect the shape of the distribution curve but will still provide an accurate estimate of the residence time. Furthermore, as stated above, the vapor/ liquid equilibrium component is the major factor driving residence time distribution of components.

The discussion in this paper provided a lot of emphasis on the liquid phase in the stripping section of the tower. However similar analyses can be conducted for the rectifying section.

Simulating tracer experiment is not intended to replace radiotracer techniques done on real equipment. Results from radiotracers distribution test provide valuable information on the actual tower behavior. The simulation methodology may however complement the actual tracer test by providing a theoretical baseline of the system. Comparison of theoretical and actual distribution will provide better insights for diagnosing distribution problems.

Nomenclature

C: Concentration
CSTR: completely stirred tank reactor
L_n: Liquid flow rate from stage N
m: Component mass flow
Q: Volumetric flow rate
Q_R: Recycle volumetric flow rate
t: Time
$\tilde{t}$: Average residence time
V: Volume
V_n: Vapor flow rate from stage N
X_n: Liquid composition of tracer from stage n
Y_n: Vapor composition of tracer from stage n

References

1. Levenspield, O., "Chemical Reaction Engineering", Second Edition, McGraw-Hill.
2. Smith, J. M., "Chemical Engineering Kinetics", Third Edition, McGraw-Hill.
3. Aspen Plus® Dynamics 11.1 Reference Guide, Aspentech.
4. Jones, T.L., "Investigation of problems in distillation column operation using radioisotope. technology", Inst. Chem. Eng. Symp. Ser., Volume 2, Issue 128. 1992.
5. Kister, H. Z., "Distillation Design", McGraw-Hill.
6. Bernard, A., Pickett, T. M., Manek, M. B., Frye, D. K., "Flux Oil Stream Import to Quench System Risk and Impacts", 23rd Ethylene Producer's Conference, AIChE 2011 Spring Meeting.

(130f) Dynamic Simulation for Distillation Column Start-Ups in an Ethylene Plant

Shujing Zhang, Jian Zhang and Qiang Xu, Dan F. Smith Department of Chemical Engineering, Lamar University, Beaumont, TX

The start-up operation of a distillation column normally presents a challenging task for a chemical plant, because complex heat and mass transfers are coupled in the transition period associated with nonlinear responses/controls such as pressure, temperature, and product compositions. The reduction of distillation column start-up time has important merits to a plant in terms of raw material and energy savings, susceptible to process disturbances, and flare and downtime reduction. To effectively reduce start-up time of an distillation column, a common way is to set up the total recycle first before the feed is directed to the column; then ramp up feed input associated with adjustment of controls of reflux rate, heating and cooling duties, and etc. The shifting from the operating status with total reflux to normal operation status under various disturbances, so as to produce on-spec products, is the most important and also the slowest step during the entire startup operation, which needs extensive studies.

In this paper, rigorous dynamic simulations are employed to study startups of two distillation columns in an ethylene plant. The first case is about LP DeC_3 (low-pressure depropanizer) start-up, where the significance (time savings compared with the plant historian start-up operation) of smooth transition from the total reflux status to normal operation status is demonstrated. In the second case, the start-up optimization of a C3 splitter is investigated, where the operating pressure, heating and cooling capacities are well adjusted under realistic constraints. Both case studies help understand start-up dynamic behaviors of distillation columns and result in improved start-up performances.

DISTILLATION COLUMN STEADY-STATE IDENTIFIER

Paper No. 142a

Anand N. Vennavelli and M. R. Resetarits, Fractionation Research, Inc.
Ting Huang and R. Russell Rhinehart, Oklahoma State University

INTRODUCTION

Identification of both steady state and transient state in noisy process signals is important. Steady state, for example, triggers the collection of data for process model adjustment, process analysis, fault detection, data reconciliation, or neural network training. A transient state, for example, triggers the collection of data for dynamic modeling, or recognition of points of change.

Often, engineers run a sequence of experiments to collect data throughout a range of operating conditions, and process operators sequence the next stage of a process. Each sampling event is initiated when the operator observes that steady conditions are met, and then the operator implements the new set of operating conditions. However, this visual method of triggering requires continual human attention, and it is subject to human error in the recognition of steady state when measurements are noisy, or process changes are slow, or there are multiple dynamics. An automated, online, real-time, steady state identifier would be useful.

The Steady State Identification (SSID) and Transient State Identification (TSID) approach here uses an easily implemented statistical method [1] with defined critical values [2, 3]. It was extended to multivariable processes and demonstrated on lab-scale and pilot-scale processes to automatically trigger an experimental sequence [4]. It was also demonstrated as a convergence criterion in nonlinear regression optimization [5].

This study reports on an application to trigger sampling and transitions in a commercial scale distillation unit.

Fractionation Research Inc. (FRI) [6] is a non-profit research consortium supported by memberships, which include many petroleum, chemical, and engineering companies. FRI is the only independent, commercial-scale, distillation experimentation program operating with hydrocarbon systems at pressures ranging from deep vacuum to 500 psia.

Fractionation Research, Inc. collects hydraulic and mass-transfer data on distillation column internals in its 4-foot diameter low-pressure and high-pressure distillation columns in Stillwater, OK using several hydrocarbon systems. Almost all of these data are collected at steady-state conditions.

A typical FRI test involves several transitions between steady-state conditions ranging from the maximum operating capacity (flood point) to the minimum operating capacity. In addition, column dynamics are influenced by the operating pressure (deep vacuum to 500 psia), internals (trays or packing), and the system physical properties.

The steady-state identifier described in this paper was implemented at the FRI unit for real-time steady-state detection. The data from the FRI distillation unit was used to demonstrate that this SS and TS identification method works in an industrial process. The steady-state identifier methodology and preliminary results are described in the paper.

METHOD

This SSID and TSID method uses the R-statistic method, a ratio of two variances, as measured on the same set of data by two methods [1]. In order to reduce computational effort, exponentially weighted moving average and variances are calculated in place of the conventional average or variance. Figure 1 illustrates the concept. The dotted line represents the true trend of a process. The value starts at 15, ramps to a value of 10 at a time of 50, and then holds steady. The diamond markers about that trend represent the measured data. The dashed line, the true trend is unknowable, only the measurements can be known and they are infected with noise-like fluctuations, masking the truth.

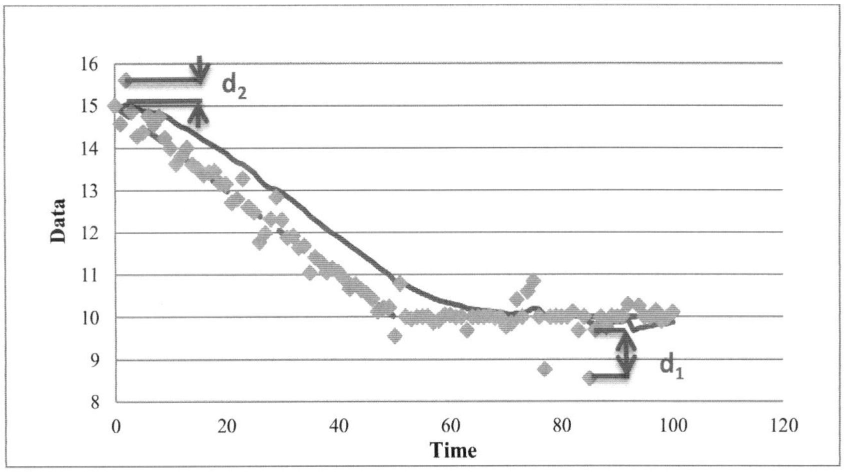

Figure 1: Illustration of concepts

The SSID/TSID approach first calculates a filtered value of the process measurement, indicated by the curved line that lags behind the data. Then the variance in the data is measured by two methods. The deviation indicated by d_2 in the upper left of the figure is the difference between measurement and the filtered trend. The deviation indicated by d_1 in the lower right is the difference between sequential data measurements.

If the process is at SS, as illustrated in the 80 to 100 time period, X_f is almost the middle of the data. Then a process variance σ^2, estimated by d_2 will ideally be equal to σ^2 estimated by d_1. Then the ratio of the variances, $R = \frac{\sigma^2{}_{d_2}}{\sigma^2{}_{d_1}}$, will be approximately equal to unity. $R \cong 1$.

Alternately, if the process is in a TS, then X_f is not the middle of data, the filtered value lags behind, and the variance as estimated by d_2 will be much larger than the variance as estimated by d_1, and ratio will be much greater than unity. $R >> 1$. Figure 2 reveals the flow chart to describe the R-statistic method calculation.

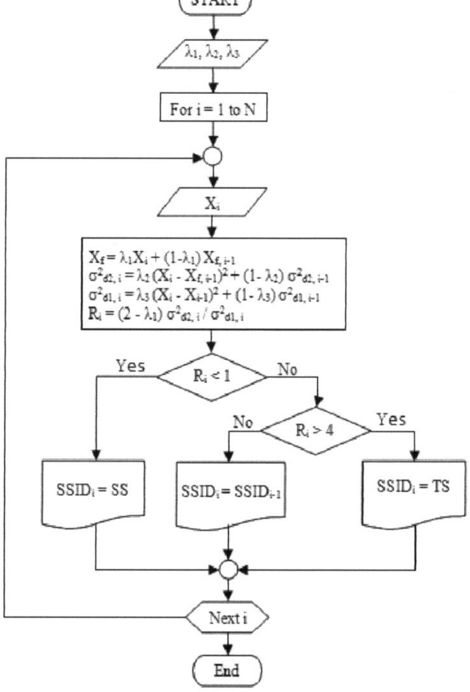

Figure 2: Flow chart

EXPERIMENTAL

The experimental data were recorded through a Yokogawa Exaquantum™ historian at the FRI control room. The data collection was performed as follows.

The steam to the reboiler of the 4-foot diameter FRI distillation unit was set at a constant pre-determined rate. No other changes to the unit were made until the operator determined that a steady state was achieved. The unit was held at this steady state for a sufficiently long time to study the performance of the internals. The steam rate was then changed to a new value, and the process was repeated several times over a wide operating range.

Several process variables were used to establish the steady state by the operator. These variables were broadly classified as hydraulic (flows, levels), pneumatic (pressures, differential pressures), and thermal (temperatures).

IMPLEMENTATION OF THE SSID PROGRAM

The implementation of the SSID program was done in three phases. In the first phase, offline data from the historian were collected, and the SSID program was implemented on the historical data as a preliminary check on the working of the program.

Figure 3 shows the result of the application of the SSID program on the offline data. An SSID of zero indicates transient state, and an SSID of one indicates steady state. The steady state and the transient state times predicted by the program agreed with visual observations.

The offline analysis also confirmed that the sampling interval chosen for the SSID program was appropriate. Based on the offline analysis results, it was observed that the steady state was detected by the SSID program with a 95% confidence approximately one hour after a change in the reboiler duty. The one-hour window could have easily been shortened at the expense of false positives, but was deemed acceptable and no changes were made.

In the second phase, the SSID program was implemented online. The SSID program was run simultaneously in a Microsoft Excel™ VBA program monitoring a subset of variables tracked by the operator. When one or more variables correlated with each other, the number of variables monitored by the SSID program was limited to one of the correlating variables. For example, not all temperatures along the column were included for monitoring as the temperatures all tend to move together. Such a simplification was expected to improve the accuracy of the SSID program. Depending on the nature of the internals in the distillation column, typically 6-12 variables were monitored for a given test.

During the second phase, online real-time implementation of the program was effected. The online implementation results were first monitored by the authors and were deemed accurate enough to make the program available to the operators for the next phase. Figure 4 shows the results from the online implementation of the SSID program.

In the third phase, the real-time implementation results were made available to operators. The program consistently detected steady state and transient state and was in perfect agreement with the operator-determined times for steady state. Figure 5 shows the results of the third phase of implementation. The operator determined steady state times, as shown in Figure 5, clearly agree with the steady state predictions of the program.

Based on several runs of the program on real-time FRI data, the authors concluded that the program was accurate enough to predict steady state and transient state every time, and at the same time, robust enough to deal with the industrial process data.

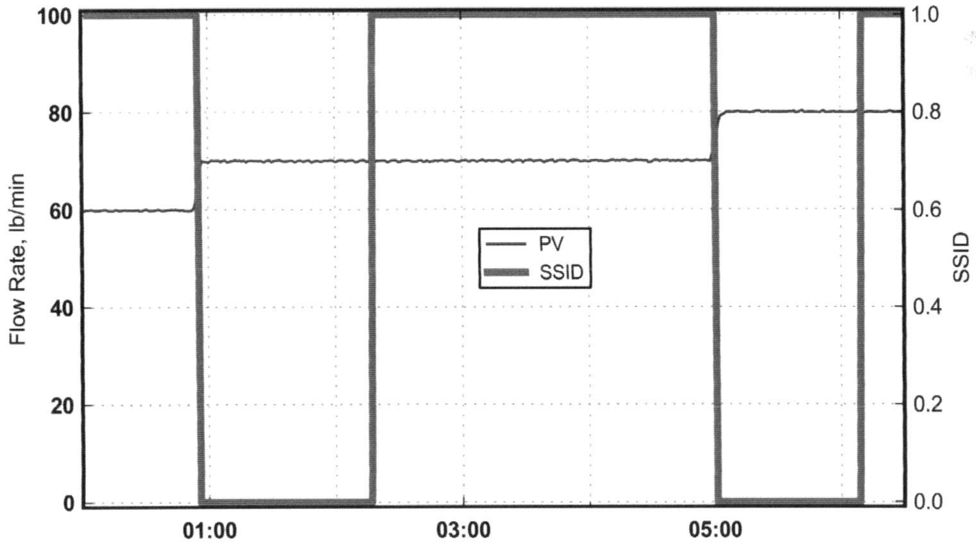

Figure 3: SSID implementation on offline data

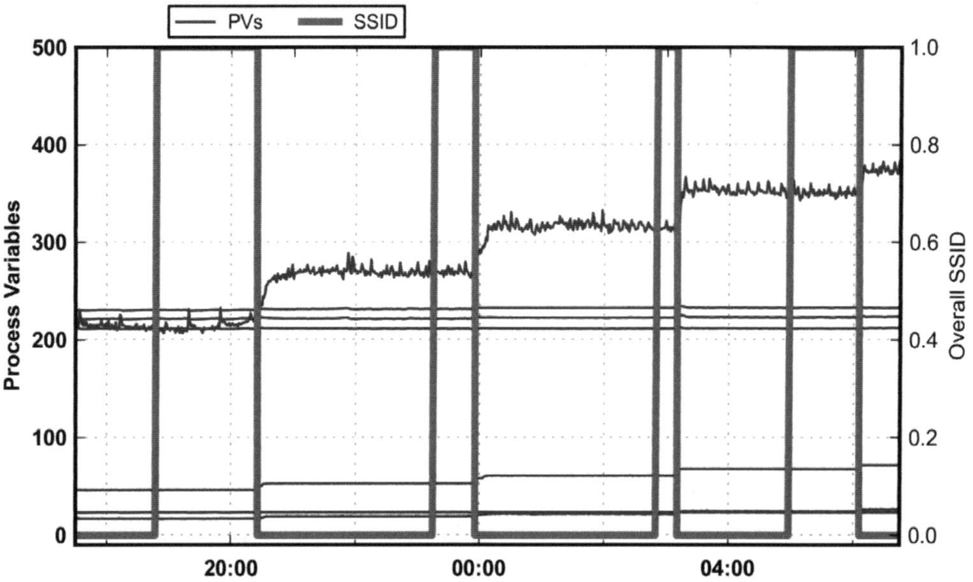

Figure 4: SSID implementation on real-time data. Program results agree with visual observations

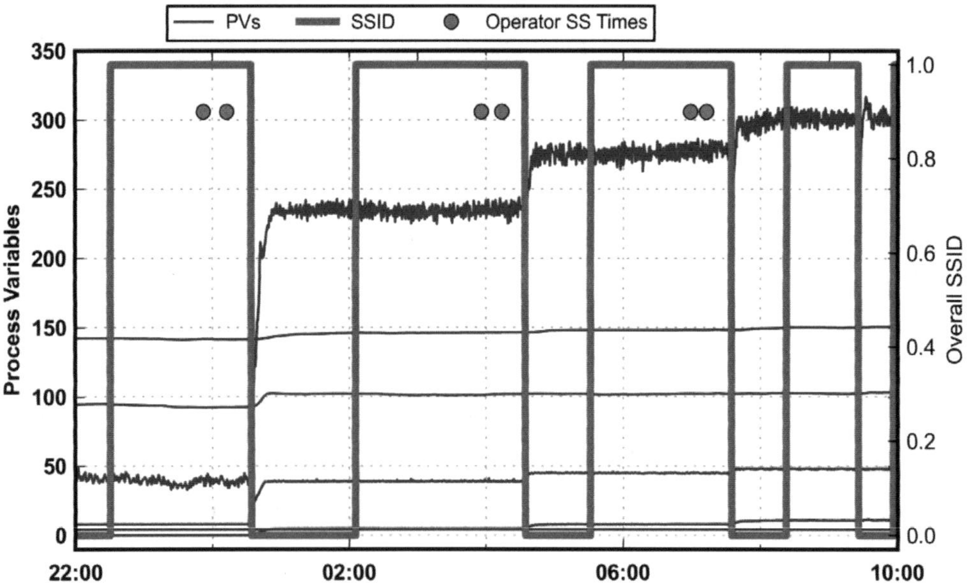

Figure 5: Online implementation on real time data. Corroboration with operator observed steady states.

CONCLUSION

The SSID method described in this paper was successfully used to detect steady state and transient states in offline and online modes at the FRI distillation unit. The steady state times detected by the method were consistent with the operator-determined steady state times.

The authors contend that the successful implementation of the program at FRI provides proof that the SSID method described in this paper has the potential to be easily implemented in an industrial process.

REFERENCES

1. Cao, S., and R. R. Rhinehart, "An Efficient Method for On-Line Identification of Steady-State," Journal of Process Control, Vol. 5, No 6, 1995, pp. 363-374.
2. Cao, S., and R. R. Rhinehart, "Critical Values for a Steady-State Identifier," Journal of Process Control, Vol. 7, No. 2, 1997, pp. 149-152.
3. Shrowti, N., K. Vilankar, and R. R. Rhinehart, "Type-II Critical Values for a Steady-State Identifier", Journal of Process Control, Vol. 20, No. 7, pp 885-890, 2010.
4. Brown, P. R.; and R. Russell Rhinehart, "Demonstration of a Method for Automated Steady-State Identification in Multivariable Systems", Hydrocarbon Processing, Vol. 79, No. 9, 2000, pp 79-83.
5. Iyer, M. S., and R. R. Rhinehart, "A Novel Method to Stop Neural Network Training," Proceedings of the 2000 American Control Conference, June 28-30, 2000, Chicago, IL, Paper WM17-3, pp 929-933
6. http://www.fri.org/

Minimum Remixing Effect in Distillation Secuences

Josué Santos-Méndez, L. Vladimir Amador-Mata and Salvador Hernández*
Universidad de Guanajuato
Facultad de Química
Noria Alta s/n
Guanajuato, Gto., 36050
México

Abstract

In this paper the retrofit of five conventional distillation trains for the separation of quaternary mixtures of hydrocarbons for feed compositions with high or low content of intermediate components is presented. The retrofit only requires the introduction of liquid or vapour recycle streams between the conventional distillation columns. Each recycle stream removes one condenser or one reboiler. The introduction of thermal links can lower the energy consumption up to 40% in contrast to conventional distillation trains widely used in the industry. This efficiency in the use of energy is achieved because the recycle streams, introduced in the conventional distillation trains, reduce the remixing associated with higher energy consumption.

Keywords: Distillation trains, energy savings, thermal link, remixing *Corresponding author, E-mail: hernasa@quijote.ugto.mx (S. Hernández).

Introduction

Conventional distillation trains are characterised by their large demands of energy in the reboilers[1]; as a result, researchers and engineers are interested in developing new distillation arrangements that can reduce energy consumption. One method that has been used for this purpose, is the use of thermally coupled distillation schemes which in the case of the separation of ternary mixtures can decrease the energy requirements around 30% in contrast to conventional distillation trains[2,3,4,5,6,7,8,9,10,11,12]. The thermal link reduces the remixing in the conventional distillation sequences which contributes to the energy savings achieved in the thermally coupled distillation sequences (Triantafyllou and Smith[4] and Hernández et al.[13]). Mixtures with more than three components have not been studied to the same degree as ternary mixtures have[14,15]. As a result, in this paper we present a study of the retrofit of conventional distillation sequences for the separation of quaternary mixtures of hydrocarbons (A,B,C,D) with low or high content of intermediate components because in the case of ternary mixtures, the energy savings depend strongly on the amount of the intermediate component.

Retrofit of Conventional Distillation Trains

It is well known that the separation of a quaternary mixture can be done using the five conventional distillation trains[16] shown in Figure 1. Figures 1A and 1B show two classical distillation trains: the direct distillation train in which the components are removed one by one in the overheads and the indirect distillation train in which the components are obtained one by one in the bottoms products. These sequences are widely used in the industry. Conventional distillation trains can be changed into thermally coupled distillation sequences by removing a condenser and introducing a liquid recycle stream or eliminating a reboiler and adding a vapour recycle stream to the bottom of the column (Figure 2). For example, in the case of the conventional direct distillation train (Figure 1A), the reboiler of the first column is replaced by a recycle stream of vapour from the second column, and the reboiler of the second column is eliminated with a vapour recycle stream from the third distillation column. At this stage, recycle streams have been introduced, but it is necessary to optimise their flowrates in order to guarantee optimal energy consumption. The recycle streams are varied until the minimum energy requirement is detected. The number of stages remains fixed in both types of distillation arrangements because in the retrofit process only recycle streams are introduced and condensers and reboilers are eliminated.

Cases of Study

We have explored the energy consumption and the remixing effect in the five conventional distillation sequences shown in Figure 1 and the corresponding thermally coupled distillation schemes of Figure 2, for the separation of two quaternary mixtures of hydrocarbons (Table 1) and two feed compositions (Table 2). The study was carried out via rigorous simulations by using the process simulator, Aspen Plus V7.2.

Discussion of Results

Energy

In the first stage of the analysis, energy consumptions in the conventional distillation trains were obtained and compared (Table 3). For the case of mixture M1 and the two feed compositions, the conventional distillation train CDS I (Figure 1A) was the most energy efficient, savings in energy of 37% and 11% for feed F1 and F2 respectively were obtained in comparison with the worst conventional distillation train. When the mixture M2 was considered, the sequences CDS V (Figure 1E) and CDS IV (Figure 1D) were the optimum options for the separation of feeds F1 and F2 respectively with savings of 58% and 41%.
In the second stage of the study, thermal links were introduced in the conventional distillation trains. As stated before, in order to obtain the optimum design in the use of energy, the interconnecting flows were varied until the minimum demand of energy was achieved. This task was completed with the process simulator, Aspen Plus V7.2. In the

optimisation procedure the interconnecting streams were changed one by one and the optimisation curves were obtained. The optimisation curves shown in Figure 3 were obtained when the mixture M1 for the separation of feed F1 was considered. In Figure 3, it can be noted that the energy consumption depends strongly on the values assigned to the recycle streams.

The optimum energy consumption found for the thermally coupled distillation trains is shown in Table 4. It can be observed, for the case of the separation of mixture M1 and the two feed compositions, that the TCDS I (direct thermally coupled distillation train) train of Figure 2A was the most efficient and presented energy savings of 27% in comparison to the CDS I scheme (direct distillation train). It is important to note that in these cases, the conventional direct scheme was the optimum sequence for both feed compositions when thermal links were introduced.
For the separation of the mixture M2 and the two feed compositions, the optimum thermally coupled distillation trains presented savings of 40% and 30% in contrast to the conventional distillation sequences. For feed F1 when thermal links were added to the CDS V train (Figure 1E), the corresponding TCDS V sequence (Figure 2E) also presented the minimum energy demand.
In summary, the introduction of thermal links in the conventional distillation trains produced savings in energy of up to 40% in comparison with the optimum conventional distillation train.

Remixing

The remixing, presented naturally in the conventional distillation trains, causes inefficiencies in the use of energy. These inefficiencies increase the energy required in the reboilers for the separation. In figure 4 we observe the composition profiles for the four components in the first column of the CDS I for the separation of the mixture M1 and feed F1, as a typical example, the composition of component B increases until a maximum of 0.57 and then declines to 0.28 in the bottom of the column, this is the remixing for component B. A similar effect is presented in the composition of component C in the second column of the conventional distillation train CDS I. According to Figure 5, the component C reaches a maximum value of 0.55 and a value of 0.4 in the bottoms product of the second column. As we can observe significant remixing is presented. When the two recycle streams were introduced in the distillation sequence, the remixing effects in components B (Figure 6) and C (Figure 7) were reduced significantly and this caused the energy savings obtained previously.

Having analysed all the mixtures and feed compositions, it was found that any conventional distillation column with at least one multicomponent product will have remixing.

Conclusions

The retrofit of conventional distillation trains was studied introducing thermal links for the separation of quaternary mixtures of hydrocarbons. Energy savings of up to 40% were obtained because the retrofit reduced the inefficiencies of the conventional distillation trains. This result is important because the retrofit involved only the introduction of recycle material streams. When a recycle stream of vapour was added one reboiler was eliminated; similarly the introduction of a liquid recycle stream eliminated a condenser. The recycle streams diminished significantly the remixing effects presented in the conventional distillation trains, associated with higher energy demands.

Acknowledgement

J. Santos-Méndez was supported through a scholarship from the Universidad de Guanajuato, México, during the development of this project.

References

1. Tedder, D. W., and Rudd, D.F., 1978, Parametric Studies in Industrial Distillation: Part I. Design Comparisons, AIChE J., 24:303.

2. Glinos, K., and Malone, F., 1988, Optimality Regions for Complex Column Alternatives in Distillation Systems, Chem. Eng. Res. Des., 66:229.

3. Fidkowski, Z., and Krolikowski, L., 1990, Energy Requirements of Nonconventional Distillation Systems, AIChE J., 36:1275.

4. Triantafyllou, C., and Smith, R., 1992, The Design and Optimisation of Fully Thermally Coupled Distillation Columns, Trans. Inst. Chem. Eng., 70:118.

5. Finn, A. J., 1993, Consider Thermally Coupled Distillation, Chem. Eng. Progress, October:41.

6. Wolff, E. A., and Skogestad, S., 1995, Operation of Integrated Three – Product (Petlyuk) Distillation Columns. Ind. Eng. Chem. Res., 34, 2094-2103.

7. Hernández, S., and Jiménez, A., 1996, Design of Optimal Thermally-Coupled Distillation Systems Using a Dynamic Model, Trans Inst. Chem. Eng., 74:357.

8. Hernández, S., and Jiménez, A., 1999a, Design of Energy-Efficient Petlyuk Systems, Comput. Chem. Eng., 23:1005.

9. Hernández, S., and Jiménez, A., 1999b, Controllability Analysis of Thermally

Coupled Distillation Systems, Ind. Eng. Chem. Res., 38:3957.

10. Shah, P. B., 2002, Squeeze More Out of Complex Columns. Chem. Eng. Progress, July:46.

11. Schultz, M. A., Stewart, D. G., Harris, J. M., Rosenblum, S. P., Shakur, M. S., and O'Brien, D. E., 2002, Reduce Costs with Dividing-Wall Columns, Chem. Eng. Progress, May:64.

12. Rong, B.G., Kraslawski, A., and Turunen, I., 2003, Synthesis of Functionally Distinct Thermally Coupled Configurations for Quaternary Distillations, Ind. Eng. Chem. Res., 42:1204.

13. Hernández, S., Pereira-Pech, S., Jiménez, A., and Rico-Ramírez, V., 2003, Energy Efficiency of an Indirect Thermally Coupled Distillation Sequence, The Can. J. Chem. Eng., in press.

14. Christiansen, A., Skogestad, S., and Lien K., 1997, Complex Distillation Arrangements: Extending the Petlyuk Ideas, Comput. and Chem. Eng., 21:S237.

15. Rong, B.G., and Kraslawski, A., 2002, Optimal Design of Distillation Flowsheets with a Lower Number of Thermal Couplings for Multicomponent Separations, Ind. Eng. Chem. Res., 41:5716. 16. King, C. J., 1980, Separation Processes.,2nd edition, McGraw-Hill, NewYork, USA

List of Figures

List of Tables

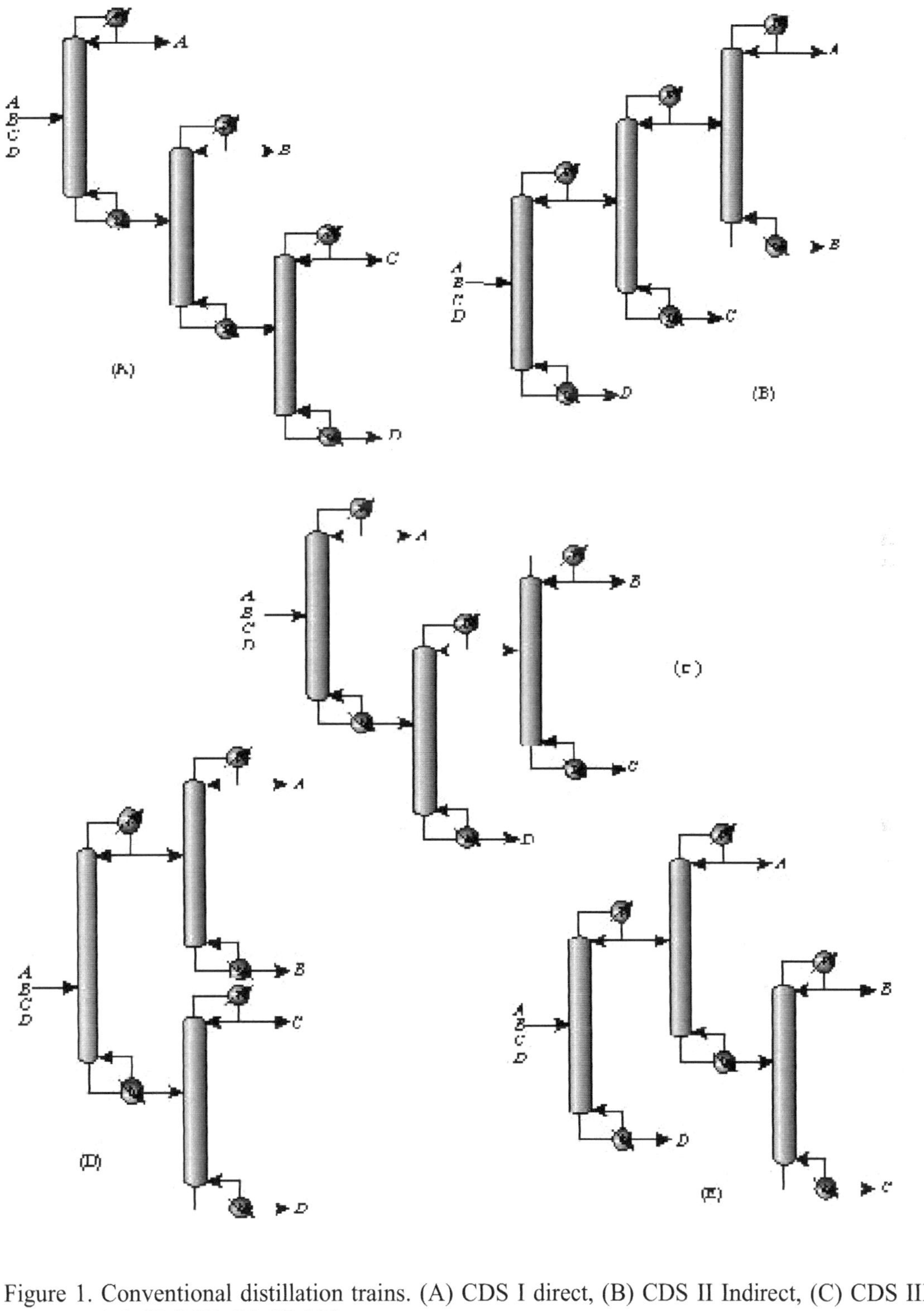

Figure 1. Conventional distillation trains. (A) CDS I direct, (B) CDS II Indirect, (C) CDS III, (D) CDS IV, (E) CDS V.

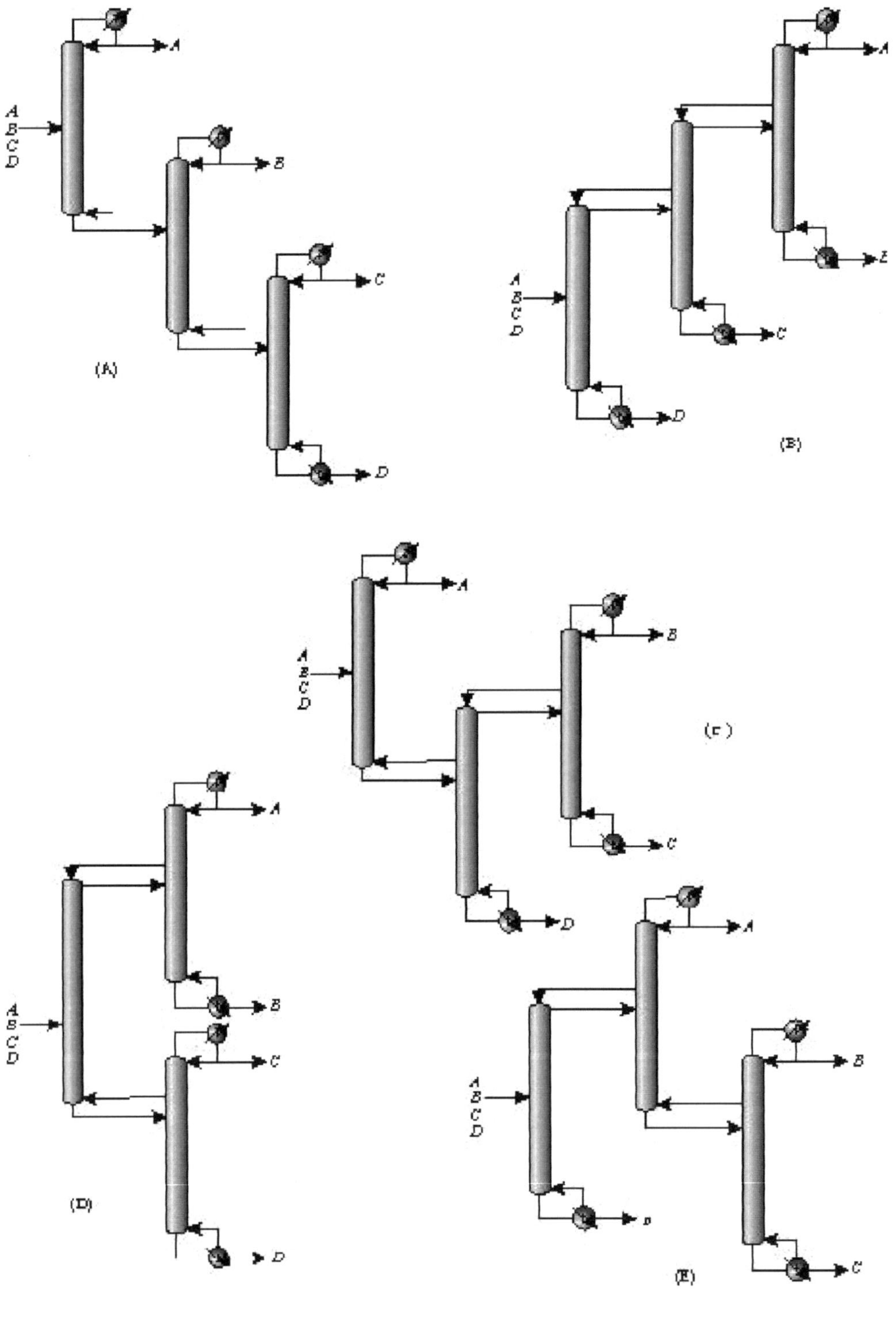

Figure 2. Thermally Coupled Distillation Sequences. (A) TCDS I direct, (B) TCDS II Indirect, (C)TCDS III, (D) TCDS IV and (E) TCDS V.

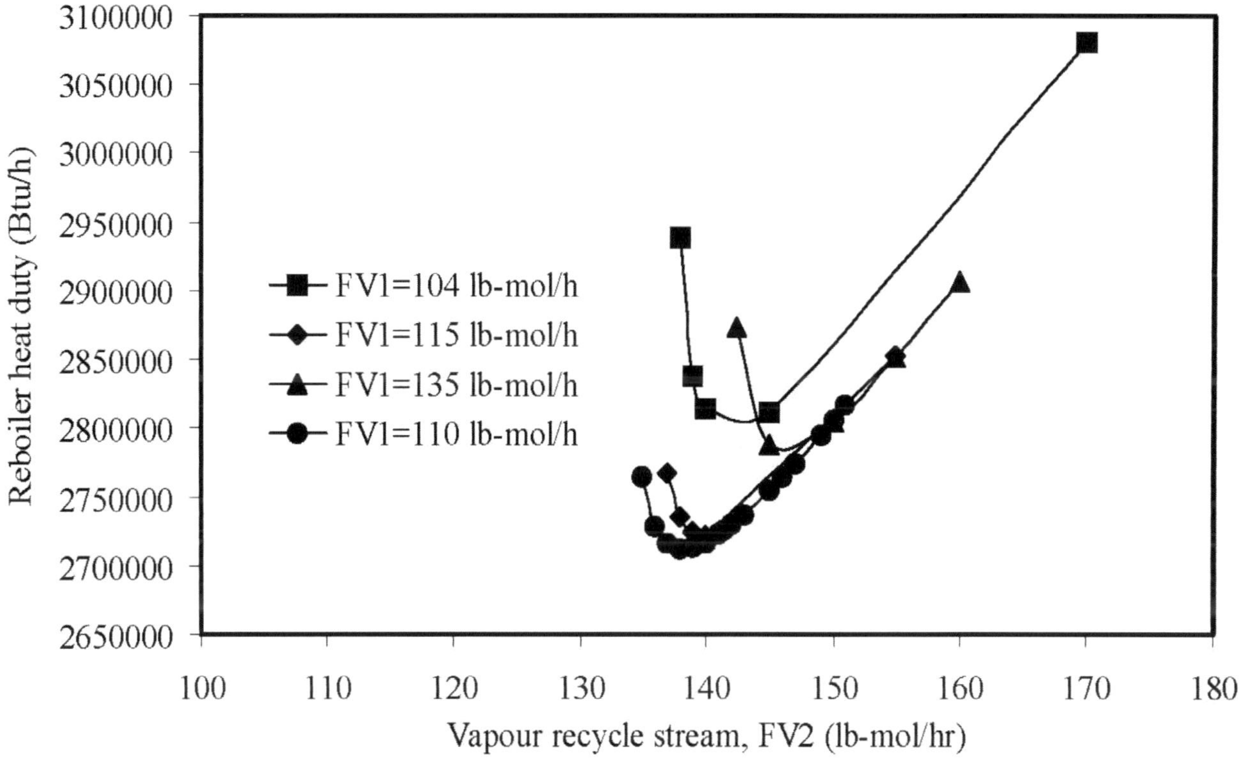

Figure 3. Optimisation curves for the TCDS I, Mixture M1, Feed F1.

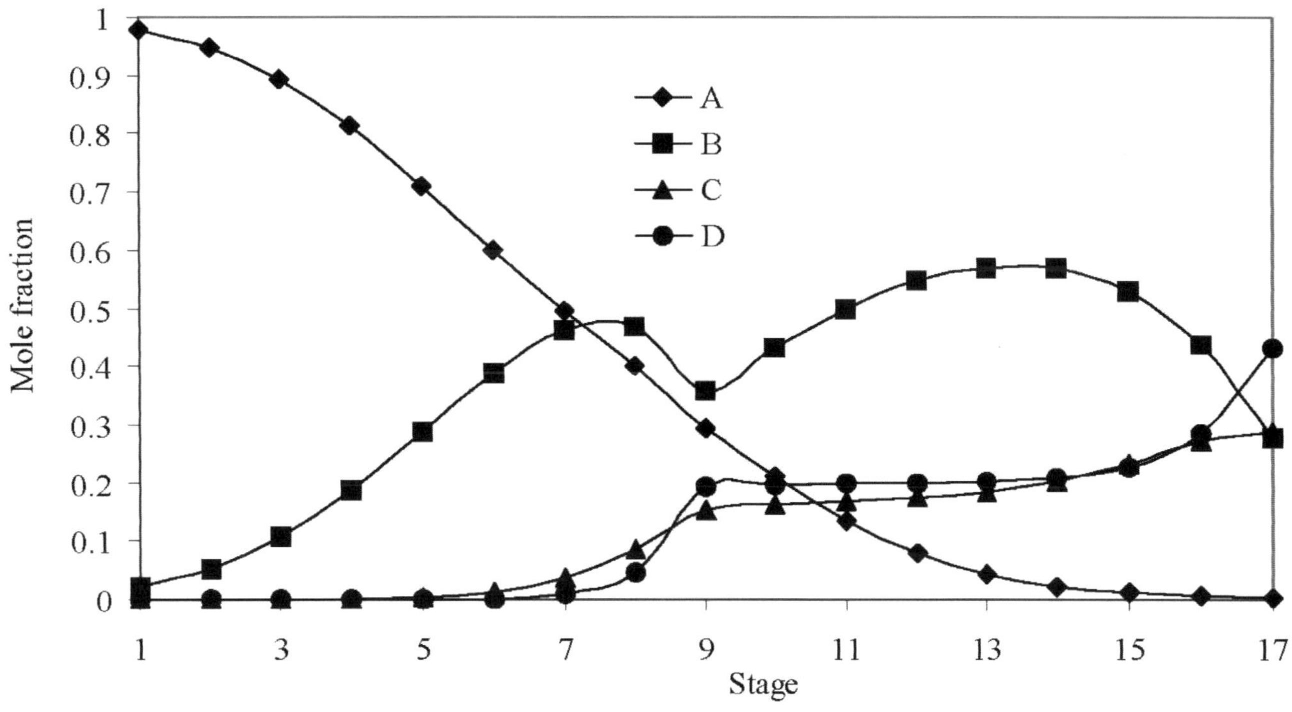

Figure 4. Composition profiles in the first column of the CDS I, Mixture M1, Feed F1.

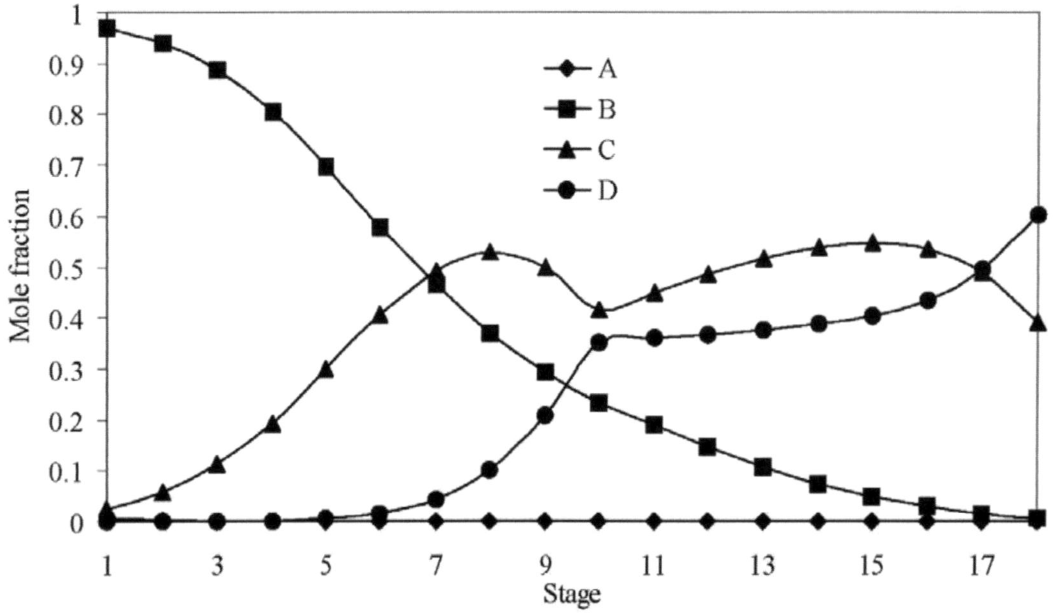

Figure 5. Composition profiles in the second column of the CDS I, Mixture M1, Feed F1.

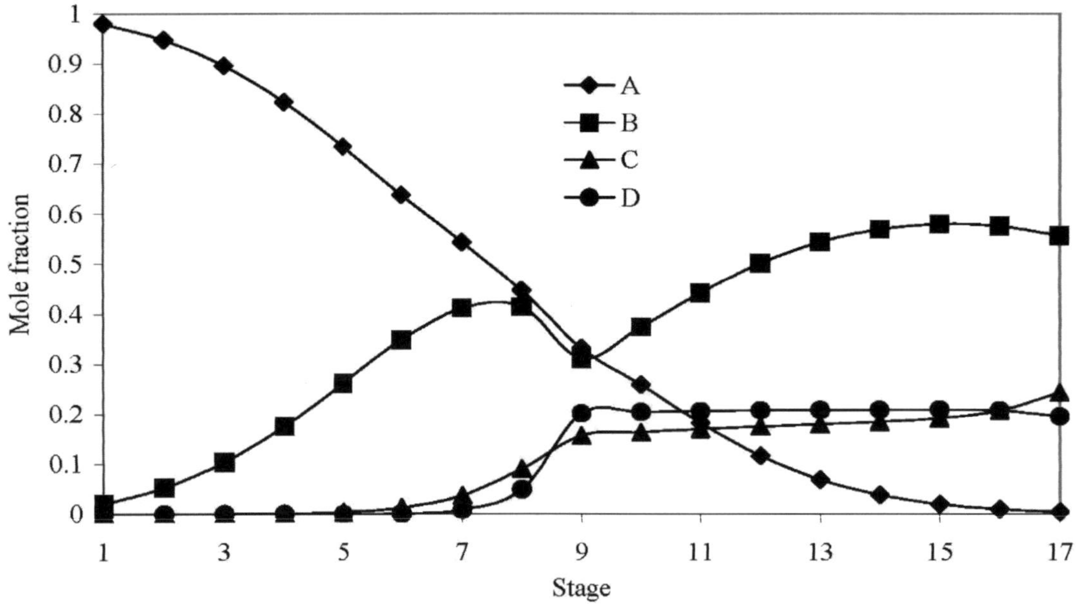

Figure 6. Composition profiles in the first column of the TCDS I, Mixture M1, Feed F1.

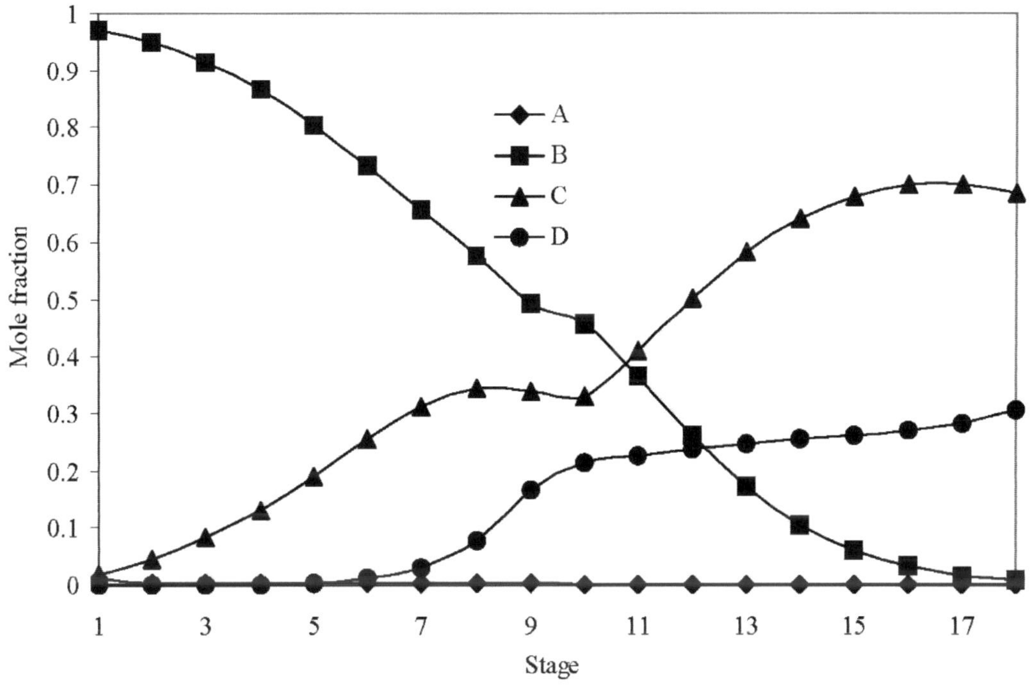

Figure 7. Composition profiles in the second column of the TCDS I, Mixture M1, Feed F1.

Table 1. Mixtures analyzed.

Mixture	**Components (A,B,C,D)**
M1	n-pentane, n-hexane, n-heptane, n-octane
M2	n-butane, i-pentane-n-pentane, n-hexane

Table 2. Feed flowrates for each feed (lb-mol/h).

Feed	**Flowrate (A/B/C/D)**
F1	30/20/20/30
F2	20/30/30/20

Table 3. Energy consumptions (Btu/h) of the conventional distillation trains.

	TCDS I	TCDS II	TCDS III	TCDS IV	TCDS V
			Mixture M1		
Feed F1	**2712669.4**	3070826.3	3688361.6	2788522.1	3141563.5
Feed F2	**2860660.9**	3640303.3	3382957.9	2925223.3	3205683.6
			Mixture M2		
Feed F1	4318728.7	4596249.2	5544802	292522.3	**2624437.5**
Feed F2	**4400424.2**	6935670.6	4946987.2	4448687.7	5714816.2

IMPROVING DISTILLATION ENERGY NETWORK USING HIGH PERFORMANCE DISTILLATION EQUIPMENT

Soun Ho Lee, GTC Technology, Irving, Texas, United States
Kwang Gil Min, GS Caltex Corporation, Yeosu, Korea

Presented for AIChE 2013 Spring Meeting, Abstract ID # 293319

INTRODUCTION

Energy costs are the largest percentage of a hydrocarbon plant's operating expenditures. This is especially true of the distillation process which requires substantial energy consumption. Concerns over recent high costs and economic pressures continually emphasize the need for efficient distillation design and operation without a loss of performance.

This paper will illustrate how energy efficient design can be applied in a distillation unit through optimization between the distillation column and heat network system. Through a case study, a successful retrofit of an aromatics distillation unit will be discussed. Detailed retrofit activities including complex heat network evaluation, process simulation modeling and energy-friendly, high-performance distillation equipment implementation are described.

DISTILLATION ENERGY NETWORK IMPROVEMENT

As continuous distillation requires simultaneous heat input and removal (thus requiring significant energy consumption), complex heat integration becomes more common for modern distillation units to improve unit energy efficiency. Since a distillation column's degree of separation and enthalpy balance influence each other, it is critical to evaluate and optimize the distillation column and heat exchanger networks together in order to maximize plant economics.

There are numerous strategies to improve the energy efficiency of distillation processes, with the amount of improvement through each strategy varied according to process condition. The following are common strategies that can be applied to practical energy improvement projects.

Feed temperature

Feed temperature is a major factor in influencing the overall heat balance of a distillation column system. Increments in the feed enthalpy can help reduce the required energy input from the reboiler at the same degree of separation. Installing a feed pre-heater is a very common process option to minimize reboiler heat duty. If the feed preheater can be integrated with other valuable process streams (as a heating medium), overall energy efficiency of the distillation system can be improved further. However, increasing feed temperature does not always improve the overall energy efficiency of a distillation unit. Excessive feed temperature increments can cause a significant amount of flash of heavy key and heavy non-key components at the distillation column feed zone. In this case, a higher amount of reflux stream is necessary to maintain required overhead distillate purities. This augmented reflux ratio thus requires a higher boil up ratio. Overall energy efficiency is eventually aggravated. [1] Therefore, careful review of feed temperature and phase is critical to minimize overall energy consumption of the distillation unit.

Feed location

Improper feed location of a distillation column can also increase reflux/boil up ratio and energy consumption. An ideal feed location is a section of the distillation column where column internal liquid traffic composition is similar to feed stream composition. In this case, the composition gradient between feed stream and distillation internal fluids is minimized. In actual distillation column operation, actual feed compositions often are changed from the original design conditions. In case of significant deviation, discrepancy between column internal liquid composition and feed stream composition can increase and results in a non-optimum feed location. Therefore, evaluating feed location is one of essential steps for a successful distillation unit energy improvement.

Inter-condensers and inter-reboilers

Adding inter-condensers and/or inter-reboilers can help improve overall energy efficiency. "Pumparound", one of inter-condenser concepts has been widely applied to numerous petroleum multi-product fractionators. On the other side, implementing an intermediate reboiler can reduces the main reboiler duty. As the required temperature of an intermediate reboiler is lower than the main reboiler, this strategy may allow heat integration with other valuable heat sources that are less costly or not fully utilized in a plant.

Column operating pressure

Relaxation of column top operating pressure decreases the distillation column temperature profile and results in lower reboiler duty. It has been observed that numerous commercial distillation columns have been operated with lower operating pressures than the original design values. However, this strategy is not applicable to columns operated under an atmospheric pressure range. Column overhead circuit pressure drop and condenser temperature approaches both heavily influence feasibility. In addition, column pressure reduction expands vapor traffic and pushes the limits of existing distillation equipment.

Column pressure drop

Reducing column pressure drop can lower reboiler duty at the same degree of separation. The amount of reboiler duty saving relies on operating pressure and enthalpy balance. This strategy is generally feasible when the distillation column is operated under vacuum pressure range. Meanwhile, pressure drop improvement does not often provide noticeable energy savings in high pressure range distillation service.

Column efficiency improvement

Column efficiency improvement can reduce the reflux/boil up ratio at a given degree of separation. This strategy can be delivered by increasing the number of theoretical stages and/or enhancing the distillation equipment efficiency. The feasibility can be gauged by a dedicated sensitivity analysis. Constructing a column efficiency curve with a reflux ratio is one of core tools for the sensitivity analysis. A typical curve is displayed in the Figure 1. This curve visualizes column efficiency sensitivity and energy saving gain. The curve can be categorized by three district zones: steep, moderate, and flat sensitivity zones.[2]

Column efficiency improvement is usually very feasible when the reflux ratio falls in the steep sensitivity zone and at ratios considerably in excess of the minimum reflux ratio. In this scenario, even the small addition of stages, or an increase in distillation equipment efficiency, can enhance overall column separation with significant energy reductions.

Improvement gain is diluted in the moderate sensitivity zone. Further detailed feasibility study is necessary through economic analysis. The magnitude of energy savings is negligible when reflux ratio variation follows flat motion in the remaining zone.

Figure 1: Typical Column Efficiency vs. Reflux Ratio Curve

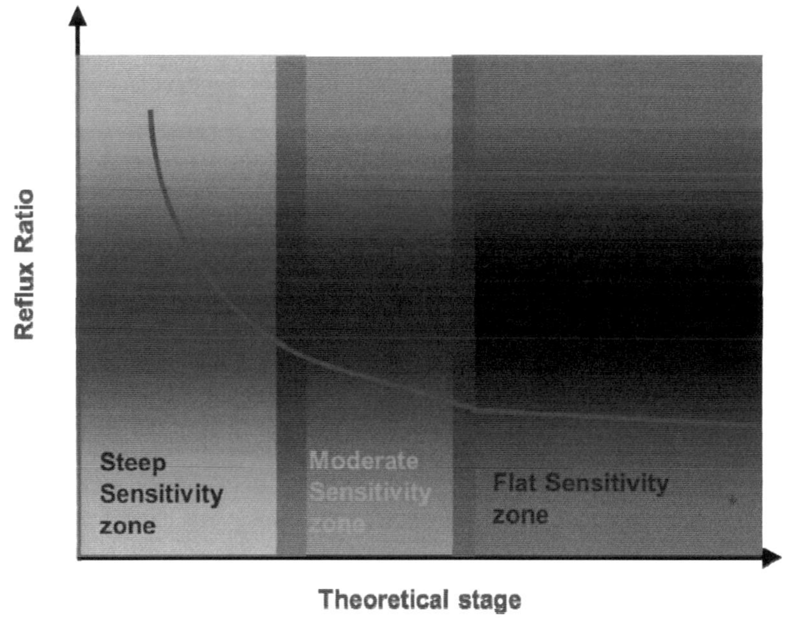

CASE STUDY: UNIT DESCRIPTION

The following discussion is a xylene mixture separation unit revamp case study that demonstrates well thought out proven design practices and selection of the correct high efficiency distillation equipment to fulfill the energy efficiency improvement. Figure 2 illustrates the xylene mixture separation unit configurations under discussions. This schematic reveals that original distillation units have implemented the full heat integration network for energy efficient operation.

The function of the xylene column is to separate the feed mixture to xylene components and heavier C9+ components. This column has two different feed sources. The reformate splitter bottom stream and the toluene column bottom stream (which belongs to the aromatic extraction unit) are introduced as the xylene column bottom feed stream. The bottom feed stream is split equally and charged to two different feed trays. The reformate splitter bottom stream is treated at the clay towers to eliminate traced olefin components before charging to the xylene column.

Meanwhile the deheptanizer bottom stream (from the xylene isomerization unit) is charged as the xylene column top feed stock. This stream is also split and introduced to three different feed trays.

The xylene column overhead vapor stream is split to three parallel streams. Two vapor streams are utilized as the heat source of the extract column reboiler and the raffinate column reboiler respectively. The condensed xylene column overhead liquid streams are returned to the xylene column receiver. The other vapor stream is supplying heat to the xylene column overhead steam generator which produces #250 steam. The condensed overhead liquid stream is also returned to the xylene column receiver. The overhead distillate of the xylene column is sent to the para-xylene recovery unit.

In the xylene column reboiler circuit, the xylene column bottom reboiler inlet stream is first transported to the other two distillation column reboilers as heating mediums. After providing heat to these reboilers, the xylene column bottom streams are combined and introduced to the furnace-type xylene column reboiler.

In the para-xylene recovery unit, the xylene components from the xylene column are separated through the adsorption process. The pre-separated extract stream from the adsorption process is charged to the extract column in order to separate para-xylene from the desorbent. At the same time, the pre-separated raffinate stream from the adsorption process is charged to the raffinate column to rectify the raffinate components (meta-xylene, ortho-xylene and ethyl-benzene) as side cut product.[3] A pasteurizing section is arranged at the top of the raffinate column to remove moistures from the side cut product [4].

Figure 2: Xylene Column Heat Network Configuration

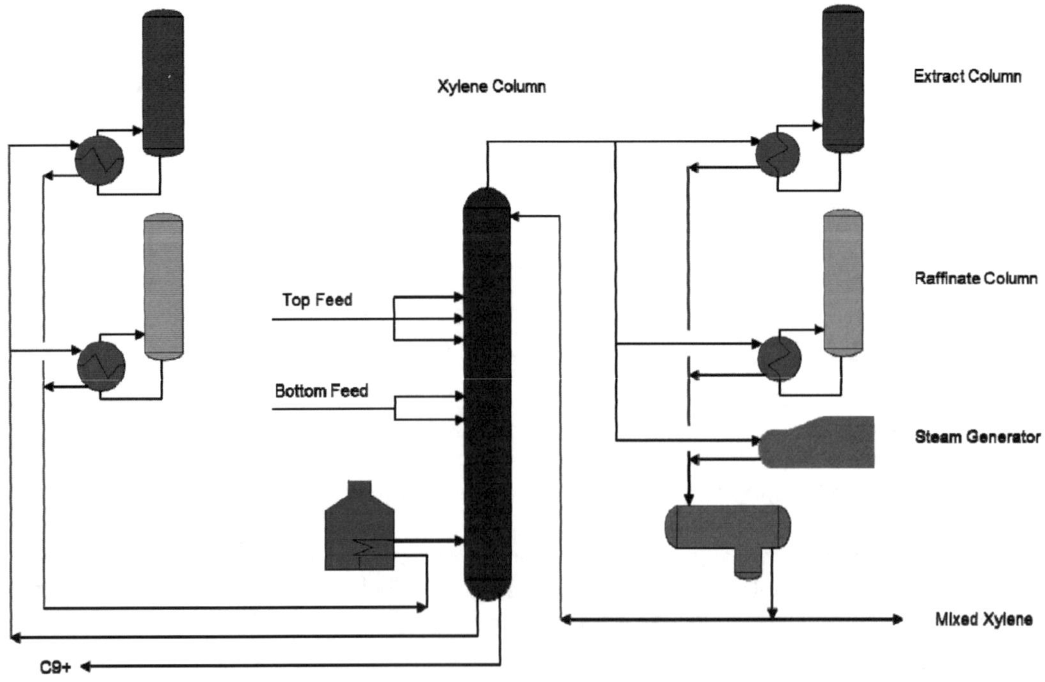

CASE STUDY: PROCESS EVALUATION FOR ENERGY EFFICIENCY IMPROVEMENT

To achieve additional energy efficiency gain, a dedicated process evaluation was conducted for the unit. Column operating conditions were first compared to the original design conditions. This comparison helps comprehend deviations between the original design and the actual operational environment. It was observed that actual product purities of the xylene column were higher than aromatic rundown products requirements. Relaxing the degree of separation of the xylene column can reduce the reflux and boil up ratio and save fuel consumptions for the xylene column furnace reboiler. However, the xylene column overhead vapor streams are utilized as the raffinate and extract column reboiler heating mediums and contribute steam production in the current unit energy network. Lower reflux / boil up ratios of the xylene column decreases the amount of xylene overhead vapor used as heating medium for the extract/raffinate column and/or steam generation.

Process simulation modeling was utilized as part of the process evaluation activities to quantify and predict gains in energy efficiency. Equilibrium base simulation software was utilized for the modeling. Base simulation modeling was first constructed through pertinent unit test run data. Gathered major process stream flow rates were verified via flow meter orifice calculations. Regular stream composition analysis reported bulk compositions for non-key components such as non-aromatic and C9+ component groups. Preliminary simulation modeling showed that component assumptions for these component groups varied simulation results significantly. To improve accuracy of simulation, detailed component analysis was specially arranged for the test run. Detailed component analysis was utilized for rigorous simulation modeling. Key component balances closures for the extract and raffinate column were less than 3%. Reconstructed feed compositions using products were applied for the extract and raffinate column simulation. For the xylene column, given overall mass balance closure was off by 5%. It was found that measured feed rates were more reliable than product rates and bottom product rate was less reliable through overall unit mass balance investigation. Based on this investigation, simulation modeling for the xylene column focused on matching feed and overhead distillate rates.

Reasonable matching reflux temperature as well as rate is critical to quantify reliable column internal traffic conditions. It has been observed that matching reflux temperature is often overlooked in simulation modeling. Reflux rate is usually metered at a flow meter located on the reflux piping. When the external reflux rate is recycled back to the column, the internal reflux rate will vary depending upon the external reflux temperature. Therefore a poor matching reflux temperature in the simulation will not predict the actual internal column traffic accurately. This can result in erroneous efficiency assumptions of the existing column, or provide a misleading, incorrect result.

Instrumentations for the extract and raffinate column pressure drop measurement were not pertinent. Matching pressure drop was not ignored in simulation modeling; instead matching column temperature data was focused. Temperature profiles are more important to predict distillation column energy consumptions. Tables 1 summarizes base model simulation results and compares test run data for the three columns. This table depicts that base model results were reasonably matched to the test run data. As mentioned earlier, the xylene column bottom rate was not matched. Through various sensitivity analyses, tray efficiencies of the columns were quantified.[5]

Table 1: Pre-revamp test run and base simulation result comparison

Case Parameter	*Unit*	*Pre-revamp Test run*	*Simulation Results*
Extract Column			
Column top temperature	^oF	*BASE*	*+ Δ0.9 ^oF*
Column bottom temperature	^oF	*BASE*	*+ Δ 1.1 ^oF*
Reboiler return temperature	^oF	*BASE*	*Δ 0 ^oF*
Reflux ratio (to overhead distillate)	*Volume*	*BASE*	*+1.1 %*
Reflux temperature	^oF	*BASE*	*Δ 0*
Overhead distillate rate	*BPD*	*BASE*	*Δ 0*
Bottom rate	*BPD*	*BASE*	*Δ 0*
p-DEB impurity in overhead distillate	*wt ppm*	*BASE*	*Δ 0*
Xylene impurity in bottom	*wt ppm*	*BASE*	*Δ 0*
Raffinate Column			
Column top temperature	^oF	*BASE*	*+ Δ 6.7 ^oF*
Side cut draw temperature	^oF	*BASE*	*- Δ 4.3 ^oF*
Column bottom temperature	^oF	*BASE*	*- Δ 0.5 ^oF*
Reboiler return temperature	^oF	*BASE*	*- Δ 2.9 ^oF*
Reflux ratio (to side cut)	*Volume*	*BASE*	*- 0.58 %*
Reflux temperature	^oF	*BASE*	*Δ 0*
Overhead distillate rate	*BPD*	*BASE*	*Δ 0*
Bottom rate	*BPD*	*BASE*	*Δ 0*
p-DEB impurity in side cut stream	*wt ppm*	*BASE*	*Δ 0*
Xylene impurity in bottom	*wt ppm*	*BASE*	*Δ 0*
Xylene Column			
Column top temperature	^oF	*BASE*	*- Δ 0.9 ^oF*
Column bottom temperature	^oF	*BASE*	*Δ 0 ^oF*
Reboiler return temperature	^oF	*BASE*	*- Δ 2.25 ^oF*
Reflux ratio (to overhead distillate)	*BPD.*	*BASE*	*+ 0.04 %*
Reflux temperature	^oF	*BASE*	*Δ 0 ^oF*
Overhead distillate rate	*BPD*	*BASE*	*- 0.1 %*
Bottom rate	*BPD*	*BASE*	*- 5 %*
C9+ impurity in overhead distillate	*wt ppm*	*BASE*	*Δ 0*
Xylene impurity in bottom	*wt ppm*	*BASE*	*Δ 0*

Extensive case studies were performed for feasibility of energy efficiency improvement of the unit. The case studies focused on energy consumptions in the extract and raffinate columns. Energy improvements in the other two distillation column reboilers in the xylene column reboiler circuit also help reduce fuel consumptions of the xylene column furnace reboiler but magnitude was not significant.

Since both columns are operated under atmospheric pressure, reducing column operating pressure is not applicable. In addition, feed streams of both of the columns are preheated by bottom product streams. Adding independent feed preheaters was not feasible due to limited plot and poor economics. The study showed that reducing column pressure drop using the low pressure drop nature of trays does not deliver reboiler duty savings in both of the columns.

Case study results for column efficiency improvement showed that energy efficiency improvement was feasible. The column efficiency curves were constructed using simulated reflux ratio and theoretical stage values and base modeling and improved column efficiency points were plotted. Figures 3 and 4 display these curves. An improved column efficiency point was predicted through an increased number of trays scenario. Simulated tray efficiency values of the base modeling were maintained for improved column efficiency case study. Extra individual tray efficiency improvement was not considered. Original trays were arranged with 600 mm (~24") regular tray spacing. The increased number of tray was predicted through reduced tray spacing scenario: 450 mm (~18") regular tray spacing. As a higher tray count can increases the column pressure drop, increased column pressure drop values were applied for column efficiency improvement case studies. The charts in Figures 3 and 4 show that the reflux ratios of the columns were positioned in the steep sensitivity zone and that enhancing column efficiencies are beneficial to improve energy consumptions of both columns. Reduced reboiler duties of the extract and raffinate columns contribute to the xylene column furnace reboiler duty in the unit energy network.

Figure 3: Extract Column Sensitivity Analysis (Revamp Design)

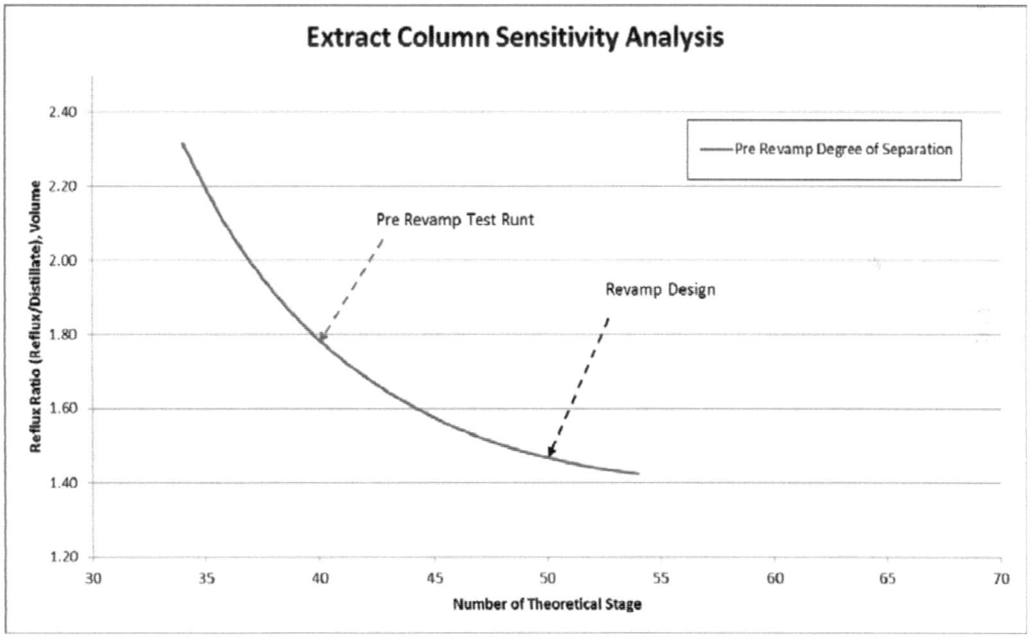

Figure 4: Raffinate Column Sensitivity Analysis (Revamp Design)

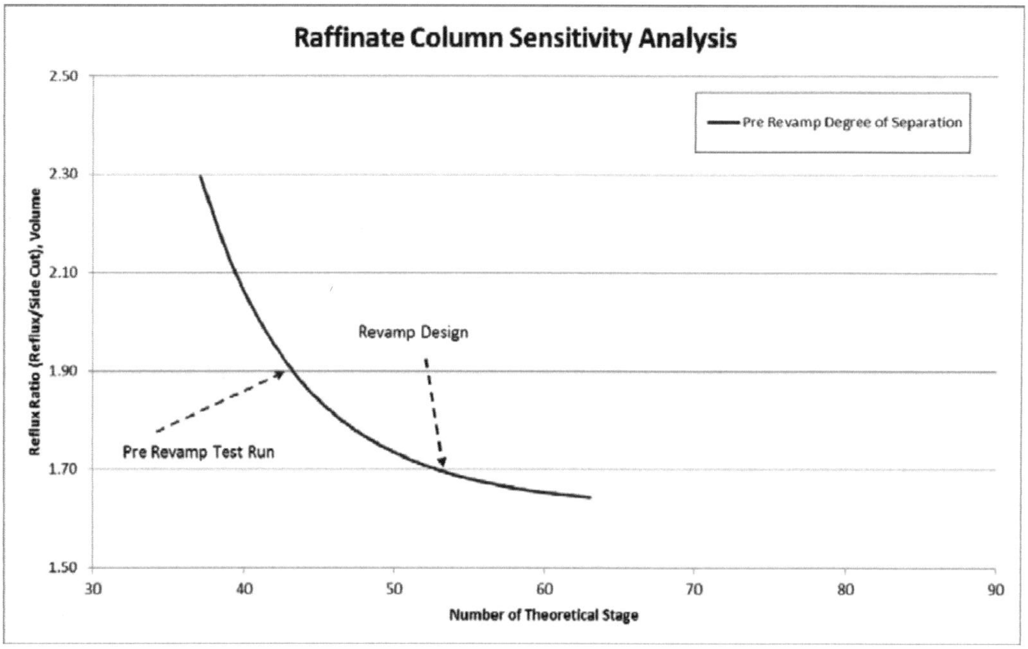

COLUMN MODIFICATION

Based on the aforementioned case study results, the extract and raffinate columns were modified. The number of trays was increased in both of the columns. At a given column shell height, a higher number of trays requires short tray spacing causing tray capacity loss. To prevent column capacity reductions, GT-OPTIM™ high performance trays were implemented in both of the columns. The higher capacity nature of the GT-OPTIM tray maintains the desired column capacity with shorter tray spacing. In addition, efficiency enhancement features of GT-OPTIM trays can contribute to maximizing column efficiency. Various performance enhancing features adapted in GT-OPTIM trays improves the vapor-liquid contact mechanism and enhances tray efficiency. These include specialized shaped downcomers, liquid inlet momentum breakers, tray inlet vapor/liquid contact initiation devices and directional valves positioned in the tray periphery area. Tray pressure drop was optimized to prevent a "too low" tray vapor velocity that can downgrade tray efficiency due to insufficient vapor/liquid contact volume.

Original trays for the pasteurization section of the raffinate column were designed with three pass. A chimney tray was positioned between three pass pasteurization section trays and two pass rectification section trays. It is inherently difficult to achieve a uniform liquid-to-vapor traffic ratio in each section of the three pass trays. Moreover, the original pasteurization section trays and the chimney tray did not equip any feature for proper flow ratio balancing. Each chimney pass open area ratio was not matched to the neighboring three pass tray pass open area ratio. In order to improve flow ratio balancing, the number of passes for the pasteurization section trays was changed from three to two and a new chimney tray was installed as per pass change.

CASE STUDY: POST REVAMP OPERATION REVIEW

The pre- and post-revamp performances are summarized and compared in Table 2. As the overall aromatic unit capacity has been expanded, the column charge rates are also increased. Post-revamp operation verifies that reflux ratios of the extract and raffinate column are reduced and these reduced reflux ratios eventually contribute to energy savings in the Xylene Column reboiler furnace. Measured furnace fuel consumption per the feed rate is substantially improved. Since the Para-Xylene recovery unit adsorbent upgrade also contributes furnace fuel consumption saving, net energy saving contribution through the column modifications was simulated and included in Table 2.

As product quality specifications are little relaxed to maximize the energy saving and column feed structures are changed, it is necessary to re-evaluate the extract and raffinate column performances with post-revamp operating conditions. The column efficiency curves are constructed using post-revamp operation simulation modeling and compared to the pre revamp base modeling curves. Degree of separation lines between pre and post revamp operations are displayed in Figures 5-6.

Table 2: Test Run Data Comparison

Case Parameter	Unit	Pre-revamp Test run	Post-revamp Test run
Extract Column			
Feed rate	BPD	BASE	+ 18 %
Overhead distillate rate	BPD	BASE	+ 36 %
Feed temperature	^oF	BASE	+ Δ 2.3 ^oF
Reflux temperature	^oF	BASE	+ Δ 16.9 ^oF
Column top pressure	psi	BASE	+ Δ 1.3 psi
Reflux ratio (to overhead distillate)	volume	BASE	- 28%
p-DEB impurity in overhead distillate	wt ppm	BASE	- Δ 1 ppm
Xylene impurity in bottom	wt ppm	BASE	+ Δ 16 ppm
Raffinate Column			
Feed rate	BPD	BASE	+ 14 %
Side cut product rate	BPD	BASE	+ 28 %
Feed temperature	^oF	BASE	+ Δ 5.6 ^oF
Reflux temperature	^oF	BASE	+ Δ 0.1 ^oF
Column top pressure	psi	BASE	+ Δ 1.4 psi
Reflux ratio (to side cut)	volume	BASE	- 6%
p-DEB impurity in side cut	wt ppm	BASE	+ Δ 23 ppm
Xylene impurity in bottom	wt ppm	BASE	+ Δ 4 ppm
Xylene Column			
Feed rate [note1]	BPD	BASE	+ 30 %
Unit reboiler fuel consumption [note2]	EFO BPD/BPD	BASE	- 22 %
Unit reboiler fuel consumption [note3]	EFO BPD/BPD	BASE	- 9 %
250# steam generation	Lb/hr	BASE	+ 26 %

Note

1. *Total feed rate*
2. *Equivalent furnace fuel oil consumption rate per feed charge rate*
3. *Simulated fuel consumption saving through the column efficiency improvement*

Figure 5: Extract Column Sensitivity Analysis (Pre and Post Revamp)

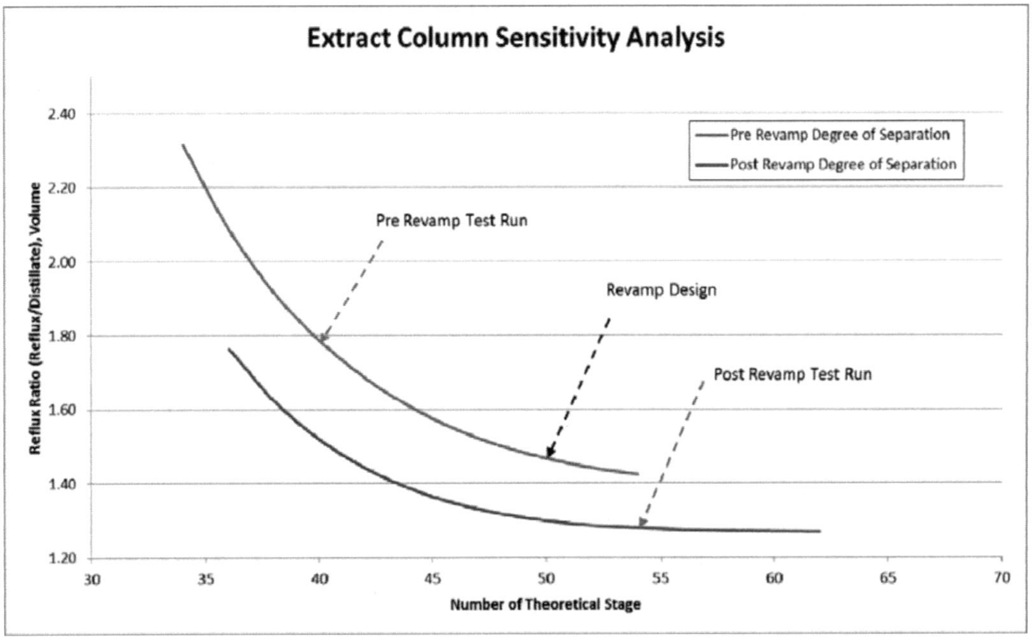

Figure 6: Raffinate Column Sensitivity Analysis (Pre and Post Revamp)

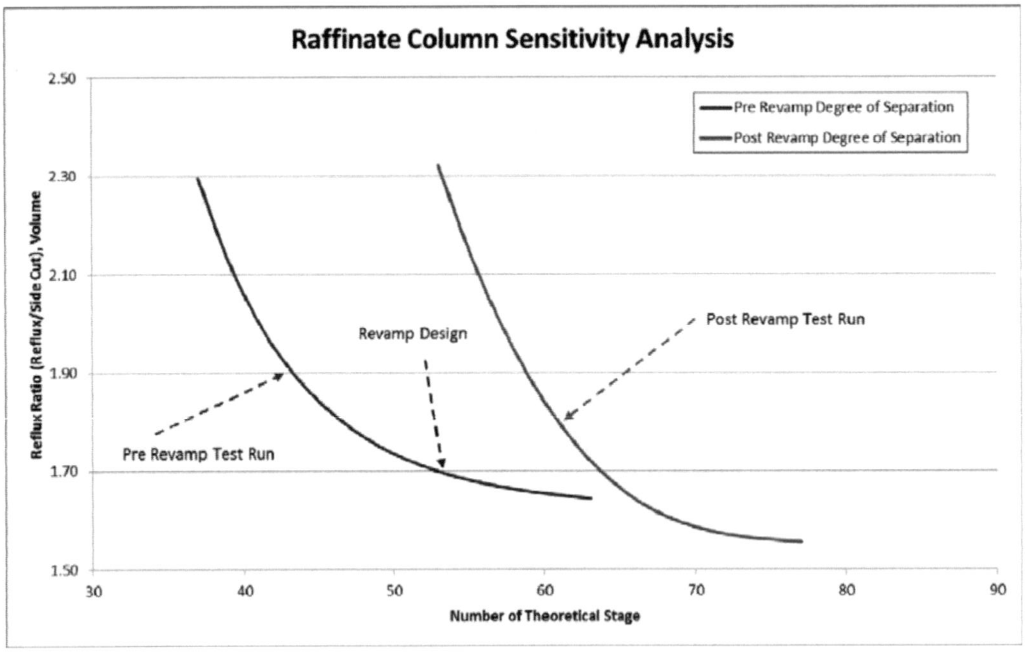

It is found that the raffinate column feed structure is changed in the post-revamp operating mode. Sensitivity analysis through simulation modeling shows that post-revamp degree of separation line is substantially shifted by the changed feed compositions and the pre-revamp degree of separation line is no longer applicable. Figure 6 visualizes post-revamp degree of the separation curve for the raffinate column is significantly shifted by the new feed composition.

The post revamp economic evaluations are updated and compared to the revamp target evaluations.[6] The evaluations are based on 3% inflation, 10% weighted average cost of capital, 15% depreciation, 1% of the total investment for maintenance, 22% tax bracket and Year 2012 average fuel price. Profitability indexes are expressed with regards to Payback Period, Net Present Value (NPV) and Internal Rate of Return (IRR). These Indexes are charted in Figures 7-9. These charts show actual revamp profitability is better than expected.

Figure 7: Profitability index – Payback Period

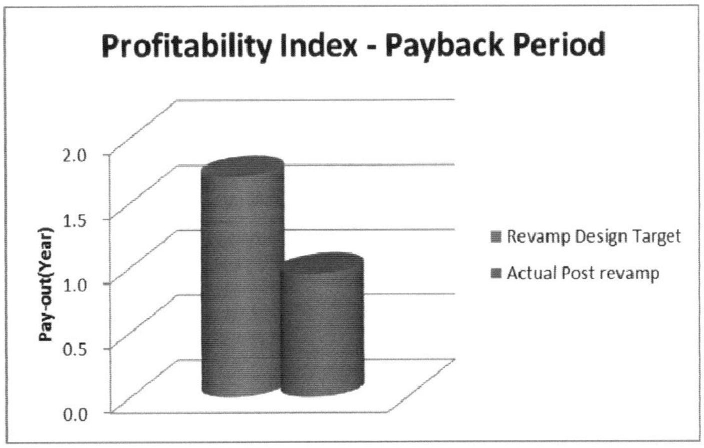

Figure 8: Profitability index – Net Present Value

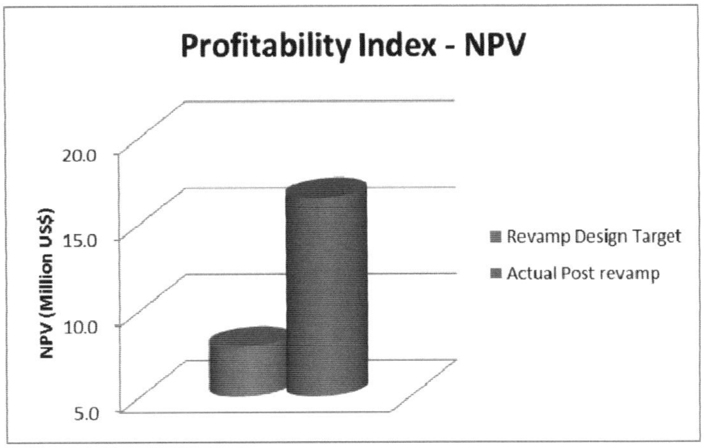

Figure 9: Profitability index – Internal rate of Return

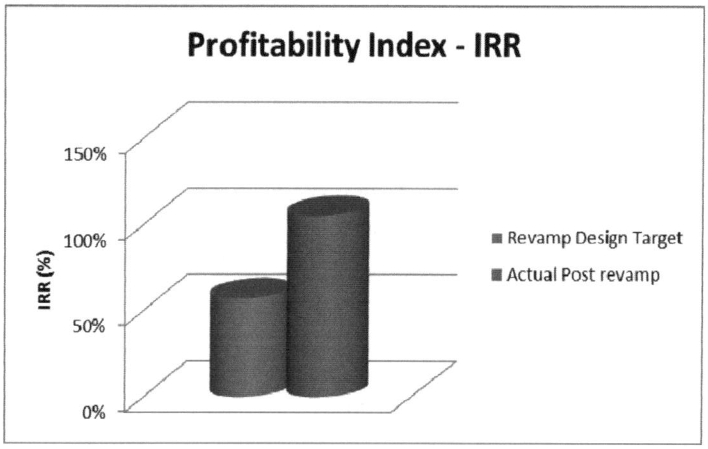

REFERENCES

1. Lee S.H, *et al*, "Optimize Design for Distillation Feed", Hydrocarbon Processing June, 2011.
2. Hanson D, *et al*, "High capacity distillation revamps", PTQ Autumn 2001.
3. Meyers R, "Handbook of Petroleum Refining Processes", McGraw-Hill Company, 1986.
4. Moczek J.S, *et al*, "Control of a Distillation Column for Producing High-purity Overheads and Bottom Streams", I&EC Process Design and Development, 1963
5. Kister H, *et al*, "Sensitivity analysis is key to successful DC5 simulation" Hydrocarbon Processing October, 1998
6. Largeteau D, *et al*, "Challenges and opportunities of 10 ppm sulfur gasoline: part 2", PTQ Q4 2012.

(142d) Development of Internal Reflux Splitter for Dividing Wall Column

Joonho Shin[1], Jongku Lee[1], Sungkyu Lee[1], Byungkyum Lee[2] and Moonyong Lee[3], (1)Corporate R&D, LG Chem, Taejon, South Korea, (2)AMT Pacific Co. Ltd, Seoul, South Korea, (3)School of Chemical Engineering, Yeungnam University, Gyeongsan, South Korea

Dividing wall column (DWC) technology has been successfully implemented in petrochemical processes. It is well known that the DWC can save the energy usage by removing a remixing phenomenon within the conventional two column system for a ternary mixture and minimize the capital investment intensifying a conventional two column system into a single column with a simple dividing wall. One of the major components in a DWC is a reflux splitter, which splits the internal liquid from the column overhead into the two liquid streams, one to the pre-section and other to the main column within the dividing wall section. Optimizing and controlling the liquid split rate is a crucial issue for successful design and operation of the DWCs. Several reflux splitters has been adopted in the industrial applications. It has been found that a more reliable reflux splitting system should be developed to make the entire column system stable and to minimize the process down time. In this article, several types of reflux splitter will be reviewed and compared. A novel development of the reflux splitting system is discussed with its structural details and benefits of the newly devised splitting system. The results from its real application to commercial DWCs show that the developed reflux splitter satisfies all of the technical requirements or specifications and even reduces the costs related to fabrication and installation.

DISTILLATION 2013 - LEARNING FROM PAST LESSONS, IMPROVING PERFORMANCE, AND ADVANCING TECHNOLOGY
2013 Spring AIChE Meeting
San Antonio, Texas
April 28 – May 2, 2013

OLEFIN-PARAFFIN SEPARATION BY REVERSIBLE REACTIVE DISTILLATION (RRD)

Steven G. Schon, P.E. (Arkema Inc.)
Hang Kuen Lau (Drexel University)
Aye Aye Min (Drexel University)
Merlin Thomas (Drexel University)
Laura Wu (Drexel University)

ABSTRACT

Olefins, especially ethylene and propylene, are large-volume industrial chemicals that are produced commercially by the separation from petroleum refining, natural gas liquids, and on-purpose chemical syntheses. A key step is the separation of the corresponding (same carbon-number) paraffins from the olefins, conventionally done by fractional distillation in "splitter" columns, e.g. ethylene from ethane ("C2 Splitter"), propylene from propane ("C3 Splitter"), etc. Due to the very close relative volatilities of the olefins and their corresponding paraffins, separation by distillation, while widely practiced industrially, is energy- and capital- intensive: They are carried out in high-pressure distillation columns, often requiring more than 150 theoretical stages and high reflux ratios. The latter results in large energy requirements, and large-diameter, heavy-walled columns, because of the large internal hydraulic traffic. It is therefore highly desirable to find an alternative olefin-paraffin separation scheme that requires substantially less energy and preferably smaller, less expensive equipment as well.

It is proposed to more efficiently separate olefins from paraffins by means of reversible reactive distillation process. An olefin-paraffin mixture and suitable reactants are fed to a first reactive distillation column, operating at suitable conditions to convert the olefin to a high degree to a high-boiling adduct, taking the paraffin as the overhead distillate product, and taking the adduct as the bottoms product. The olefin-adduct is fed to a second reactive distillation column, operating at suitable conditions to back-convert the adduct to the original olefin, which is taken overhead as the distillate product, and the original reactants, which are taken as the bottom and/or side-draw product and recycled to the first distillation column. If the olefin distillate from the second column contains residual amounts of the adduct and/or reactants, it may be sent to another separation unit to purify the olefin and recover and recycle the adduct and/or reactant back to the reactive distillation system.

A conceptual design study examined the feasibility of olefin-paraffin separation by reversible reactive distillation, for the specific case of C3 (propylene-propane) splitting. The calculated total reboiler duty per pound of propylene product was 35 – 45 % lower for the reversible reactive distillation process compared to a conventional C3 Splitter. Even though 3 columns are required in place of a single conventional C3 Splitter, in aggregate, they require ~40 % fewer total theoretical separation stages, ~ 85% less total tray area, and ~80% less total column volume. In addition to the energy savings, the projected capital investment is ~20% lower for the reversible reactive distillation system compared to a conventional steam-reboiled C3 splitter.

INTRODUCTION

As part of their graduation requirements, senior chemical engineering students at Drexel University do a process plant design study, including process calculations, sizing equipment, developing cost estimates, and preparing economic evaluations to assess the business feasibility of the chosen project.

One of the authors (Schon), who is employee of Arkema Inc, is an industrial advisor and mentor to several of the Drexel student groups. Based on a presentation he saw by Prof. Aspi Kolah of Michigan State University, on reactive distillation for separating cyclohexane from cyclohexanol, he came up with the idea to separate olefins form paraffins using a reactive distillation. Arkema graciously allowed him to submit the concept as design topic for his Drexel students. While he provided process design guidance in the subject study, the bulk of the work, especially the details, simulation, and optimization of the reactive distillation units, was done by the students, who are the co-authors of this paper.

BACKGROUND

Olefins, especially ethylene and propylene, are large-volume industrial chemicals that are produced commercially by the separation from petroleum refining, natural gas liquids, and on-purpose chemical syntheses. The separations are conventionally performed by a series of fractional distillations. Often, the corresponding paraffins of the same carbon number are present and co-distill with the olefins, as the olefins and paraffins of the same carbon number have similar relative volatilities. This necessitates an additional distillation ("splitter") step, to separate an olefin form its corresponding paraffin, e.g. ethylene form ethane ("C2 Splitter"), propylene from propane ("C3 Splitter"), etc.

Because of the close relative volatilities of the olefin/paraffin pairs, separation by distillation, while widely practiced industrially, is energy and capital intensive: They are carried out in high-pressure distillation columns, often requiring more than 150 theoretical stages and high reflux ratios. The latter results in large energy requirements, and relatively large-diameter heavy-walled columns because of the large internal hydraulic traffic (volumetric flows).

It is highly desirable to find an alternative olefin-paraffin separation scheme that requires substantially less energy and preferably smaller, less expensive equipment as well. One alternative is to chemically react the olefin to a compound (adduct) that has a much lower relative volatility compared to the paraffin, allowing easy separation of the paraffin, followed by reaction of the adduct to recover the olefin.

This can be achieved using an equilibrium-limited reaction, selecting suitable conditions of concentration and temperature to drive the reaction forward to form the adduct, and subsequently, after the paraffin removal step, select suitable different conditions of temperature and concentration to drive the reverse reaction, converting the adduct back to the original olefin.

However, with equilibrium reactions, by their very nature, high degrees of conversion cannot be achieved in a single step; if a single-stage reactor is used for the forward reaction, a significant portion of the olefin would remain un-reacted, with its attendant difficulty in separation from the paraffin. If stand-alone reactors are used, multiple reactors or reaction stages are needed to drive the conversion of the olefin high enough to facilitate easy separation of the paraffin. Similar difficulties are encountered with the reverse reaction – with a single stage reaction, equilibrium limits the back-conversion, leaving substantial fractions of the adduct un-converted, again necessitating multiple reactors or reaction stages to achieve high recoveries of the desired olefin.

In a well-established technique for working with equilibrium-limited reactions, high conversion can be achieved by reactive distillation (RD), in which a distillation column has a reaction zone in which the equilibrium reaction takes place. The distillation continuously removes the reaction product from the reaction zone, allowing the reaction to be driven further towards completion (by Le Chatelier's principle). The reactive distillations are configured so that the reaction zone effectively functions as a series of counter-current reactive stages, offering the advantage of maximal chemical driving forces and multi-stage reaction.

It is proposed to separate olefins from paraffins by carrying out the reversible equilibrium olefin reactions in reactive distillation systems. This would facilitate the distillative separation of the paraffin with reduced energy (reflux) requirements (due the much larger relative volatility between the adduct and the paraffin, compared to the paraffin-olefin separation), followed by the recovery and distillative purification of the olefin with low energy requirements during the reverse reaction.

PROPOSED REVERSIBLE REACTIVE DISTILLATION (RRD) PROCESS

Olefins are separated and purified from their corresponding paraffins by a reversible reactive distillation process. An olefin-paraffin mixture and suitable reactant(s) are fed to a first reactive distillation column, operating at suitable conditions to convert the olefin to a high degree to a high-boiling adduct, taking the paraffin as the overhead distillate product, and taking the adduct as the bottoms product. The olefin-adduct is fed to a second reactive distillation column, operating at suitable conditions to back-convert the adduct to the original olefin, which is taken overhead as the distillate product, and the original reactant(s), which are taken as the bottom and/or side-draw product(s) and recycles to the first distillation column. If the olefin distillate from the second column contains residual amounts of the adduct and/or reactant(s), it can be sent to another separation unit (e.g. fractional distillation, crystallization, membrane separation, etc.) to purify the olefin and recover and recycle the adduct and/or reactant(s) back to the reactive distillation system.

CONCEPTUAL DESIGN STUDY

A conceptual design study examined the feasibility of olefin-paraffin separation by reversible reactive distillation (RRD), for the specific case of C3 (propylene-propane) splitting.

A base case design assumed a conventional steam-reboiled C3 splitter. The subject study involved the design of a suitable RRD system to achieve the same propylene production rate and purity as the conventional C3 splitter, assuming an identical mixed-C3 feedstock. The scope of the study made the flowing assumptions:

- Production capacity of 1.2 billion pounds propylene per year (100% basis).
- Mixed C3 feedstock: 43.5% Propane / 56.5% Propylene (typical C3 splitter feed composition from a propylene synthesis plant).
- Final propylene product: >99.5% pure (polymer grade propylene)
- RRD Unit would be retro-fitted into an existing facility, to replace a conventional C3 splitter.
- Both the conventional C3 splitter base case and the RRD system assume identical outside battery limits (OSBL) facilities and utilities, and identical operating labor requirements

Base Case – Conventional C3 Splitter

270,000 pounds per hour of a crude propylene mixture containing 56.5% propylene and 43.5% propane, are fed to a conventional distillation column ("C3 Splitter"),

For a given number of theoretical stages, the feed location was varied, to determine the minimum reboiler heat input, corresponding to the optimum feed location. This procedure was repeated for various numbers of theoretical stages. The required reflux ratio (at the optimal feed location) vs. the number of theoretical stages is shown in Figure 1. From the asymptotic shape of the curve, this separation requires at least 75 theoretical stages (as the reboiler duty approaches infinity), and alternatively, a minimum of reflux ratio of 13 (as the number of theoretical stages approaches infinity).

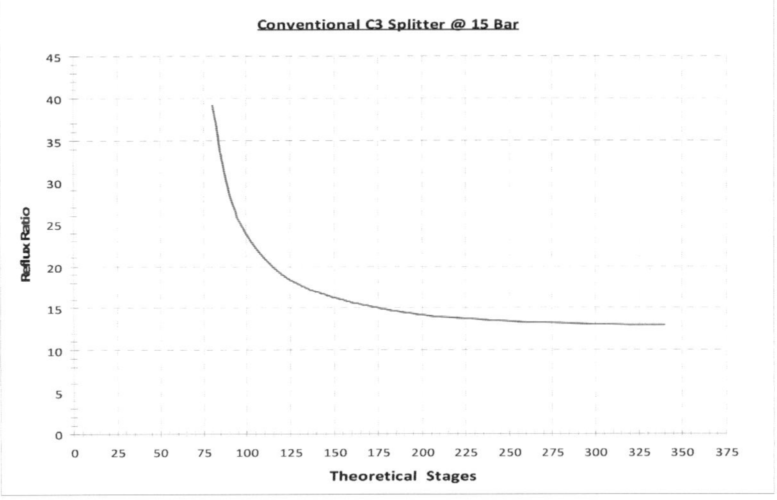

Figure 1

Using a widely accepted heuristic of 1.1 times the minimum reflux ratio as the design point for the number of theoretical stages, this results in a conventional C3 Splitter with 195 theoretical stages (including the condenser and reboiler). Assuming an overall column efficiency of 80 %, this corresponds to 241 actual trays, the required column diameter is calculated to be 27.5 feet, assumes 4-pass valve trays. The corresponding reboiler heat input is 2930 BTU/lb propylene product.

RRD Process Synthesis Approach

The process synthesis approach used the following sequence:

1. Develop a conceptual RRD process flow scheme to convert the (propylene) into a high-boiling adduct and distill-off the propane, then back-convert the adduct to the original olefin and distill it away form the recovered original reactants, which are recycled to form more adduct.
2. Identify suitable reversible reactions to make the olefin adduct. Criteria include large relative volatility differences between propane and the adduct, and propylene and the reactants, as well as availability of suitable catalysts, and minimizing the potential for safety or environmental risks.
3. Select a candidate chemistry and catalyst system, based on availability of equilibrium & kinetic data, as well as safety and cost considerations.
4. Use process simulations to determine practical design and operating conditions, and to size the equipment required for both the conventional C3 splitter base case and the RRD process.

5. Optimize the process to minimize capital and operating (energy) requirements, including opportunities for heat integration with the parent facility.

Selected RRD Process For Evaluation

The selected class of reversible reactions was esterification, since propyl esters are well known and produced commercially. The target propylene adduct was propyl acetate (mixed iso-propyl and n-propyl isomers, since the both with de-esterify back to propylene).

While propylene can be directly esterified with acetic acid to isopropyl acetate, this reaction route uses boron trifluoride (BF_3) and hydrofluoric acid (HF) as homogeneous catalysts. In addition to being extremely hazardous materials to work with, homogeneous catalysts are difficult to integrate into reactive distillation columns, if the reaction zone is above the reboiler.

The literature showed that propylene can be converted to propyl esters by the reaction with acetic acid in the presence of water and a strong acid catalyst. This is achieved by simultaneous reactions:

- Propylene + Water ⇔ Isopropanol (dominant pathway)
- Propylene + Water ⇔ Propanol (secondary pathway)
- Isopropanol + Acetic Acid ⇔ Isopropyl Acetate + Water
- Propanol + Acetic Acid ⇔ Propyl Acetate + Water

It was therefore decided to evaluate a 2-step esterification process. Based on patent literature, these reactions can be carried out using solid hetero-polyacid (HPA) catalysts. The specific catalyst chosen in this study is used in the commercial production of ethyl acetate; the patent taught that it can be used to make isopropyl acetate as well.

Process simulations indicated that because of the multiple components present and the non-idealities associated with the polar species (water, alcohol, acid, and ester), the propylene distilling out of the second (de-esterification) RD column would still contain some isopropyl acetate. This distillate was therefore sent to conventional distillation column, to separate and purify the propylene as the distillate, and recycling the acetate bottoms back to the de-esterification RD column.

The final 3-column process flow sheet is shown in Figure 2:

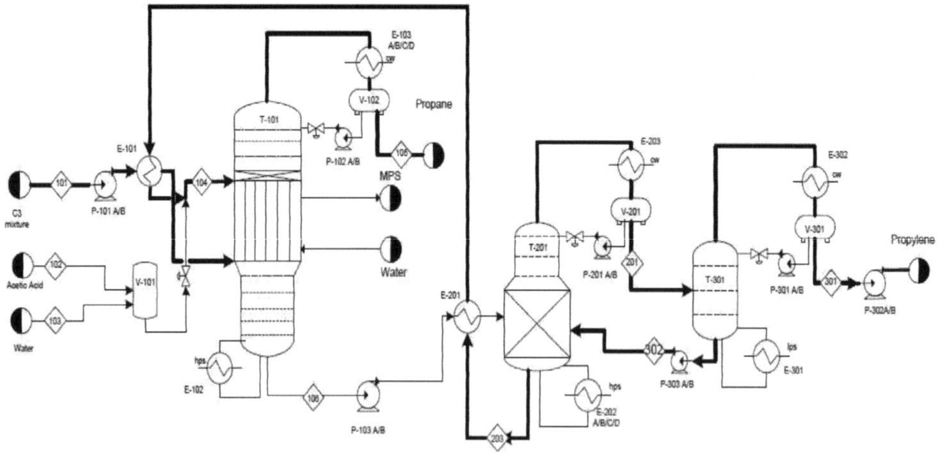

Figure 2 RRD Flowsheet

Key process variable for the RRD design, in particular to establish the equilibrium reaction conditions and component concentrations, are reaction temperature, pressure, acetic acid : propylene feed ratio, the water : acetic acid ratio, and the catalyst loading and number of reaction stages in the RD columns. The following figures illustrate the effect of T, P, and feed ratios on the equilibrium conversions:

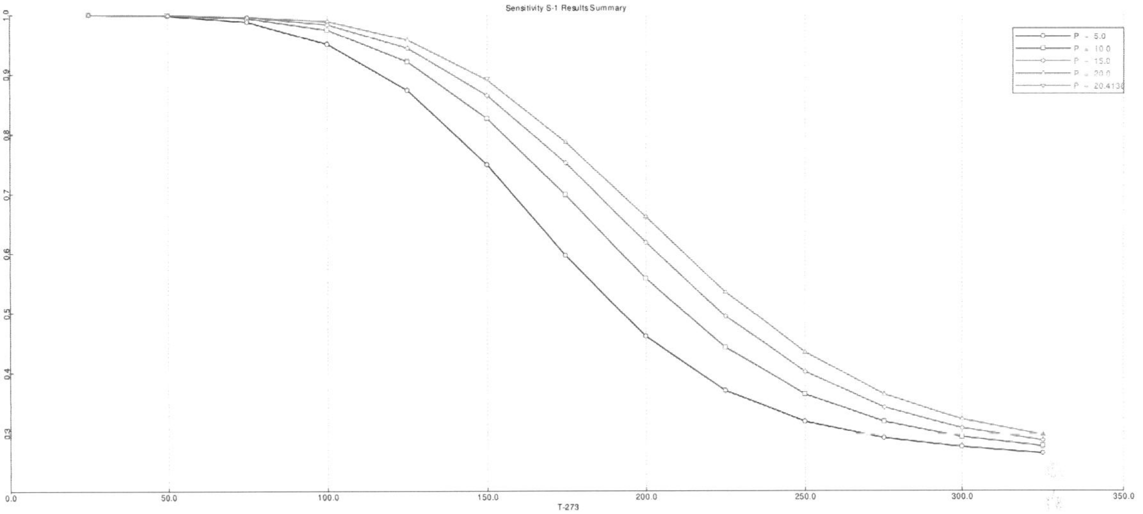

Figure 3 Equilibrium C3= Conversion vs. Temperature and Pressure

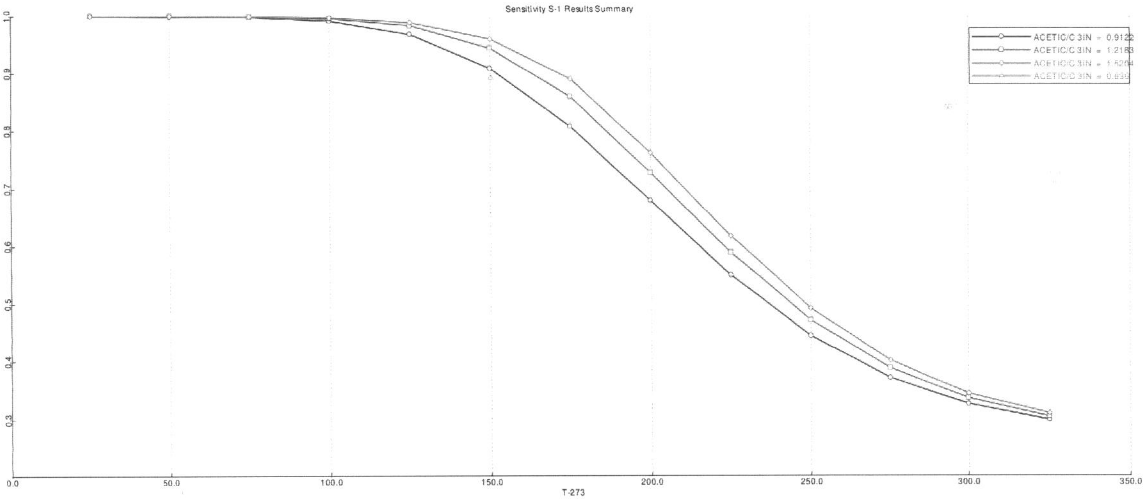

Figure 4 Equilibrium C3= Conversion vs. Temperature and Acetic Acid : C3= Feed Ratio

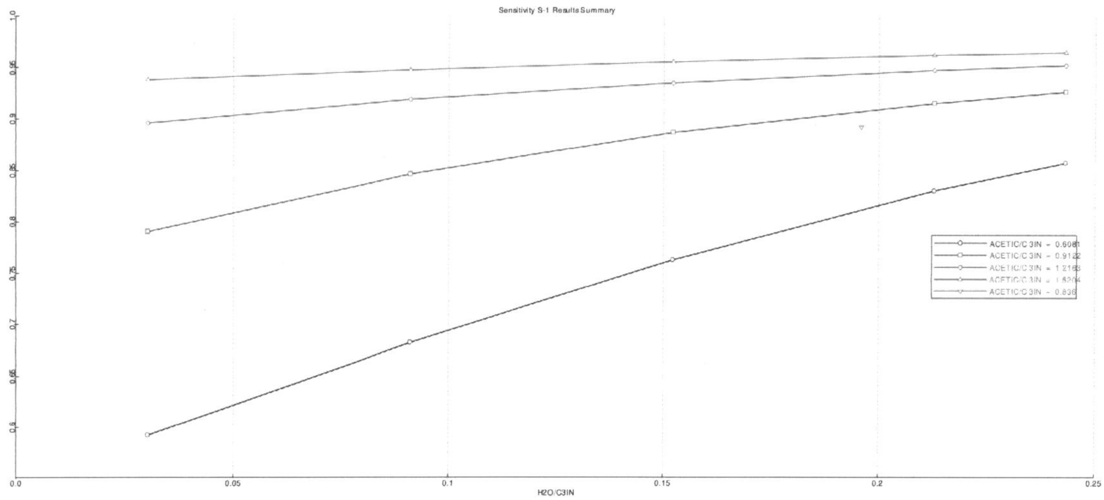

Figure 5 Equilibrium C3= Conversion vs. Acetic Acid and Water : Propylene Feed Ratios

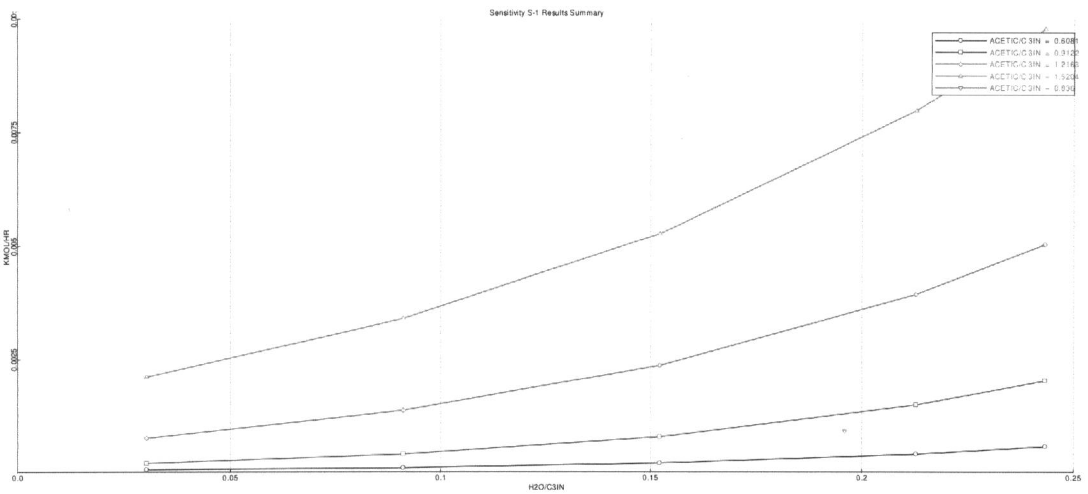

Figure 6 Acetic Acid and Water:C3= Feed Ratios Effect on Equilibrium Water Concentration

Based on these equilibrium curves, the selected RRD process consisted of the following:

- A first reactive distillation column (RD1) with a total of 50 theoretical stages (including the reboiler and condenser), of which the top rectification section has 11 ordinary fractionation stages, a reaction zone that has 27 stages in which fractionation and equilibrium reaction occur simultaneously, and a bottom stripping section that has 12 ordinary fractionation stages. The reaction zone stages are provided with bales of the HPA solid-phase catalyst.
- A second reactive distillation column (RD2) with a total of 36 theoretical stages (including the reboiler and condenser), of which the top rectification section has 14 ordinary fractionation stages, a reaction zone that has 19 stages in which fractionation and equilibrium reaction occur simultaneously, and a bottom stripping section that has 3 ordinary fractionation stages. The reaction zone stages are provided with bales of the same HPA solid-phase catalyst used in RD1.
- A purification column (C3) is a conventional distillation column comprising a total of 36 theoretical stages (including the reboiler and condenser), of which the top rectification section has 23 fractionation stages, a bottom stripping section that has 13 ordinary fractionation stages.

RRD Process Description

The propylene-propane feed mixture is preheated (by cross exchange with the bottoms of RD2) and fed immediately below the reaction zone of RD1. RD2 bottoms, which is an acetic acid stream containing 1.0 wt% water and up to 4.4 wt % (total) of propyl acetate and isopropyl acetate is fed 3 theoretical stages below the top of the reaction zone of RD1.

In the RD1 reaction zone, propylene is converted to propyl acetate and isopropyl acetate, which goes to the bottom of RD1, along with unconverted excess acetic acid and small amounts of propyl and isopropyl alcohols and the unrecovered propane.

RD1 operates at 15 bar (top pressure) to allow condensation for the reflux using cooling water, as well as establishing the temperatures in the forward reaction zone. RD1 uses a partial condenser, where the vapor distillate is propane containing 2.0 wt% propylene.

The calculated temperature and composition profiles in RD1 are shown in Figures 7, and 8, respectively. A surprising result is that the temperatures of several fractionation stages below the reaction zone are cooler than the (reactive) fractionation stages above. This is attributable to the shift in composition to a higher fraction of more-volatile materials in vicinity of the propylene-propone feed point (note the propylene concentration bulge between stages 38 and 44).

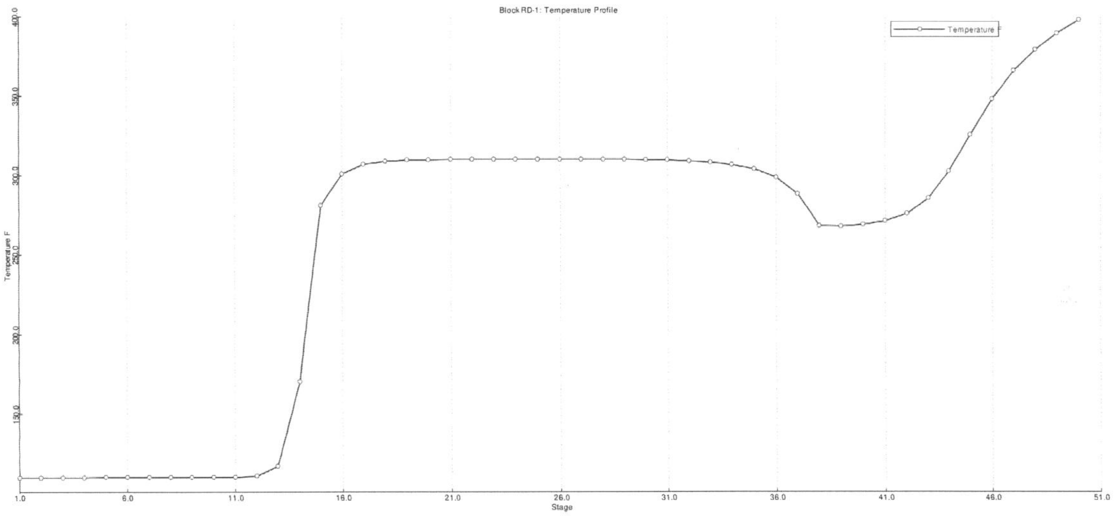

Figure 7 RD1 Temperature Profile

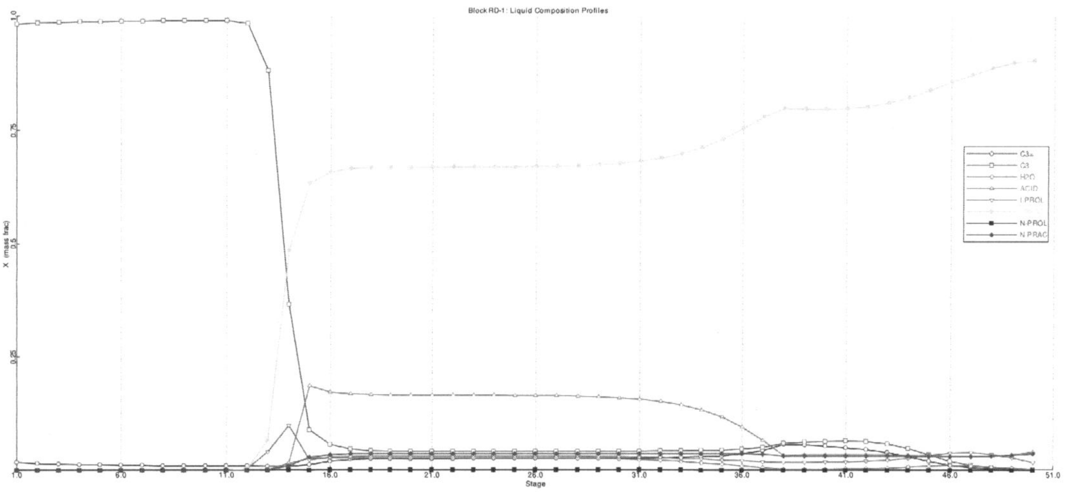

Figure 8 RD1 Composition Profile

The bottoms of RD1 is preheated (by cross-exchange with the bottoms of RD2) and fed 7 theoretical stages below the top of the reaction section of RD2. Recycle n-propyl and isopropyl acetates from the bottoms of C3 are fed 2 theoretical stages above the bottom of the reaction section of RD2.

In the RD2 reaction zone, which operates at higher temperatures than the RD1 bottoms, the propyl esters are back-converted to propylene, which goes to overheads distillation of RD2, along with some unconverted propyl esters, small amount of propyl and isopropyl alcohols and water, and the unrecovered propane. The acetic acid and water regenerated by the de-esterification, along with the remaining un-converted propyl esters, are withdrawn from the bottom of RD2 and recycled (via the RD2 and RD2 feed preheaters) as the reactant feed to RD1.

RD2 operates at 30 bar to maintain the elevated temperatures in the reverse-reaction zone. The calculated temperature and composition profiles in RD1 are shown in Figures 9, and 10, respectively.

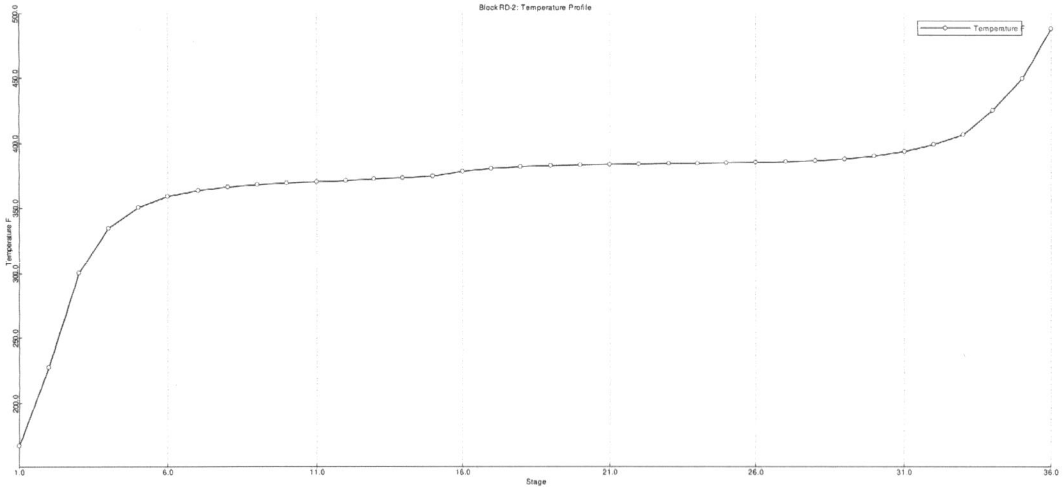

Figure 9 RD2 Temperature Profile

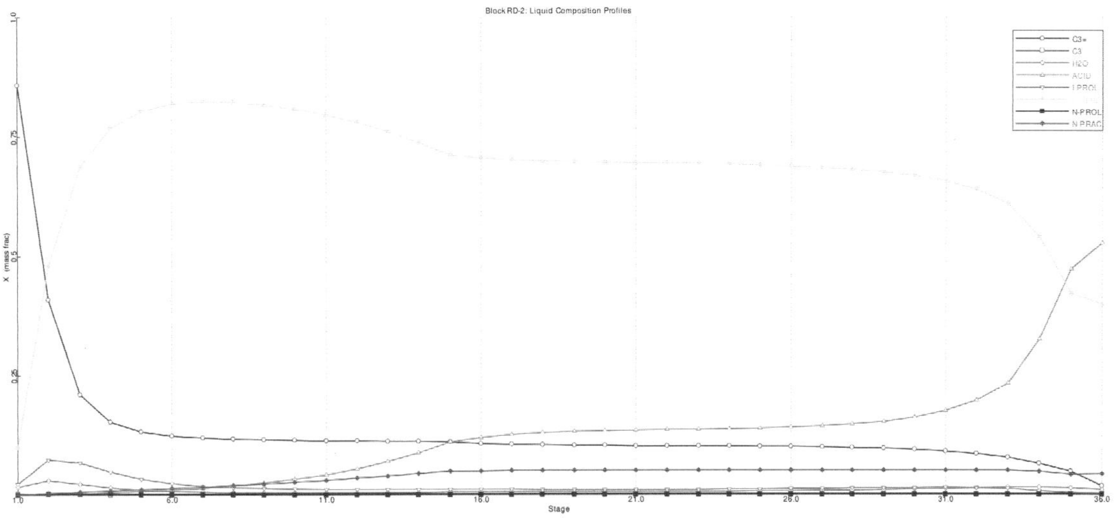

Figure 10 RD2 Composition Profile

The distillate of RD2 is fed between the rectifying and stripping sections of C3. C3 operates at 19 bar to condense the reflux using cooling water. It uses a partial condenser, where the vapor distillate is the propylene product at a purity of 99.7 wt%. The separated propyl acetates and water are recycled from the C3 bottoms back to the lower part of the reverse-reaction zone of RD2.

Comparative Column Sizes – RRD Vs. Conventional C3 Splitter

The diameter of RD 1 is of 17.5 feet in the rectification section, 15.5 feet in the forward reaction section, and 13 feet in the stripping section. The diameter of RD2 is 11 feet in the rectification section, and 13 feet in the reverse reaction and stripping sections. The diameter of C3 is 8 feet. The combined reboiler heat duties of RD1, RD2, and C3 is 1847 BTU/lb propylene product.

By comparison to the conventional C3 splitter, the RRD process requires 37 % less heat input per pound of propylene product than the conventional C3 Splitter. Even though 3 columns are required in place of a single C3 Splitter, in aggregate, they require considerably less material for the column vessels and fractionation internals, hence likely to require less investment. As Table 1 shows, RRD requires 40 % fewer total theoretical separation stages (not including their respective reboilers and condensers), 87% less total tray area, and 80% less total column volume than a conventional C3 splitter.

Table 1

	Diameter	Theor Stages	Trays @ 80% Efficiency	Tray Spacing	Tray Area (ft2)	"Active" Column Volume (ft3)
			Conventional Distillation			
C3 Splitter	**27**	**193**	**241**	1	**138129**	**138129**
			Reversible Reactive Distillation Process			
RD-1 Rectification	**18**	10	13	1	3007	3007
RD-1 Reaction	**16**	27	34	2	6368	12737
RD-1 Stripping	**13**	11	14	1	1825	1825
RD-2 Rectification	**11**	13	16	1	1544	1544
RD-2 Reaction	**13**	19	24	2	3152	6305
RD-2 Stripping	**13**	2	3	1	332	332
C-3	**8**	34	43	1	2136	2136
	Totals	116	145		18365	27886
	vs. Conventionial	**60%**	**60%**		**13%**	**20%**
			Reversible Reactive Distillation Process w Heat Recovery			
RD-1 Rectification	**15**	10	13	1	2064	2064
RD-1 Reaction	**16**	27	34	2	6368	12737
RD-1 Stripping	**12**	11	14	1	1555	1555
RD-2 Rectification	**11**	13	16	1	1544	1544
RD-2 Reaction	**13**	19	24	2	3152	6305
RD-2 Stripping	**13**	2	3	1	332	332
C-3	**8**	34	43	1	2136	2136
	Totals	116	145		17152	26673
	vs. Conventionial	**60%**	**60%**		**12%**	**19%**

Opportunity For Heat Recovery

The forward esterification reaction is exothermic, and operates at an elevated temperature. This suggested that the heat of reaction could be recovered, by generating steam, rather than rejecting it via the condenser of RD1.

RD1 was modified, replacing the reaction zone trays, with a multi-tubular heat exchanger packed with catalyst to a depth providing the equivalent of the specified number of theoretical stages. Cooling with boiler feed water in the shell of the reaction zones generates moderate-pressure (25 – 40 psig) steam, which can be exported to use elsewhere on the site (although it is not hot enough to reboil RD2, which requires higher-pressure steam).

The heat removed from the forward reaction zone of RD1 decreases the condenser cooling duty (and associated condenser size) of RD1 35 %. The associated reduction in RD1 vapor traffic allows an 18 % reduction in the diameter of the rectification section. These reductions in condenser and rectification section sizes further reduces the capital investment

Taking the credit for the energy value of the generated steam, the net total heat input requirement into the process (i.e. the combined reboiler duties of RD1, RD2, and C3, less steam generator duty) is reduced to 1581 BTU/lb propylene product, improving the RRD process energy savings to 46 % less heat input per pound of propylene product than the conventional C3 Splitter.

Comparative Economics

The capital costs of the conventional steam-reboiled C3 Splitter and the RRD systems were estimated with a software tool suitable for conceptual-level cost estimates (+/- 50 % accuracy). As shown in Figure 11, even though the RRD system requires more equipment items and upgraded metallurgy (stainless-steel clad carbon steel to resist corrosion by the acetic acid, vs. carbon steel for the C3 Splitter), it is projected to have ~ 20% lower ISBL investment cost, because of the much smaller column sizes.

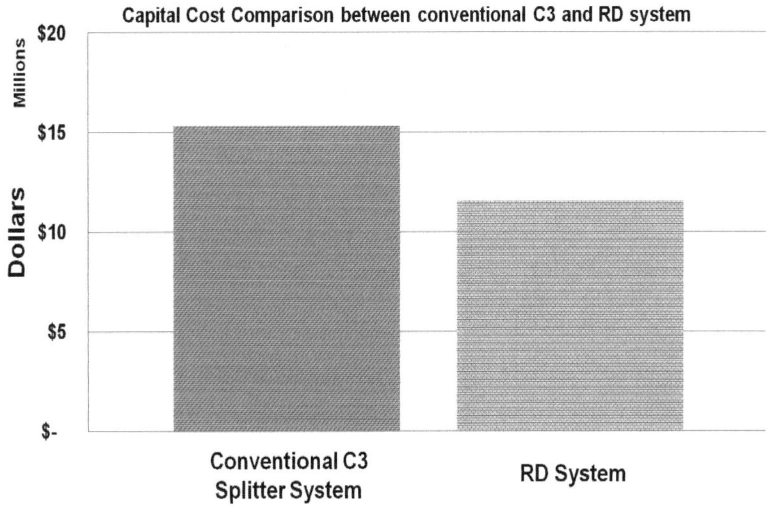

Figure 11.

The assumed operating cost bases are given in Tables 2 and 3:

Table 2

Utility	Cost	Unit
Fuel	4.50	\$/MMBtu
Cooling Water	0.40	\$/MMBtu
Electricity	19.71	\$/MMBtu

Table 3

Materials	Cost	Unit
Catalyst	9.00	\$/lb
Acetic Acid	0.42	\$/lb
Process Water	0.12	\$/gal

Using the estimated capital cost, and with the assumed operating costs, shutting down a conventional steam-reboiled C3 Splitter and installing an RRD system to reduce the energy consumption would have an internal rate of return (IRR) of about 27%

CONCLUSIONS

- An RRD system for separating propylene from propane has the potential to reduce the separation energy requirement by 35 – 45 % compared to a conventional steam-reboiled C3 Splitter.

- Despite the increased complexity, RRD requires significantly smaller equipment than a conventional C3 Splitter, resulting in lower overall capital costs. RRD requires 60% fewer trays, 85% less tray area, and combined 80% less column volume.

- RRD appears to be an attractive alternative to conventional splitter towers: The energy savings may justify replacing conventionally reboiled / condensed splitter in an existing process. RRD looks especially attractive for a new installation. In addition to lower energy costs, it has the potential to have lower capital as well.

ISSUES

- The RRD process has not yet been demonstrated at any scale. The present study is a "paper" process, intended to explore whether RRD merits further study.

- RRD has limited applicability to other olefins. Because double bond formation (on reverse reaction) favors secondary position, RRD can only be used for olefins whose adducts re-generate the same (original) double bond as the starting olefin, regardless of which isomer adducts are formed. Unless there are steric hindrances, it appears that this "symmetry" requirement limits RRD to the separation of ethylene, propylene, and iso-butylene from their corresponding C, C3, and C4 paraffins.

- Catalyst durability / life is unknown. This study assumes 2 – 3 years between change-outs of the HPA catalyst, based on typical experience with this catalyst in ethyl acetate production.

- Materials of construction requirements. This study assumes that RRD equipment wetted by streams containing acetic acid or water will use stainless-steel-clad carbon steel. However, the elevated temperatures associated with the forward and reverse esterification reactions may make the liquid more aggressive, requiring more expensive materials of construction.

- This study assumed as a base case C3 Splitters using steam-heated reboilers and water-cooled condenser. However, it is a common practice to use vapor-recompression systems to upgrade the rejected heat from condensers in narrow-boiling-range towers, to provide most of the reboiler heat duty. This could potentially eliminate the energy savings advantage of an RRD system, albeit at the cost of a large compressor, and the potential reliability issues associated with complex moving machinery.

PATH FORWARD

- The RRD process should be evaluated for other suitable olefins, i.e., ethylene (vs. C2 Splitters) and iso-butylene (vs. C4 Splitters)

- Alternative reversible olefin reaction chemistries or adducts (other than acetates or even esters in general) should be investigated, to determine if more favorable operating conditions or milder materials of construction can be used with RRD.

- The economics of the RRD process should be compared against Splitter towers using vapor recompression for reboiling.

- If the conceptual design studies show that RRD remain attractive, significant process development work will be required to bring this concept to commercial fruition. Major activities include demonstrating "proof of concept" in a working system (e.g. bench scale operation), catalyst selection and life studies, pilot plant operations with extended runs in closed-loop recycle mode, and materials of construction validation.

(142f) A New "Spin" on Trays

Izak Nieuwoudt, Koch-Glitsch, Wichita, KS and **Neil A. Sandford**, Koch-Glitsch LP, Wichita, KS

The capacity of any conventional cross-flow tray is limited by gravity due to the settling velocity of the droplets. In addition, the efficiency of a cross-flow tray is reduced as the liquid rate is reduced. Cross-flow trays also struggle to deal with the pitch & roll that are experienced on some off-shore applications. In this presentation, information on two types of trays to deal with all of these issues will be presented. In both of these trays, centrifugal motion is used to overcome the gravity limitation. The liquid flow patterns are managed to ensure that the nature of the froth maintains good efficiency. In the case of off-shore applications, baffles are used to negate the influence of pitch & roll. Large scale test data on these trays will be presented.

(142g) Evaluation and Comparison of Internal /External Heat Integration Configurations for the Separation of Propane and Propylene

Dawei Chen, Lianghua Xu and Xi-Gang Yuan, State Key Laboratory of Chemical Engineering, Tianjin University, Tianjin, China

The separation of propane and propylene by distillation consumes vast amount of energy owing to the close-boiling feature of the binary mixture. Various energy saving schemes have been proposed, including the state-of-the-art heat pump configuration and the promising internally heat integrated distillation column (HIDiC). More recently, the idea of simplifying HIDiC by several external heat exchangers has come into notice and was claimed to be a more practical configuration with greatly reduced capital cost. To investigate the energy and economical performances of heat pump assisted distillation along with HIDiC and its various simplifications, comparisons based on simulation was made in the present work. Compressing ratios and total numbers of stages were optimized to achieve an optimal configuration in terms of total annualized costs for each scheme before compared with others. For HIDiC and its simplifications, the impacts of different heat distribution patterns were also considered. It was found that the result greatly depended on the criterion selected for comparison, and some discussions and remarks were also presented.

AUTHOR INDEX

AUTHOR INDEX